WAITING FOR ROBOTS

The France Chicago Collection

A series of books translated with the generous support of the University of Chicago's France Chicago Center

The Hired Hands of Automation

WAITING FOR ROBOTS

ANTONIO A. CASILLI

Translated by Saskia Brown
With a foreword by Sarah T. Roberts

THE UNIVERSITY OF CHICAGO PRESS
CHICAGO AND LONDON

The University of Chicago Press, Chicago 60637
The University of Chicago Press, Ltd., London
© 2025 by Saskia Brown
Foreword © 2025 by The University of Chicago
All rights reserved. No part of this book may be used or reproduced in any manner
whatsoever without written permission, except in the case of brief quotations in
critical articles and reviews. For more information, contact the University of Chicago
Press, 1427 E. 60th St., Chicago, IL 60637.
Published 2025
Printed and bound by CPI Group (UK) Ltd, Croydon, CR0 4YY

34 33 32 31 30 29 28 27 26 25   1 2 3 4 5

ISBN-13: 978-0-226-83707-9 (cloth)
ISBN-13: 978-0-226-82095-8 (paper)
ISBN-13: 978-0-226-82096-5 (e-book)
DOI: https://doi.org/10.7208/chicago/9780226820965.001.0001

Originally published as Antonio A. Casilli, *En attendant les robots. Enquête sur le travail
du clic* © Paris, Éditions du Seuil, 2019.

www.centrenationaldulivre.fr

The University of Chicago Press gratefully acknowledges the generous support of the
France Chicago Center toward the translation and publication of this book.

Library of Congress Cataloging-in-Publication Data

Names: Casilli, Antonio A., 1972– author. | Brown, Saskia, translator. | Roberts, Sarah
T. (Professor of information studies), writer of foreword.
Title: Waiting for robots : the hired hands of automation / Antonio A. Casilli ;
translated by Saskia Brown ; with a foreword by Sarah T. Roberts.
Other titles: En attendant les robots. English (Brown) | France Chicago collection.
Description: Chicago : The University of Chicago Press, 2025. | Series: France Chicago
collection | Includes bibliographical references and index.
Identifiers: LCCN 2024017903 | ISBN 9780226837079 (cloth) | ISBN 9780226820958
(paperback) | ISBN 9780226820965 (ebook)
Subjects: LCSH: Labor supply—Effect of technological innovations on. | Labor
supply—Effect of automation on. | Automation—Social aspects. | Information society.
Classification: LCC HD6331 .C2913 2025 | DDC 331.25/6—dc23/eng/20240514
LC record available at https://lccn.loc.gov/2024017903

♾ This paper meets the requirements of ANSI/NISO Z39.48-1992
(Permanence of Paper).

CONTENTS

Foreword vii

Preface xi

INTRODUCTION *1*

Part 1: What Automation?

1 · WILL HUMANS REPLACE ROBOTS? *17*

2 · WHAT'S IN A DIGITAL PLATFORM? *37*

Part 2: Three Types of Digital Labor

3 · ON-DEMAND DIGITAL LABOR *57*

4 · MICROWORK *78*

5 · SOCIAL MEDIA LABOR *109*

Part 3: The Horizons of Digital Labor

6 · WORK OUTSIDE WORK *151*

7 · HOW DO WE CLASSIFY DIGITAL LABOR? *167*

8 · SUBJECTIVITY AT WORK,
GLOBALIZATION, AND AUTOMATION *189*

CONCLUSION: WHAT IS TO BE DONE? *214*

Acknowledgments 229

Notes 231

Index 291

FOREWORD

The title of this book in its original French, as published by Seuil, is *En attendant les robots*. The English title, *Waiting for Robots*, is a direct translation, but our pronunciation obscures the pun. Typical of its author Antonio Casilli's dry humor, this is a nod to *En attendant Godot* (*Waiting for Godot*), the mid-twentieth-century absurdist play by Irish expatriate Samuel Beckett in which, quite famously, and decades before it served as the main thrust for the entirety of *Seinfeld*, nothing of consequence happens. The conceit of the play is that the two main characters wait for the arrival of the titular Godot—an arrival that, unbeknownst to the two hapless protagonists, will never transpire. The dialogue between the characters, and the nonevents of the play itself, are therefore focused on an exercise in futility and uselessness; it is much consternation and handwringing over a state (Godot's arrival) that will never come to pass. (Fond of both puns and literary references, Casilli entitled an earlier monograph *Les Liaisons numériques*, or *Digital Liaisons*, after the scandalous Choderlos de Laclos epistolary novel featuring the libertine and malicious machinations of eighteenth-century ancien régime courtiers.)

Unlike *Godot*, however, this book arrives just in time: in a moment in which the collective has focused its greatest dreams, or deepest anxieties, depending on where you sit, on the imminent reordering of society under the regime of large language model–fueled natural language processing tools such as ChatGPT and other AI-based automation of all sorts. No matter where one may fall in the mainstream debate underway about these computational tools and what they portend for economic growth, for the workforce and for society as a whole, it is the razzle-dazzle promise of their quasi-magical abilities to transform work as we know it, and, with it, the economic conditions of most workers, that is typically, breathlessly at the fore. Both proponents and detractors claim these tools, once adopted, will have the ability to replace legions of humans, eliminating the need for their waged work. It is an unprecedented situation, we are told, that is always-already on its way. Much like our midcentury main characters, all we can—and must—do is wait.

And yet.

Since the late 1960s, a variety of theories have emerged, suggesting significant changes in work and society, challenging traditional practices dating back to the Industrial Revolution. These post-Fordist theories insist upon an imminent rupture with mass manufacturing and factory arrangements prevalent through to the mid- to late twentieth century in the Global North. Although those contexts were, of course, deeply intertwined with many technological advancements that permitted the exponential growth in production in manufacturing sectors, these new theories focused on the changes to be wrought by digital technologies, in general, and the widescale introduction of the computer (and later, the personal computer) into all sorts of workplace tasks. In these new workplace arrangements, the focus would shift from the manufacturing of goods (see Daniel Bell's *The Coming of Post-Industrial Society*) to a form of mindwork—analytical work, advertising and marketing, dealing with data, and so on—all of which would be undertaken in symbiosis with, and greatly facilitated by, computers.

It was this latter characteristic, claimed many sociologists and cultural critics of the past three decades (see work by Manuel Castells, Nicholas Negroponte, Henry Jenkins, the entire run of *Wired* magazine, and anyone who has ever called himself a "futurist" since the 1990s) that would herald, in turn, a new age of less work and more leisure. Rhizomatic social arrangements would free us from those that trapped society into hierarchies of class (if less attentive to concerns over the constraints of race and gender) and allow for new and creative points of connection. All of these changes would yield, through the use of computers and an increasingly digitally networked world, a liberatory outcome. Can't we all just taste the freedom?

Today's rhetoric is not too different. AI proponents and detractors alike are predicting the elimination of work as we know it, and the disappearance of employees from the scene. Yet, as Casilli shows us through the many rich examples peppered throughout *Waiting for Robots*, the case is likely much less one of the *elimination* of work than its atomization, its fracturing into meaningless parts, and, ultimately, its *degradation*, to invoke the subtitle to Harry Braverman's canonical 1974 text, *Labor and Monopoly Capital: The Degradation of Work in the 20th Century*.

But as the longer historical and theoretical context Casilli provides within shows, it turns out that the truth beyond the rhetoric is both far more complex and simpler at the same time. These tools, robots by another name, are invariably described as *just on the verge of* taking over all aspects of our work lives and cultural output, but they simply cannot exist without the humans that inform them. As Casilli demonstrates time and again, workers in the digital and AI-backed economies do not disappear. They are simply moved out of sight in a sleight of hand that places them elsewhere in the production chain, thereby creating, maintaining, and activating the tools that use their extracted human intelligence at levels of remove, and validating a logic that devalues their contributions. After all, Casilli says, "today, digital platform workers are themselves the cogs in the machine that threatens to replace them. In other words, the fragmentation of their work is not a consequence, but a precondition of automation." So, as we await the near future for the robots' arrival, the changes necessary to allow for the continued degradation of human work take place in real time, and under our very noses. If the robots never arrive as described (and they never will), it won't matter; it was the cheapening and reordering of humans' contributions that was the real project all along.

A professor of sociology at the Institut Polytechnique de Paris—designated as a *grande école* in the French higher education system—and a now by-fellow at Churchill College, Cambridge, Antonio Casilli is a leading researcher and intellectual in contemporary Europe who holds expertise on a wide range of issues at the nexus of digital technologies and society. He is also well known to the French general public; a welcome and frequent presence on radio and TV, his signature quick wit and polymath's grasp of world history and culture make him a critical voice advocating for nuanced and evidence-based labor policy both nationally and within the EU. As a sociologist, he is keenly concerned with the structural impacts of digital technologies and AI, working in an advisory capacity to the French government, as well as a consultant with NGOs, labor unions, and grassroots organizations to address the negative role of surveillance and computational automation on people's lives.

To this end, he has contributed a prolific and important body of work (in French, in Italian, in English, as well as in translation) to the many debates around the role of computational automation, AI, and digital technologies as they proliferate and find their way into every aspect of personal and collective/civic life. Indeed, his is a key voice defining the parameters of the ongoing debates regarding the social role of AI, and has greatly helped to move the needle, in Europe and beyond, to terms that are more reality based and technically informed than the general discourse or accepted wisdom, while also providing pointed, insightful, and desperately needed criticism of these technologies, their implementation, and their true end-to-end social costs. As Casilli reminds us, the "digit" of "digital labor" refers to numbers, yes, but it also refers to the digits of the hand, leaving their fingerprints wherever they touch these technologies, as long as we know where, and how, to look.

Working alongside a global cohort of colleagues such as, but not limited to: Mary Gray, Rafael Grohmann, Mark Graham, Lilly Irani, Karen Gregory, Jen Schradie, Julian Posada, Nikos Smyrnaios, Jack Linchuan Qiu, Ursula Huws, Jérôme Denis, Trebor Scholz, and so many others (myself included), Antonio Casilli has created community and given space to those of us who share his commitments and concerns and invited us to join him. He has engaged with us intellectually, invited us to lecture at his university, and co-constituted research communities (first with the European Network on Digital Labour, followed by the International Network on Digital Labour). Indeed, he is a rigorous mentor, an exacting teacher, a generous peer, an incredible colleague, and a dear friend. It is my great privilege and honor to introduce Antonio Casilli, and *Waiting for Robots*, to the wide English-language audience it deserves.

Sarah T. Roberts

PREFACE

This book, originally published in French in 2019, had an epigraph. It was from a speech delivered by Alexis de Tocqueville to the French Parliament on January 27, 1848. That day, deputies could see the gathering crowd of Parisian revolutionaries from the windows of the National Assembly. Yet they seemed oblivious to the "social question" these citizens raised. A moderate politician, Tocqueville issued his famous warning, "Gentlemen, we are sleeping on a volcano."

Readers of the French edition of this book interpreted the quotation in several ways. For some, the volcano was a symbol of digital technologies threatening to erupt without AI regulation. Others saw it as an allegory—like lava burning through resistance, ideologically diverse waves of political unrest would characterize the early 2020s—the yellow vests in Europe, the anti-lockdown riots in China, the Black Lives Matter movement in the United States. There are even those who saw the volcano as foreshadowing the COVID-19 pandemic, the economic recession, the geopolitical tensions, or the global climate crisis.

Prognostication has never been my cup of tea. I meant for readers to understand the quote less as a grand allegory and more as a simple analogy. In my mind, the volcano on which we sat was the social awareness raised by workers—those Uber drivers and Amazon microtaskers, people we refer to as "platform workers"—who produce and distribute goods and services hosted by big tech companies.

The cultural and political struggle to recognize digital labor as "real" work has taken decades, and we are not done. There have been gains for logistics, transportation, and online workers. Food delivery couriers working for digital platforms are now considered employees in Spain under the 2021 "riders' law." All digital laborers are presumed to be employed by platforms under a new 2024 European directive. Crowdworkers working for Ixia, a Brazilian platform, cannot be hired as freelancers, following a 2022 decision of the labor court in São Paulo.

Today few (at least among those who would read a book with this title) would argue that gig workers are given exciting opportunities or experience fair work environments on online platforms. We are aware of the important role that ride-hailing drivers, internet content moderators, and microtask service providers play. But we are also likely to decry the precarity, algorithmic surveillance, and harsh working conditions that these platform workers endure.

Though it's been five years since I wrote the French version of this book and I have updated these pages throughout, my work's underlying message is the same. We're still decades away from protecting the work of AI's hidden helpers—the unseen people behind the machines who achieve "automation," including the unpaid users whose data feeds the system.

As a society, we can now more easily identify capitalist exploitation. We see through the promises of "being your own boss" and "working the hours you want." But technological fantasies are still pervasive, even among the most informed.

The purpose of this book is to demonstrate the presence of inconspicuous labor in AI solutions and digital services. I offer a framework for detecting continuities of exploitation between observable platform labor (like delivery, logistics, and transportation), and other forms of not-yet-discovered labor (online crowdwork, click farming, etc.). I argue that labor does not exist in a cultural and regulatory vacuum, and that workers are not passive recipients of decisions made by platforms. Workers have agency, albeit a conflictual one. Disputes between them and corporations, governments, and sometimes even users are the result of the workers' desire for autonomy over the digital tools they use and the artificial intelligence they shape.

I hope to steer between two extreme views of AI—as a destroyer of worlds, both economic and environmental, and as a benevolent human caretaker, the panacea to our social and political problems. AI is a technological process that isn't actually artificial; it reflects the values of the people who produce it.

It is difficult to achieve this understanding. This book was first published as part of a wave of scholarship—including works by Mary Gray and Sid Suri,[1] Kate Crawford,[2] Kylie Jarrett,[3] and Sarah Roberts[4]—that helped us learn more about this subject. More people are paying attention now, and they are focusing on the large data extracted for machine learning, and the humans who train it.

During this training, systems are taught how to operate based on examples from datasets that are not easy to produce. Both users and workers generate, annotate, and refine the data to make it usable. Before anyone first interacted with Siri or Alexa, workers had to listen to hours of real-life recordings of mundane conversations. Before one can appreciate ChatGPT-generated sonnets in the style of Shakespeare, workers had to evaluate tomes of similar generated text.

But once the AI is trained and launched on the market, humans—underpaid or unpaid—continue to contribute. Using AI is like owning a car that needs endless maintenance. Among tech companies, this mainte-nance is often referred to as "verification." Updating and troubleshooting an AI system is like having someone fixing your car's brakes or check your engine's oil. As users, we do these verification tasks too, whether we realize it or not. Contributing your personal data to customize AI services is a lot like filling up the tank of your car.

Finally, and most disturbingly, AI sometimes needs workers to "imper-sonate" algorithms. These "fake" artificial intelligences might seem the result of unethical companies scamming users. But this practice occurs among more than a few bad apples. Having "humans in the loop" has been a feature of computing since the late twentieth century. The only differ-ence is that today it is more prevalent and less visible. Those interested will find stories of hidden workers simulating recommendation algorithms, surveillance cameras, voice assistants, and chatbots—and I'll explore some of these examples in this book.

Developing truly powerful technologies is a difficult task. Our world is populated by faulty AI systems, glitchy software, and weak image filters. In the minds of their creators, technologies are given a lot of leeway, anthro-pomorphized, and considered more efficient than they actually are. Users also personify and confer power on digital objects. They complain when systems don't "obey"; they worry about the "influence" of algorithms on their choices; and they express desires about "convenient" technologies. They see themselves reflected in these tools and treat them like animated beings. However unwillingly, when users overestimate the intelligence of their software or devices, they fail to recognize how their fellow humans contribute to their efficiency.

Despite the mounting evidence that AI's production requires human labor, our contemporaries worry that Artificial General Intelligence (AGI) will eventually surpass human intelligence. Whether propagated by artists, media moguls, or tech companies themselves, these myths are so easy to imagine. Can you picture that translucent robot in a thinking pose? You've seen this image in endless magazine covers, documentaries, YouTube thumbnails.

The myth that machines displace jobs is one of the most difficult to dis-pel. The previous edition of this book already included a critical appraisal of this theory—which I had to thoroughly update for this new version.

Sadly, doomsayers of a labor apocalypse continue to appear. Economists continue to bet their careers on how many jobs will be destroyed by the technologies developed by the capitalists who fund them.

We can assume that those who support this theory have an inadequate understanding of the amount of work required to manufacture our digital artifacts. Yet, there are people among them who know intuitively that a huge human workforce operates remotely both from low- and high-income countries to train and operate automated systems but worry about a "great robot replacement" anyway. As long as fear is concerned about robots taking jobs from US workers, reason cannot accept that AI is made up of outsourced digital labor.

While debunking this myth, we must not fall into the ideological trap of blaming foreign workers and not robots as the source of "great replacements." Rejecting this proposition instead allows us to focus on another vital goal of organized labor: redistributing value and power among workers.

Our goal should be to recognize and compensate digital labor collectively. I discuss this in the conclusion of this revised edition, where I outline some of the latest initiatives to overcome today's political stances around digital labor. Platforms are more than just technological paradigms. They have a long political history. By reactivating some of their original values, such as solidarity, resistance to exploitation, and civic engagement, it is possible to formulate policies that are radically progressive and antiauthoritarian. These policies have been discussed extensively in James Muldoon's *Platform Socialism*, published in 2022. The book explores the transformative political potential that platforms offer.[5] In Muldoon's analysis, platforms are used to challenge power in a manner that is resolutely anti-capitalist. This represents a substantial departure from both the leftist fixation with "resistance" as the only way to express political disagreement, and a liberal political philosophy that promotes regulation to restore market power and consumer rights but doesn't question how information, labor, and the environment get commodified.

The results in my book's first edition are still relevant. That's a sad testament to how AI still continues to cause inequality and confusion. It's also a testament to how important these analytical categories remain for describing the labor-digital nexus.

Naturally, this new English edition contains updates and revisions. More than its previous version, this book also incorporates empirical data collected over the past years with my colleagues of the DiPLab research team. DiPLab (which stands for Digital Platform Labor) was co-founded with Ulrich Laitenberger and Paola Tubaro at the Institut Polytechnique de Paris to explore the relationship between technologies and work. To date, it has conducted research projects in more than twenty countries and collected the testimonies of more than three thousand workers. The empirical vignettes that complement this book's theoretical framework originate from fieldwork conducted in Europe, Africa, and Latin America between 2018 and 2023.

With DiPLab, I have made significant acquaintances in Egypt, Bolivia, Madagascar, Nigeria, and Venezuela, as well as in European countries like France and Germany. Our fieldwork has given me fresh theoretical insights, which I have elaborated on in this edition. I've also gained a stronger understanding of the reality that workers all over the world face today.

However, my core argument has not changed. AI is not only produced in technological campuses outside rich cities in the US; it can also be found in cybercafes in Africa, in family homes in high-crime areas in South America, and in studio apartments in middle-class neighborhoods in Europe's peripheries. Move away from the fantasy of the "entrepreneurial garage" that has dominated American startup culture for decades.[6] Rather, picture a preindustrial cottage industry where AI is built by low-wage workers of all races and genders, not only by ambitious data scientists and software engineers.

In addition to the pages that depict the circumstances of workers' lives directly, the stories of these "hired hands of AI" are also linked to the scholarly literature that I reference. In addition to studies that my collaborators and I have published, this research comes from the scholarship developed by the community of academics, activists, and policy experts gathered around the International Network on Digital Labor (INDL), that I am a cofounding member of. As part of an interdisciplinary research effort, I've also included references from other fields, spanning from computer science to history and philosophy. I hope these resources will be useful for readers interested in learning more about platforms, artificial intelligence, and digital labor.

There's still a long way to go until we gain a sober understanding of our technologies. Since the arrival on the scene of big data and algorithms, we have experienced a perpetual "AI summer." Both panic and enthusiasm are fueled by the notion of unstoppable progress. While our society has changed over the years, AI still uses the same positivistic arguments that have been around for centuries.

Already in Thomas Jefferson's Monticello mansion, so-called dumbwaiters—platforms connected to ropes and pulleys that delivered food and drinks to rich diners from poor kitchen staff below—appeared to operate without human intervention. Of course those mechanical lifts and hidden doors required the unseen labor of enslaved people. Cooking, arranging, and delivering food using pulleys were the first requirements of these workers. Another was to be as discrete as possible. We can even imagine Jefferson dining with abolitionists and avoiding the appearance of hypocrisy.

Whether we are talking about a dumbwaiter or artificial intelligence, technology hides subjugation. Human workers make automation possible. Recognizing the value of their contributions to these digital infrastructures is the first step toward worker reparation and rebalancing the current power dynamics.

Antonio Casilli, June 2024
Listening to Ebba Grön, "Staten och kapitalet"

INTRODUCTION

Simon and AIA (Artificial Intelligence from Antananarivo)

I interviewed Simon in 2017. That is not his real name, and SuggEst is not the real name of the startup where he was an intern at the end of his master's degree at a business school. The company was one of the rising stars of the European AI innovation sector. Its business model consisted of selling data on wealthy customers to luxury brands. Are you a politician, an athlete, or an actor? The SuggEst app promised to give you "100 percent personalized offers from the most famous French brands and most talented designers in the luxury market on the most advantageous terms." "Thanks to machine learning," the app detects the preferences of these rich potential customers to predict their choices. The traces they leave digitally—on their social media profiles; at public events they attend; and in selfies with friends, fans, and relatives—are automatically assembled using artificial intelligence. The results are aggregated and analyzed, and a product suggestion is made.

Yet behind this AI lurks a very different reality. Three days into his internship, during a coffee break, Simon grew suspicious. He asked why the company seemed not to employ a single software engineer or data scientist. "You don't need software engineers if you don't have any AI software," one of the company's founders admitted. AI was given credit for work being done by human workers abroad. Rather than have AI harvest information and an algorithm offer results, the app transferred users' requests to Antananarivo, Madagascar.

The app relied not on software in Europe but on humans acting as AI on an African island nation. And what work were they doing? The platform would send the workers an alert along with the name of one of the app's users. By dredging social media and the internet, they would collect all information available: texts, photos, videos, financial transactions, and website visits. They were doing the work of a bot, a program for data aggregation, all by hand. After investigating these individuals on social media, sometimes by creating fake accounts, they would write summaries of the person's preferences, and send them to SuggEst. The company would then compile the information and sell it to brands in the luxury sector, which in turn would suggest products.

How many hired hands are working as artificial intelligence in the world to-day? Nobody knows for sure, but a conservative estimate puts the figure at several million workers. And how much are they paid? Barely a few cents per click, often with no contract and no job security. And where do they work? Internet cafés in the Philippines, homes in India, college computer rooms in Kenya, abandoned factories in China. And why do they accept this work? Well, doubtless for the pay, especially in countries where the average salary of a worker is rarely higher than a few tens of dollars per month.

Simon's fellow interns confirm that this practice is standard. Whole city districts and whole villages in Mozambique and Uganda are put to work in this way, clicking on images and transcribing segments of text. The ultimate purpose, Simon learns, is to "train the algorithms," that is, to teach the machines to perform their automated tasks. But up to now, African remote workers have simply been generating data manually. This data will eventually feed an AI. When will the machine learning process be completed? It's difficult to tell. SuggEst app users are always changing, and new users want new offers. The machine needs to evolve at a pace that workers find difficult to keep up with. The platform continues to channel more work to more outsourced workers in Africa. "Ideally," another of the founders confided, "we would produce the AI. But, at this stage, we have so many requests from clients that it is best for us to concentrate on the platform for our remote workers, to make it more efficient and profitable." The interns, too, work to generate data, sometimes. So, just like the others, Simon spends his "off-peak hours" training an AI—or rather playing at being one.

Apart from what Simon calls "misleading advertising" (the company sells an AI solution that isn't one) and from the data collection carried out under nontransparent conditions, there is also the small problem of SuggEst's links to large companies in the digital sector. The company is part of an ecosystem that includes leading businesses in the field. It is a pioneer in artificial intelligence, having received accolades from the press. How much, Simon wonders, do tech giants know about the subcontracting chain that stretches all the way from a startup in Europe to the outskirts of a city in Madagascar? Will they admit that the artificial intelligence contribution of this company is only a handful of interns and freelancing African workers? And are they likely to acknowledge that, if the salaries of these workers are cheaper than the cost of a team of computing experts paid to devise automated solutions, there is no good incentive, economically speaking, for the startup to ever develop the AI?

Forget about losing jobs to automation. Remarkably, the reality is that humans steal the jobs of robots.

Understanding Digital Labor: A User's Guide

This account provides only a taste of what sociologists learn when they venture into the technology sector and explore the rhetoric and reality of automation. From their attempts to identify what AI experts call the "human in the loop," it becomes clear that today's technological fantasy emphasizes scientists in white coats, startup founders in hoodies, venture capitalists in sneakers and fleece vests, and banks of high-tech equipment. But these common representations omit the reality of many other people who work from all sorts of places, especially from home. Even before the COVID-19 pandemic introduced a growing number of workers to online platforms and apps, this distributed workforce was providing the backbone for technology. Indeed, AI often means that behind one white-collar worker, millions of blue-collar ones exist as well.

This book attempts to make sense of what Simon, the pseudonymous intern, experienced and to provide answers to the question that his story leaves hanging: Is this startup simply an isolated case of "AI washing," or do its methods reveal a more general tendency to conceal the labor behind its facade of automation? We shall have to pry off the mask of machine learning to get an answer, guided by other questions: who makes automation happen? By what concrete means? Within what social configurations? And with what political consequences? More generally, what is the underlying bond between human work and this new organization of our technical environment?

The book is divided into three parts: the first ("What Automation?") analyzes the economic and cultural links between artificial intelligence as a scientific project and the techno-economic paradigm of the digital platform; the second ("Three Types of Digital Labor"), provides a series of examples, ranging from Uber to Amazon and from Facebook to Google, that shed light on the variety of forms of work that have emerged out of the economic models geared to smart solutions; the third part ("The Horizons of Digital Labor") provides theoretical tools for addressing the overexploitation and economic asymmetry resulting from the restructuring of labor markets in the digital age. In the conclusion ("What Is to Be Done?"), I suggest some avenues for countering or getting beyond these effects.

The contemporary infatuation with artificial intelligence is our starting point for the analysis of chapter 1 ("So, Will Humans Replace Robots?"). Reading the 2023 edition of the AI Index report that is published annually by Stanford University, the impression one gets is that the California gold rush has nothing on our pursuit of AI. Since the end of the 2010s, AI usage has burgeoned. With the largest number of newly funded startups, the United States leads all regions, followed by China and the United Kingdom. The percentage of companies using AI in at least one department grew from 20 percent in 2017 to more than 50 percent in the following five years.[1] However, the impact of these technologies on the labor market, and the delirium it elicits among investors and consumers, raise thorny issues. How do we view human productive work, especially given the difficulty of distinguishing work from the tasks that constitute it? If we confuse the two, we end up thinking that the automation of certain elements of a job inevitably leads to the complete disappearance of that human occupation. As a result, the "great technological replacement" hypothesis has come to dominate intellectual debate for several decades now. You'll see that I argue this is both a dystopian fantasy for workers, and a wet dream for capitalists.

What is new about the current situation is not the claim that automation has destructive effects on employment. Prophecies about the "end of work" have existed since the dawn of the industrial era. Rather, the problem is first to acknowledge and evaluate the quantity of human work that is necessary to produce automation, so that we can then understand how this technology, in turn, affects labor. For this, some economic and statistical indicators may quickly shake us from our artificial intelligence illusions.

Contemporary anxieties about the disappearance of work are actually a symptom of what is really occurring today: work is going digital. Human productive acts, reduced to underpaid or unpaid micro-operations in this large-scale technological and social transformation, fuel an economy based on information. The latter is reliant essentially on data extraction, a process in which the productive tasks delegated to humans are massively devalued, considered to be too small, too inconspicuous, too much like play, and too menial to count.

Another point to come from these preliminary remarks is that today's rhetoric of automation conceals the proliferation of digital platforms. These platforms combine a particular technological structure with a specific economic organization in an original way. Without a "core business" in the strict sense, platforms function by serving as intermediaries in information transfer between other economic players. Technologists' dreams of intelligent robots are fueled by the profits reaped by these vast digital oligopolies. In chapter 2 ("What's in a Digital Platform?") we will look closer at the technical paradigm of platformization, which today affects both technology companies and other sectors (insofar as the latter are undergoing a "digital transition"). First, we will establish a genealogy of the concept of *platform*, showing its continuity with certain notions in seventeenth-century political theology (a "platform" as a "political program," but also the doctrine of a congregation). Today's digital platforms distort some of the values that were initially attached to the idea, such as the pooling of resources, the abolition of private property, and the refusal of wage labor. When recuperated by a capitalist discourse, these values emerge in techno-economic structures as advocacy for "sharing" goods, for "liberating" work, and for "opening up" information resources.

Because platforms depend on algorithmic methods that require large amounts of data to function, they end up disrupting traditional markets, especially the labor market. They capture the value generated by their producers, suppliers, and consumers. The "users' work" is necessary to produce different types of value. First, there is the value derived from the work of *qualification* necessary for the platform's daily functioning; users add quality to online items by sorting information and by commenting on and rating goods and services, thus directly contributing to the platform themselves. Second is the value derived from *monetization* (taking commissions, or selling the data provided by users to other operators), which increases the platform's short-term liquidity. The third and last value is the *automation* value (redeploying users' data and content to train artificial intelligence systems). Automation ensures the platform's longer-term growth.

Platforms are not specialized in the production of a single good or service. Rather, they bring together quite distinct activities and business models. In the chapters that make up the second part of the book, I will consider three of these models: platforms for "on-demand" services such as transportation app Uber or food delivery service Postmates; for microwork like Amazon Mechanical Turk or Scale; and social media platforms such as Facebook or TikTok. The tasks from which digital platforms manage to extract value vary, as some of these platforms produce personal services, others offer content and manage information, and still others (like dating and friendship apps) market social relations themselves. Each category of platform requires different types of workers, making it possible to classify them according to a variety of criteria, such as working arrangements, geographical range, payment methods, and conflicts around the extraction of value.

Chapter 3 ("On-Demand Digital Labor") discusses platforms, such as Uber, Airbnb, Deliveroo, or TaskRabbit, that connect, in real time, seekers and potential suppliers of a material service in a specific geographical location. The visible nature of these services should not deceive us, since what is involved is primarily an activity of data production. I will focus on the case of Uber drivers and their connected daily lives, which unfold not so much behind the wheel as on their smartphones, as they carry out information management tasks such as clicking, adding to GPS routes, filling out schedules, sending messages, and managing their reputation score. I will then show how passengers also generate data on their trips. We will then be able to examine in detail how Uber's pricing algorithm works.

This case study of Uber will shed light on two points. The first is the gap between the ideal image of a "collaborative" economy and the reality of on-demand digital labor. The ethos of sharing and connecting, which underpins some of these services, would seem to justify the contingent work done by the platform's users. The forms of labor discipline, and the conflicts between service providers and owners of the technological infrastructures, resemble the struggles sparked by industrial manufacturing over the last few hundred years. The second point concerns the use of big data extracted from the activity of Uber's drivers and passengers to create a particular type of intelligent robot—the autonomous vehicle. We will look at how it really functions and show that this celebrated "autonomy" isn't all it seems. Driverless cars move around with a "vehicle operator" on board who can take back control at any moment. Additionally, and contrary to all expectations, these cars shift the responsibility of driving onto the passengers and require remote analysis by human image recognition operators. These individuals are the "annotators," who help the vehicle's AI to interpret the signage, or who correct the routes calculated by its GPS.

But who are the annotators? Not engineers, or cartographers, as the Uber platform calls them. As we will see in chapter 4 ("Microwork"), they are dubbed "human robots," that is to say, workers paid to carry out or support the work of AI. At light years' distance from the fantasies of robot cars and delivery drones that investors and media personalities gush over, myriads of nonspecialized clickworkers perform the tasks necessary for selecting and improving data, and for making it interpretable. We will illustrate this point by studying the case of Amazon's Mechanical Turk, a service for recruiting hundreds of thousands of microtaskers located all over the world to perform the tasks that machines are unable to handle—filtering videos, tagging images, and transcribing documents. These Amazon workers, called "Turkers," are paid barely a few cents of a dollar for each task. Yet this digital labor is essential for producing artificial intelligence applications that are, when it comes down to it, largely handmade.

Today, the microwork market is expanding. As I will discuss in chapter 4, the workforce for these platforms is estimated at between forty million and several hundred million individuals worldwide. Since it is difficult to identify the human and technical components of these activities, this figure is necessarily imprecise. More often than not, microtaskers are invisible to users in the Global North, both because they are not mentioned by the tech giants and because they usually reside in Asian, Latin American, and African countries. By contrast, the client companies of clickworker services are mostly located in Western countries such as United States and Europe, or in industrial giants in Asia, like China, India, and Japan. Thus, the global geography of microwork comes to look very similar to other recognizable historico-political tensions and asymmetries. A "new international division of labor" has emerged that is even more unequal than the divisions denounced by critical thinkers of the second half of the twentieth century.[2] The global chains of labor relocation associated with microtasking show automation in a new light: Human workers are not being replaced by sophisticated and precise artificial intelligence applications, but by other humans, who are hidden from view, underpaid, and facing work instability.

In most cases, microtasking is a very low-wage job, based on piecemeal payments. However, data work is also sometimes performed for free. This can occur when consumers and everyday internet users are placed squarely within the productive process of training algorithms. The best-known example of such training is probably the reCAPTCHA, which delegates to internet users the responsibility of copying out words or identifying images. For some years now, this system has enabled the digitization of books for Google Books, shaped recognition for Google Images, and improved autonomous cars operated by Google's Waymo.

This last example touches on a central point in my argument. The book stresses that digital labor is not just a productive activity, but a relationship of dependence between two categories of actors on digital platforms: the platform owners and the platform users. This relationship, which in the other chapters appears in the visible activity and direct involvement of users, manifests itself in chapter 5 ("Social Media Work") in the form of contributions that are most often provided for free by the users of major social media platforms. Their work produces content (photos, videos, and text) and data (geographical location, preferences, and links) in configurations of networked labor that primarily benefit large advertising placement companies.

Meta, the owner of Facebook and Instagram, and the largest global market of unpaid contributions, is a textbook case in this respect. There are just a few content creators who thrive on social media, yet millions never see a penny from their efforts. The controversies surrounding how users are dispossessed of the value they create have elicited contradictory reactions. On the one hand, the transformation of Facebook, formerly a website for shared interests and sympathies, into a "factory" for content and data is criticized by many. On the other hand, those who believe in users' free and mutually beneficial participation in platforms stress the variety of motivations that volunteers, amateurs, and enthusiasts have in signing up to a platform. However, this "hedonistic" vision (which can be summarized in the idea that "if it's fun, it can't be actual work") ignores the pressure to participate and the gap between the economic interests of the users and those of the owners of a service like Facebook. Above all, this vision fails to recognize that, alongside the digital labor freely carried out by users for no charge in places like the United States and the United Kingdom, there are the huge volumes of data produced by underpaid workers in places like Venezuela and Kenya. This is where "click farms" have been set up to help brands go viral and to churn out content to produce videos and texts that can make website rankings higher, and this is where commercial moderator services that filter out pornographic and violent images are often located.

On social media, the overlapping presence of moderators, click farmers, content creators, and "produsers" (producer–users) struggling to monetize their online presence shows that there is a close connection between the fun these platforms promise and income-generating opportunities. Whether such activities are contested or accepted, they reveal the flip side of the vision of social media platforms as havens for free content consumption. Users are reduced to the rank of potentially unpaid clickworkers who, like their counterparts on microwork platforms, enable intelligent systems to get off the ground. Meta has adopted the same approach as Amazon; far from hiding the fact that its artificial intelligence is "human-powered," Meta makes this into a selling point for some of its automated systems. The chatbots, moderation filters, and audience management solutions provided by Meta are extremely reliant on both paid and unpaid labor. However, it is becoming increasingly apparent that the humans who keep Facebook and Instagram going are not volunteer users, enthusiastic participants, or altruistic amateurs—but a digital proletariat.

In the last part of the book, I will review the theoretical and political questions raised by digital labor. Chapter 6 discusses the important theoretical tradition devoted to "work outside work," to which current reflection on digital labor owes so much. Research carried out from the 1950s onwards—on domestic work, on the value produced by audiences of traditional media, on the work of consumers in supermarkets, and on immaterial work—represents an important attempt to extend the concept of work to activities previously overlooked as opportunities for value generation. But to what extent can digital labor be called work? Is this labor such a radical transformation that it becomes necessary to classify it under a different heading? Some authors have proposed the notions of "playbor" or "weisure," emphasizing the playful and leisurely component of some of the activities that take place on platforms. However, these notions disregard the fact that platform work can be tedious and time-consuming. Although this tediousness can be outweighed by the enjoyment users derive from social media, it disproportionately affects microtaskers in middle- or low-income countries and on-demand workers such as delivery couriers, drivers, and personal service providers. Moreover, the free and voluntary nature of the work of users on social and entertainment platform depends on keeping all the droves of moderators and click farmers well out of sight.

Chapter 7 ("How Do We Classify Digital Labor?") focuses on corroborating, both objectively and historically, the thesis that digital labor is work. First, digital labor fulfills the objective conditions of work, that is, a contract (often under the guise of "terms of service"), obligation, and control. Second, it revives certain nineteenth-century work hire practices, from before the generalization of wage labor, while borrowing other features from the "protected subordination" that characterize modern forms of formal employment. Digital labor involves partly recognizable and conspicuous elements (delivering a meal, publishing a video online, and so forth), and partly inconspicuous work (preparing and processing information and data). This second dimension is irreducible: it is work that cannot be automated precisely because it is required in order to achieve automation. The work of a designer on Etsy, a photographer on Instagram, or even a freelance programmer on Fiverr is thus very far from the ideal of the creative "exalted workers." It has more in common with an activity that is hardly specialized at all and that has no career prospects. It leaves little room for negotiation between the workers and the platforms that demand their digital labor.

Finally, in chapter 8 ("Subjectivity at Work, Globalization, and Automation"), I examine the subjectivity fashioned by these working arrangements. Digital platform workers' lack of bargaining power hinders their awareness of this subjective dimension. Their own perception of their work is ambivalent. On the one hand, the platforms exploit them, while on the other hand they gain new opportunities. Their collective subjectivity likewise oscillates between a vision of empowerment and of exploitation. Whether they see themselves as members of a "virtual class" or as a "digital proletariat," the fate of platform workers is close to that of huge numbers of workers in our globalized markets. For a growing number of people in low-income countries in particular, platform work is an extension of the migratory experience or of forms of economic dependence with which they are already all too familiar. Some authors unequivocally define the system as imperialist, colonialist, or a type of new slavery. These categories should be handled carefully to avoid trivializing these notions and compromising the specificity of their underlying historical experiences. Still, digital labor clearly poses afresh the question of North–South inequalities.

For inhabitants of low-income countries, underpaid activities on platforms tend to be the only access to the "work of the future." But the insecurity and instability of these jobs are affecting a growing number of working people in higher-income countries too, who are constrained to provide an increasing amount of free labor. It is mainly an issue for the younger generations, who are essentialized in the platforms' discourses, reduced to the role of "digital natives" who are naturally predisposed to sharing and participating online and allegedly demand no payment for their effort and time. This way of consigning one section of the global workforce to contingent work, while subjecting the other to value-producing leisure, is part of a larger effort by capitalist platforms to undermine labor in order better to evacuate it both as a conceptual category and as an activity requiring remuneration. So while I start by showing that automation cannot liquidate work, it seems possible, paradoxically, that platformization can totally devalue it. Rather than AI eliminating jobs, big tech companies are seeking to deprive workers of their wages. The question whether this drive will be successful or not does not depend on the overdetermined developments of a technological process, but on the outcome of the struggles that are brewing.

In the conclusion ("What Is to Be Done?"), I will look at several conflicts and campaigns to get platform work recognized. The concrete actions designed to improve the working conditions and rights of platform user–producers are sometimes mediated by organizations such as unions, grassroots networks, "guilds," and regulatory bodies. In addition to the resources of labor regulation (the reclassification of workers as employees, set working hours, and negotiations on fair pay), new rights for user–workers have been won using other legal instruments. These concern primarily the right to privacy, taxation in the digital economy, and business law.

In yet other cases, labor law specialists, digital rights associations, and users have come together to create positive synergies from which new forms of organization may develop. These initiatives converge around two approaches to collective action in the age of digital labor: platform cooperatives, whereby users have property rights in the platform, as a "people's" alternative to the capitalist services; and reflection on the articulation between the paradigm of the platform and that of commons. To address and mitigate some of the limitations of both approaches, I introduce a third perspective, based on a collective redistributive income for digital labor. This approach seeks not to supplant, but to take the wage model further by recognizing and remitting the value originating from the invisible work of data producers. In this sense, it differs fundamentally from, and goes against, Silicon Valley's idea of a universal basic income.

Reindexing Work: A Mode of Action

Digital platforms act as walled gardens of social relations. They impose the task of data and information production in the way they devise mechanisms to maximize participation. By taking work as the key to understanding the nature of the society fashioned by the internet and its related technologies, we aim to display the continuities between the activity of social media users, and workers on nonstandard contracts, in insecure work, or in self-employment, workers who bear the full brunt of the "uberization" of the economy.

By analyzing concrete examples using tools drawn from sociology, political science, management, law, and computer science, we will attempt to grasp the economic and social principles that govern societies shaped by digital platforms in a bid to understand how value is produced and circulates, and how the forms of domination and the imbalances are thereby generated. Then we will attempt to envisage some possible alternative scenarios.

The theoretical approach developed in this book requires a reversal of perspective; it is not the "machines" that do the work of humans, but rather humans who are tasked to provide digital labor for the machines by training, maintaining, and imitating them. Human occupations change—they become normalized and taskified—in order to produce information in a standardized form. Automation thus alters the nature of work, but it does not destroy it.

The book is thus situated at the heart of a debate led by computer science and philosophy, a debate that explores the limits of the epistemic program of artificial intelligence. Today, a number of authors challenge the ideological narrative of complete automation as the "manifest destiny" of our current technological structures (see chapter 8). This phenomenon is sometimes referred to as the delusion of "automated autonomy."[3] This rhetoric has long promised that machines would operate without human intervention. We're assured it'll happen soon. But this endeavor has been rife with bugs, and each new solution makes full automation seem farther away. There's no such thing as an entirely self-acting "intelligent" system; artificial intelligence requires extensive development and maintenance, which involves a few highly skilled specialists and a great deal of unpaid, underpaid, and poorly recognized labor. Theorists have introduced neologisms like "heteromation,"[4] "fauxtomation,"[5] and "Potemkin AI"[6] to denounce the elements of fraud and mystification embedded in the claim that AI is labor-saving technology.

The economic and cultural advocacy of automation ignores the reality of the market for these innovations. From Uber to Google, from Amazon to Microsoft, the business models of the digital giants are not geared toward commercializing AGI (Artificial General Intelligence), but rather promoting devices that require considerable human effort to operate. Despite the grand vision of big tech companies and startups alike, AI reality is constantly scaling back: users are promised autonomous vehicles, and they get assisted driving; they're promised decision-making software, and they get a drop-down menu of options; they're promised a robot doctor, and they get a medical search engine.

Since the mid-twentieth century, the "end of work" debate has had two sides: those who see in technological unemployment the collapse of the idea that work is at the center of personal life and social organization; and those more cautious voices who emphasize that the centrality of work to human experience has remained relatively constant.[7] By reminding us that the capacity of automated devices to replace human workers is in fact limited, and that prophecies in this field have been proven wrong time and again, I am unequivocally in the second camp. However, this book also aims to add another dimension to the debate by emphasizing how work offshoring and concealment are intertwined in today's automation. Our work isn't destined for obsolescence; rather, it is being shifted and hidden, moved out of sight of citizens, analysts, and policymakers, who are all too eager to abide by the platform capitalists' storytelling.

Automation relying on human labor underpins global markets where digital labor is negotiated in exchange for monetary, symbolic, or service-based remuneration. In order to expand and innovate, platforms require the labor of human beings whom they do not manage as workers, but as "users" or as "contributors."

In light of this, we need to address the qualitative aspects of work in the context of the digital transformation of organizations. The authors engaged in this other debate are divided between those who are alarmed at the erosion of the wage system and those who emphasize the opportunities for mobility, flexibility, and, ultimately, the independence of workers in the current context, in which markets would supposedly be more respectful of workers' life choices (see chapter 7). Revealing the deep-seated tensions between work "for others" and work for oneself and for one's community, digital labor on platforms is an activity that exposes users first and foremost to the risks of insecure work and social exclusion. Platforms facilitate the assignment of productive tasks to increase the number of people *at work*. Meanwhile, they effectively place them *out of work*, i.e., outside of standard labor relations.

Workers thus find themselves squeezed between proclamations of independence and material conditions in which they receive low or nonexistent pay, are subjected to other people's rhythms and goals, and do not see the outcome of their productive acts. Since they cannot spontaneously find meaning in what they do, they must set up communities and new forms of organization to make their own labor meaningful. For platform workers to cultivate the meaningfulness of their jobs,[8] they must bend and modify the top-down rules they receive from owners and developers of opaque AI algorithms.

The last theoretical pillar of this book, and the last debate it addresses, concerns digital labor's potential as a catalyst of conflicts and a driver of social change. The emergence of a collective subjectivity from this type of labor is neither immediate nor linear. It results from conflicts over the recognizing of digital activity as work, of data as information units "produced" by users, and of automated systems as sites of negotiation and confrontation for workers. The aim of this book is also to renew (and hopefully bring to a close) a dialogue I began several years ago with Italian workerist theory. Workerism, a subset of autonomous Marxism, understands labor as a collective endeavor that involves all of society's activities ("social factory"). At various stages of my career and in various capacities, I have been fortunate to collaborate with influential Italian theorists, including Paolo Virno, Franco Berardi, and Leopoldina Fortunati. These authors argue that the wealth of contemporary society is owed to two often-overlooked dimensions: collective *production*, including creativity, social interaction and cultural activities; and collective *reproduction*, including care and domestic work. The wages workers receive are therefore only a fraction of the value generated.

By applying this lens to digital labor, we can see online platforms as new iterations of the twentieth-century "diffused factory," and their users as the next generation of "social workers."[9] Today's societies are witnessing the extension of value production beyond the workplace, a trend that the workerists anticipated. Work's encroachment on life started well before digital platforms. However, it has accelerated since then and been accompanied by another phenomenon—the casualization of employment. Importantly, this theoretical perspective doesn't just look at cognitive capitalism critically,[10] it also explores how to establish autonomy in production and help commons grow.[11]

The following chapters have been an opportunity to update and revise certain aspects of this tradition of thought, which is often overly anxious to fulfill the Marxian prophecy of the *general intellect*,[12] and, to this end, has sometimes sacrificed proper attention to the material conditions of work in the age of information technologies. It is with the concept of digital labor that the workerist approach finds a new material anchorage: the finger that clicks, the *digitus*, that performs a crucial function in contemporary technological settings. By touching a screen or a mouse, not only does this finger click—the smallest of all tasks, perfect for training AI—it also restores the etymological meaning of this activity: an occupation that can rightfully be called *digital*.

· PART 1 ·
WHAT AUTOMATION?

· 1 ·
WILL HUMANS REPLACE ROBOTS?

Machines Are Just Humans Who Calculate

In 1936, the English mathematician Alan Turing delivered his paper "On Computable Numbers" at the London Mathematical Society. The consequences of his argument could not have been greater for subsequent research on artificial intelligence. If you are trying to ascertain whether machines can think, perceive, and even desire, he said, apply exactly the same criteria to the machine as you would to a human:[1] "We may compare a man in the process of computing a real number to a machine [. . .]," he wrote.[2] Performing arbitrary computations requires manipulating symbols (1, 2, 3, · · · , +, −, . . .) and rules. At the time, in Turing's view, this held true both for the teams of people hired to compute astronomical data in an observatory, and for the automated device he designed, the machine that would later be named after him.

This approach certainly inspired generations of scientists to believe in the possibility of intelligent machines, and generations of industrialists to capitalize on this idea. But, from the outset, Turing's mechanistic view of the mind also met with scathing criticism from influential philosophers, among them Ludwig Wittgenstein. The author of the *Tractatus Logico-Philosophicus* held a diametrically opposed view, as summarized in his remark that Turing machines are in fact only "human beings, who calculate."[3] By reversing Turing's statement, Wittgenstein aimed to defuse his argument. There is no equivalence between human mind and an abstract and accurate machine. People who compute are only similar to other people who compute.

Today, given current technological advances, most would regard Wittgenstein's refusal to imagine mathematical models of the human mind as outdated. Take for example the victory of IBM's Deep Blue supercomputer against the world chess champion Garry Kasparov in 1997. Or, consider GoogLeNet's ability to diagnose skin cancer with the same degree of accuracy as a doctor's in 2017.[4] Artificial intelligence has had success after success, and the media has cheered on AI at every step. Someone who declared that machines are incapable of "thinking like humans" would not get much of an audience today.

However, perhaps that's an oversimplification of Wittgenstein's views. He was not skeptical of the ability of machines to adequately simulate human cognitive processes. Instead, he was making an argument about the nature of technological innovation. These machines, Wittgenstein pointed out, owe their very existence to the human beings who teach them how to think. Imagining him considering today's advances, he might have said that IBM's supercomputer would not have beaten the world champion if several human grandmasters hadn't supplied it with more than 700,000 games, collected in a huge database of chess openings, from which it learned the secret of their strategies. Similarly, the neural network used for medical diagnoses would never have performed so well without the million-odd images of skin cancer samples amassed in the ISIC dermoscopic archive, all produced, digitized, and enriched by hundreds of thousands of human workers.

We cannot help but look differently at artificial intelligence in the light of Wittgenstein's argument. According to a tenacious misconception, the intelligent machine's cognitive capacities make it possible to dispense with human intervention altogether. Wittgenstein stresses that the machine's autonomy is unproven. On one point, Turing would have agreed with him: artificial intelligence does not presuppose that machines have cognitive skills. A computer may, at best, "display intelligence," but this intelligence is only the effect of the mechanically executed instructions it has been given: take a variable, assign it a value, multiply it by a coefficient, and so forth. These instructions can be defined as the "atomic tasks" that comprise a program or a computational procedure.[5]

The algorithms that govern so many areas of our lives today are simply this: a series of operations performed to obtain a result. Whether we want to find the quickest route by public transport using a GPS ("calculate the starting point," "calculate the point of arrival," "superimpose the subway map on the shortest path between the two points," etc.), or a soulmate on a dating platform ("take profile A," "analyze a finite number of its characteristics," "match them with those of profile B," etc.), the processes involved are identical. Both require a computer to execute a sequence of instructions. At no point do we need to suppose that the algorithm attributes any meaning to them. If the artificial intelligence ushered in by Alan Turing is simply the mechanical execution of a series of tasks, then the philosophical puzzle of "a machine that thinks" is indeed solved.

What is artificial about artificial intelligence is precisely that these tasks require no discernment, and yet, like an emergent property, they produce a semblance of intelligence.[6] As such, regardless of whether we adopt

Turing's or Wittgenstein's position, there is a flip side to the problem. To-day's society no longer requires a machine to be intelligent, that is, capable of *inter-legere*, in the Latin sense of "reading between the lines" (in this case, reading the lines of code constituting commands for the algorithm to execute). Instead, today's AI has refashioned humans as machines that execute instructions mechanically and without challenging them. Inevita-bly, then, the scientific program of artificial intelligence involves *a certain cybernetics*, or art of controlling human beings and disciplining the execu-tion of their activities.

A Tale of Two Digital Labors

Where do we see this cybernetics of human activity in today's economy? The expression "digital labor" has been adopted in several fields but is un-derstood in very different ways. From the 2010s onwards, business consul-tants, innovators and think tank experts have used it to mean the complete automation of work thanks to the combined advances of robotics and data analysis.

A decade earlier, academics, activists, and policy analysts employed the expression for something quite different: for them, digital labor meant technologies that harness human workers. There is thus a political dimen-sion to this concept; "digital labor" stands against the attempt of technol-ogy owners to conceal and demean human work. Just as Deep Blue hid the grandmasters who taught it to play chess, automation perpetuates its myth by concealing digital labor. These two approaches reproduce the original divide between Turing's and Wittgenstein's visions of humans' role in AI systems. Those adhering to Turing's view define digital labor as "fully au-tomated," and ultimately, they worry that AI will replace human labor. Those who question how digital labor uses human work under new techni-cal and managerial conditions are on Wittgenstein's side.

The first group wishes to force or encourage humans to carry out frag-mented tasks so that machines can give the impression of thinking. They reduce the value of human work to a symbolic first step. The second group questions the effects of this devaluing of human labor. Simultaneously, they highlight its crucial role in meeting the growing need to produce data, and to carry out information management tasks.

In my research, I employ "digital labor" to mean the process of turning work into tasks (*taskification*) and into data (*datafication*), at a time when artificial intelligence and machine learning are integral to our economies and lives. As a set of practices, digital labor lies at the intersection between nonstandard forms of employment, freelancing, micropaid piecework, pro-fessionalized amateur activities, monetized leisure, and visible data production. I will explore how these disparate phenomena are connected and the impact of these technologies on human occupations.

The Fear of "The Other Great Replacement"

US Conservatives have consistently accused Big Tech of political bias, and now they worry about AI having a "woke" bias. For these ideologues, technology is synonymous with inclusivity and diversity, which makes it frightening. In other parts of the world, right-wing politicians worry about AI's impact on jobs. Far-right member of the European Parliament Jordan Bardella made headlines in 2023 when he claimed machines would replace humans. According to him, rather than migrants taking jobs, another type of "great replacement" was allegedly imminent.[7] The concept of "great replacement," popularized by French conspiracy theorist Renaud Camus, illustrates the hostility and fear of the most backward parts of our society toward migrants and people of diverse ethnicities. For them, the fear of changing demographics in the workforce and the fear of changing modes of production overlap. Although contemporary discourse is infused with fascist undertones, technological apprehension is not new.

The impact of technical devices on society has been a fear since the very dawn of our civilization. As early as the third century BCE, the Latin poet Ennius succinctly expressed this huge existential anxiety: "The machine is an immense threat [*machina multa minax*]."[8] The hexameter continues: "[The machine] poses the greatest of dangers to the city [*minitur maxima muris*]." It matters little that, for this poet, the technology in question was a weapon of war, the Trojan horse, and that the threatened polity was the actual city of Troy.

Archaic anxieties return, when machines threaten to disrupt the way we work. The fantasy of the destruction of human labor, of the "great replacement of humans by machines," is centuries old. The classical thinkers of industrialism analyzed this phenomenon. Among them was the English essayist Thomas Mortimer, who, in his *Lectures on the Elements of Commerce*, published in 1801, described two categories of machines: one "to shorten, or facilitate the labor of mankind," and another whose object is "almost totally to exclude the labor of the human race."[9] Despite his optimistic "almost totally," the author argues that all social good and judicious public policy should oppose the second type of technology.

David Ricardo, in the third edition of his *Principles of Political Economy and Taxation*, published in 1821, devotes chapter 31 ("On Machinery") to this issue. He insists that the use of mechanical solutions is purely instrumental, and not an inevitable fate. It results from the "temptation to employ machinery" that inhabits the capitalists seeking to achieve productivity gains by reducing labor costs. Automation is only one of the options available to them: they can equally replace workers by low-cost labor (obtained through "foreign trade"), or even harness the strength of animals.[10] Automation is only one option, among many, for the factory owners.

Andrew Ure, his contemporary, took this reasoning to extremes in his *Philosophy of Manufacturers* (1835), adding to the list of possible solutions "the substitution of women and children for men's work."[11] Although, according to Ure, machines aim to "supersede human labor altogether," the ultimate goal of those who use them is not the destruction of labor, but the reduction of its costs. Even when it isn't used, the mere threat of automation can lower costs; its invocation pressures workers, disciplines them, threatens their jobs and lowers wages. Every worker lives under the impression of being, potentially, superfluous.

Hence, for capitalists, automation is not just a technical solution to a technical issue. It is also a method of squashing labor conflicts. "The most perfect factory," as Ure says, can "do without the work of human hands." But mentioning this potential expendability is merely a way of controlling the work of those very human hands.

Children, foreign labor, and even animals are all equivalents, almost synonyms, of the machine. Given this ontological confusion, technology can only be defined by negation: automation is everything that "the work of human hands" is not. While confirming this definition, my goal in this book is to show that the current transformation doesn't represent the replacement of humans by machines; rather, the change is from the work of human *hands* to the work of human *fingers*—to *digital* labor.

Marooned from the Information Society

While in the nineteenth century the authors of classical economics predicted workers *could* be replaced by machines, by the late twentieth century this belief had become a radical prophecy of the end of work, shared by both doomsayers and technologists. Sociologist Daniel Bell's alarming prediction of the decline of manual jobs in post-industrial societies,[12] and the warm reception given by some government advisers who predicted a decrease in clerical work due to the advent of "telematics"[13] set the stage for futurist Jeremy Rifkin to argue that information and communication technologies would result in a jobless world, uprooting our work-based societies.

Such deterministic views have been criticized both empirically and theoretically. Philosophers have argued that the value of work remains a core tenet of our culture.[14] Economists have maintained that computing technologies will not necessarily result in an overall reduction of employment.[15] There are several factors that can influence how automation turns out. Task interdependence is one of them. The overall demand for labor will increase if a task, having been automated, depends on another that cannot be. Demand elasticity, the tendency to fluctuate in response to changes in prices, also plays a role. If demand is elastic, introducing automation decreases prices and increases production, and labor may grow as a result. In the early twentieth century, technology reduced the price of air travel, resulting in more people working in that sector. Twenty-first-century machine learning may very well lower the price of automated solutions, resulting in more spending and more work in this sector.[16]

In light of these contributions, and in order to measure the impact of information and communication technologies on contemporary developments in work, it is essential to examine the changing and asymmetrical relationships between workers and managers.

Manuel Castells's trilogy on the networked society in the information age, published at the end of the 1990s, offers perhaps the most incisive comments on automation. With the rise of economic growth models founded on information, jobs have become more flexible and fragmented. Consequently, labor markets are polarized between informed decision-makers who can "manipulate symbols," and "a disposable labor force that can be automated and/or hired/fired/offshored, depending upon market demand and labor costs."[17] Rather than ending work, automation fractures, segments, and ultimately dissolves it as a cohesive social force.[18]

The idea of a workforce split between hyperspecialized and irreplaceable jobs, and "weak" jobs undertaken by those that history has left behind has been gaining traction among economists.[19] It is true that this scenario has changed significantly over the past few years. COVID-19 has fundamentally changed companies' views of what constitutes an essential and indispensable job, emphasizing menial or manual work.[20] GPT-powered software is also accompanied by predictions that 80 percent of jobs could get "exposed" to changes, especially for people with bachelor's, master's, and professional degrees.[21] Historically, low-skilled or "lousy jobs" were characterized as repetitive activities that followed simple rules. Now that big data and artificial intelligence are supposedly capable of reproducing complex cognitive processes, creative professions with a strong intellectual and relational component are, according to this dystopian prophecy, at risk of being replaced by technology.

The "Oxford Paper" by economist Carl Benedikt Frey and computer scientist Michael Osborne, which has been at the center of intense international controversy since its first publication in 2013, is typical of this approach. The fifty-page report (not including appendices), focusing on the United States, sought to measure the number of jobs, both manual and cognitive, destroyed by machines.[22] Manufacturing, transport, trade, services, agribusiness, health, and other sectors were covered. The authors estimated the likelihood of robots or software replacing humans according to the degree of repetitiveness of the tasks involved and the level of automation already in place. The results were unambiguous: 47 percent of jobs, they claimed, were highly likely to disappear due to the wave of technological innovation based on machine learning and mobile robotics.

Other scholars have used Frey and Osborne's data, replicating, updating, or transposing their results to other contexts. These new estimates severely undermine the Oxford researchers' conclusions. The methods that they used to assess the supposed disappearance of work have also attracted much criticism, due to their limitations and biases, both conceptual and statistical. Using a reductionist "task-based approach," they argued that, if a certain percentage of job-tasks becomes automated, whole occupations will be eliminated. Moreover, the authors only examine a subsample of 10 percent of existing jobs. They do not seem to consider the possibility that the effects of displacement may be mitigated by the creation of new occupations, ones that do not yet exist or whose content has been reconfigured by technological innovation. This opens the door to the easy criticism that a growing demand for highly qualified roles such as machine learning specialists, automation experts, and human-machine interaction designers, as well as the constant new training of workers, could counter job displacement.

Following the "Oxford Paper," several *Future of Jobs* reports put out by the World Economic Forum have pointed out that technology adoption doesn't necessarily result in job losses. Employment growth and decline are driven by macroeconomic trends, as well as exogenous shocks like health crises, geopolitical instability, and environmental changes.[23] According to the business leaders gathered in Davos, several dozen million jobs will be lost globally, but they will be compensated for by the creation of 133 million new jobs.[24]

A more damaging criticism is how the Oxford researchers conceive of automation. Adopting automated solutions appears to be "frictionless"—with an easy low-cost switch from humans to machines. This is when a significant gap between the early theorists of innovation in the workplace and their late epigones becomes apparent. Ricardo, Ure, and their disciples never overlooked the inevitable labor negotiations resulting from automation. Osborne, Frey, and others didn't share this caution.

Robots or Workers?

AI experts applaud the prospect of a world without jobs. In 2020, 500 of them agreed that there's a 50 percent probability that automation will replace 90 percent of human labor within a fifteen-year timeframe.[25] In fact, the survey participants were recruited at conferences on AGI, a field of research that aims to develop autonomous systems that can perform human-level tasks. The accuracy and objectivity of this forecast may thus be skewed by respondents with an interest in promoting AI's capabilities.

In addition, expert predictions about AI have a bad track record.[26] The issue is labor productivity, which is defined as output per hour of work. It should increase as a result of technological advances in production. In reality, automation and AI are more like a version of Solow's paradox. The famous economist noted that "one can see the computer age everywhere but in the productivity statistics." Today's version of his observation would be that one can see automation affecting work everywhere—except in labor productivity statistics.

Already in 2017, the US Bureau of Labor Statistics warned that, in contrast to the previous decade, automation had increased only very slowly.[27] Productivity gains that measure the impact on workers of the introduction of automated processes have averaged less than 1 percent in the nonagricultural sector and in manufacturing.[28] By the end of the decade, concern had turned to pessimism. US labor productivity has undergone a noticeable slowdown, and many economic observers are wondering what caused it.[29] This stagnation is not restricted to the American continent; some northern countries have also had very slow or negative growth in productivity. Over the past two decades, productivity growth has declined globally, and for some G7 countries it's even been negative.[30] "This is equivalent," Dean Baker, the Director of the Center for Economic and Policy Research, quipped "to workers replacing robots: a situation where it takes more workers to produce the same amount of output."[31]

The facts contradict the proponents of the theory of the "great replacement" through automation. This is particularly visible in the robotics sector. A survey of seventeen countries between 1993 and 2007 found no significant effects of multifunctional industrial robots on overall employment in terms of total hours worked.[32] Studies on industry sectors financed by robotics companies tried to reassure a worried public. The International Federation of Robotics unequivocally emphasizes the positive impact of industrial robots on employment. Even if the labor force in one sector is shrinking, this is balanced by gains in other job types, thus leading to more jobs created worldwide thanks to these technologies.[33]

Still not convinced that automation doesn't end human work? Let's compare levels of automation and the unemployment rates of different G20 countries. The countries with the highest automation display lower unemployment rates. South Korea has 1,000 robots per 10,000 employees, and only 2.8 percent of its workforce is unemployed. Japan's robot density is comparable to Germany's (399 and 397 out of 10,000 workers, respectively) and their unemployment rates are both 2.9 percent. China and the US lag far behind when it comes to robot density and have higher unemployment rates (China's reached 5.2 percent in 2023).[34]

Whereas the robotics sector may provide a plausible approximation of automation levels in our economies, estimations of the ratio of employees to robots, correlated with the unemployment rate, give only rough indicators. Yet, robotization is not simply a matter of hardware and factories, nor is work only formal employment. Therefore, a simple correlation between these two variables cannot fully explain the phenomenon.

Popular imagination still conjures images of robots as humanlike automatons. However, in the broader context of today's technological innovation, "robot" (especially when the term is abbreviated to "bot") refers to software that interacts with humans. Although manufacturing robots are designed primarily to replace workers, they are also increasingly incorporating data-intensive features like algorithms, sensors, and voice assistants. Core academic literature on automation indicates that these solutions, which involve data, are intended to augment rather than replace human workers.[35] The fact that this substitution is not taking place thus has nothing surprising about it. Today's robots have nothing to do with the nineteenth-century automatons that anthropologist André Leroi-Gourhan described as "machines without a nervous system of their own."[36] Today, devices that sort information, compute routes, reply to chats, make purchases, etc., are robots too. It is neither their strength nor their material endurance (their hardware aspect) but their ability to execute complex informational processes (their software aspect) that define them in an industrial context.

So, despite the current wave of automation, work still seems to hold its own. This is the case both culturally, considering how central it remains to the social fabric of society, and substantially, because of its dominant role in the life paths of individuals.

This persistence reveals a new perspective. Automating parts of the tasks that constitute an occupation won't, as Frey and Osborne predicted, eliminate whole jobs. Even those jobs at higher risk of automation usually contain a large number of tasks and roles that are difficult to digitize. A twenty-one-country comparative study from the Organisation for Economic Co-operation and Development (OECD), carried out a few years after the Oxford Paper, shows that the automation potential of today's occupations has been greatly overestimated. While automation may significantly transform 50 percent of tasks, the introduction of artificial intelligence is likely to destroy only 9 percent of jobs.[37]

The Oxford Paper's ominous portents are not representative of reality. David Autor, a professor of economics at MIT, explains why human labor's obsolescence is so difficult to prove. Over the past two centuries, the job-to-population ratio has kept growing, and unemployment hasn't grown in a visible and stable way. Yet with each technological innovation the same anxieties return, and with them the same argument that "this time it is different." Today's version of this claim is that, due to the so-called disruptive digital technologies, the social order founded on work has broken away from its past. However, Autor argues, this perspective ignores the obvious complementarities between human work and automated processes. Although the dialectic between automation and work is fraught with tensions, it has led to a rise in the demand for human labor.[38]

One example of this complementarity can be found in banking, where the introduction of automatic teller machines (ATMs) from 1980 to 2010 led to the reskilling of certain categories of staff and not to the destruction of their jobs. In the United States alone, ATMs increased from 100,000 to 400,000 between the end of the twentieth century and 2010. Yet the number of clerks behind the counter remained stable, and even increased slightly from 500,000 to 550,000. The ubiquitous presence of ATMs did not put cashiers out of work because the sector's economic expansion stabilized the job numbers. The presence of ATMs enabled branches to function with fewer employees. However, this gave rise to an increase of 43 percent in the number of branches since 1990.[39]

In terms of quantitative demand for human labor, the banking sector saw little change. Qualitatively, however, the effects of automation were immense. It changed the content and even the nature of the work. The tasks of counting and handing out money, now automated, disappeared from the cashier's job description, while other duties became key, such as customer relations, assistance, and advice focused on selling investments, shares, and financial products.

This is why David Autor calls for a whole change of perspective, and for abandoning the idea that jobs are threatened with extinction, simply waiting for the death blow to come from the tiniest technological change. The focus should be on the activities that workers perform daily, which may be automated in parts, but not completely nor at the same time. Work itself does not disappear.

Replacement or Transfer?

The impact of automation on work cannot be simply described as the replacement of human workers with artificial entities like bots or intelligent systems. Instead, it involves the digitization of human tasks, which fundamentally transforms the nature of work by pushing two long-term trends to their extremes: the standardization and the outsourcing of tasks. By breaking down productive actions into standardized sequences, these tasks become compatible with digital processes. This phenomenon was observed during the Taylorist mechanization in the twentieth century, and it continues with the new Taylorism of digital platforms and smart technologies. What sets information technologies apart from their industrial predecessors is their spatial dimension. Today, production and automation can take place anywhere, extending beyond the confines of traditional company boundaries. In a sense, it occurs in multiple "elsewheres" by breaking down tasks into numerous uniform components.

To illustrate this point, let's return to the example of ATMs. The digital standardization of certain tasks does not necessarily impact employees directly. Instead, it is the users, consumers, and clients who interact with these machines. They are the ones performing transactions and handling bills, not the clerks. The same principle applies to other self-service technologies, such as airport check-in kiosks or self-service checkouts in supermarkets.[40]

To clarify, let me say this: automation technologies do not *replace* work, but rather *transfer* it, delegating a growing number of productive tasks to nonworkers (or workers who are not recognized and are not paid as such). The notion emerging here is that of "consumer work," a concept which will be pivotal in the rest of this book.[41] A corporate entity's relationship with its user is a work relationship, mediated by digital technologies with a view to producing a good or a service. This is what some authors, such as Ursula Huws, call the "unpaid labor of 'consumption work.'"[42]

By focusing on the work mediated by digital platforms, we can get away from an employment-centered approach and acknowledge the great variety of workers beyond the workplace. This connects with the unrecognized, but no less necessary, contribution made by human collectives whose access to the labor market was already difficult in the world inherited from the first Industrial Revolution—minorities, marginalized groups, and women. Consumer labor finds an echo in domestic work, for example. Both are ways for firms to exploit to the full the economic dependence of the human ecosystems that revolve around—and not within—the proverbial workplace.

Of course, "consumer labor" and "domestic labor" do not exhaust the notions that can be leveraged to describe the gamut of human productive activities in the digital age. Informal labor has a significant presence in this work. Moreover, many value-creating processes are also hidden in back offices, or miniaturized and reduced to microwork. In yet other contexts, they are made invisible, by having tasks relocated, and performed by contingent workers on the other side of the globe. And, often enough, the very nature of the activity is often denied, as it is treated not as work but as a game, a form of free participation, a means of self-realization, and so forth.

Automation or Digitization?

Digital labor is not "dead labor." In Marxist terminology, this expression refers to industrial machinery that crystallizes human labor.

Moreover, digital labor does not spell "the death of labor." Neoliberal orthodoxy associates this formula with the prophecy of complete automation of human activities.

Rather, digital labor involves outsourced and standardized productive tasks, and a reorganization of the relation between the inside and the outside of the firm, such that the proportion of value produced on the inside decreases while it increases on the outside.

However, the focus on automation should not blind us to the real issue at stake, that of digitized—or, for short, digital—labor. As discussed, I purposefully use "digital" to encompass multiple meanings, including its association with "digit," the figure, and "digitus," the finger in Latin. "Digital labor" underscores the physical activity, the movement of this finger used to count, point, click, or press the button, in opposition to the immobility of the abstract figure. By using the term digital labor, we counter the idea that digital technology is exclusively the domain of experts and specialists "of the figures." Last, it obliges us not to overlook the people who carry out the humble, ordinary, and basic tasks that underpin today's value production chains.

Additionally, is there a reason for emphasizing "labor" over "work"? The specificity of "labor" can be appreciated by looking at the history of the words employed to describe human activity. While the term "work," which finds its root in the Proto-Indo-European word *wergom* and in the Greek *ergon*, defines the relationship with the physical world, the term "labor" emphasizes the social dimension of human occupations. Due to this, its meaning is similar to terms that express subordination or dependence in other languages, such as the German term *Arbeit* and the Russian *rabota* (meaning both "slave" and "robot"). When, the individual's activity is no longer only a relation to the world of nature and need, it reveals itself as essentially a *social* relation: this is "labor."[43]

The intersubjective dimension of human occupations is a big part of what it means to live in a society. This is all the more true when the modifier "digital" is added to emphasize the technological aspects of these activities.

Sociologist Alexandra Bidet and legal scholar Jérôme Porta, for example, point out that digital labor puts the very idea of work under strain. Activities that are inseparable from the digital, such as administering one's Facebook account, creating a playlist, liking content, etc., are activities that produce wealth, but not income. The user–consumer is part of the value chain and is subjected to certain constraints [. . .]. Beyond the question of income and the equitable distribution of the value generated, the way activities carried out in the context of a wage relation are transferred (for example, when the consumer who purchases a ticket then prints it out) reveals how fragile are the conventions by which we qualify an activity as work.[44]

The concept of "digital labor" helps us to understand the technical side of human activity. We don't just think of technology as something external to society, but as something that encompasses all the material processes by which individuals operate and live in the public sphere. Insofar as work is part of a triangle whose other sides are technical mediations and social structures, it is hardly surprising that digital sociability and digital labor are interrelated. Technology has penetrated our private lives over the last few decades to such an extent that it can hardly be detached from the individual. Work mirrors these features, becoming less visible, and less linked to the expression of mechanical force.

Contemporary studies treat this transformation as though technology came only from the outside, destroying the equilibrium of life in society. The concept of digital labor, however, negates the opposition between inside and outside, since the technical environment in which work is performed must always be considered. There can be no work without a tool, as we have learned from prominent philosophers like Gilbert Simondon and anthropologists like André Leroi-Gourhan.[45]

Work needs an economic context too. So our analysis cannot stop at the invisible work of today's consumer. We must also include the phenomena of informal and occasional work, zero-hour contracts, and even those traditional forms of subcontracting that have hugely increased, and whose logic has prospered, under the influence of intelligent machines and algorithms. The objective of this book is to map both waged and unwaged digital labor to detect other trends accelerated by it, like job insecurity and poor working conditions.

Digital labor cannot be simply defined as "free work," as it delineates a continuum between unpaid, underpaid, and flexible activities. Since it is not merely about consumption either, digital labor does not represent a perspective that is entirely alien to work. It's actually a way to acknowledge the growing dependence of modern productive structures on technology that bridges work and life.

Over the past few decades, a community of digital labor scholars have made these points. My contribution takes a more general approach and shows how these tendencies are not only the result of technological processes, but also deeply rooted in the changing nature of work today. I emphasize change—diversification and fragmentation—over disappearance.

Jobs or Tasks?

To explore the boundaries between work and nonwork, we must also change our focus from jobs to tasks, as mentioned above. This does not mean simply that economists and market regulators would do well to change the conceptual categories of their analyses. It means, much more concretely, that modes of production are progressively shifting toward task-based occupations. Anthropologist and media scholar Mary Gray explained in 2016 that the shift to tasks was the cornerstone of digital labor. Fragmentation, outsourcing, and insecure jobs are inseparable: "Corporations, from the smallest startups to the largest firms, can now 'taskify' everything from scheduling meetings and debugging websites. [. . .] Forget the rise of robots and the distant threat of automation. The immediate issue is the uberizing of human labor, the fragmenting of jobs into outsourced tasks, and the dismantling of wages into micropayments."[46]

Continuing to focus on jobs while ignoring this task-centered approach to productivity is problematic for two reasons.

The first is the difficulty of thinking about "unworked" hours. How do we define working time when we deal with worker–consumers, online amateurs, people volunteering content and data, on-demand paid and unpaid microworkers?

The prism of contract-based formal employment located in an office or a factory is ill-suited to comprehending the work of individuals and human collectives that are theoretically autonomous, but who are in fact linked to productive infrastructures.

The second problem is that most analyses based on jobs have an overly narrow national framework, which compromises their impact, especially since value production today depends predominantly on global networks. The use of outsourcing to cut costs or optimize a company's manufacturing and real estate portfolio is no longer limited to multinational companies. From the biggest to the smallest firm, globalized supply chains link together suppliers and purchasers in cascading networks.

Continuing to ignore the rise of taskification leads us to relegate what are actually human productive acts to the realm of nonwork. It is because work is fragmented and defies the categories typically leveraged to analyze it that contemporary commentators are so quick to declare that work is no more: they no longer recognize it when they see it. Sure, analyzing non-standard and unstable work is a complex endeavor, much like establishing links between global influencers' monetization strategies and people in low-income countries who toil for pennies on their smartphones. But ignoring these connections can have serious consequences. Not only do we fail to discern developments in work, but we also imagine that the enormous quantities of digital labor outsourced to human communities, locally and globally, is performed "by machines." In this respect, automation is above all a spectacle, a strategy for diverting our attention from managerial decisions aimed at reducing the relative share of wages (and more generally of payment for human factors of production) compared to rewards for shareholders and investors.

The Hidden Strings of Automation

One way to understand how intelligent automation in recent years has resulted in outsourcing and taskification is to look at consumer services featuring AI solutions. From the simplest TikTok image filter to the most intricate recommendation algorithm for streaming services, these products use machine learning methods to reach a mass audience through applications, devices, and web interfaces. In contrast to business-oriented solutions, they do not try to champion artificial intelligence as a "super brain" for human-level decision-making. Instead, they promise a weak artificial intelligence, which is used in applications that help consumers manage information (like sorting songs on an online music service), that improve content (such as autocorrecting text on a smartphone), or that suggest routes (such as using a GPS for directions). Other examples include Tesla's Full Self-Driving system, which claims to automate driving on predetermined roads, slowing down if the safety distance between vehicles is not respected, and so forth. Or the voice recognition software of Apple's Siri and Amazon's Alexa, both touted as solutions to help kids with homework or parents to order groceries. These conversational technologies are generically known as "voice assistants," and they are found in smartphones and home robotics.

Human input is a major component of all artificial intelligence tools, regardless of their sophistication. Machines do not replace human beings—they assist them. Even if they help decision-making in areas as disparate as care, leisure or administrative tasks, they cannot undertake an action without involving the user, beforehand or afterward. An intelligent system must be configured and calibrated before it can function and often, after producing a result, it will be the user who verifies the quality of the results. Consider ChatGPT, and how users need to "prime" it before using it optimally, and then evaluate its performance. Priming involves supplying relevant context (for example: "you are a sci-fi writer and your assignment is to draft the plot of a space saga. Here is a 500-word sample that displays the style of the text you must emulate . . .") and instructions to shape the chatbot's responses, tailoring them to the user's needs. Once ChatGPT provides output, users can give feedback using the thumbs-down button in the interface. The AI uses this information as part of its learning process.

Chatbots can generate, locate, classify, and present information, but humans make the final choice as to whether that information conforms to their expectations. The same goes for voice assistants, that receive help to humans, as well as providing it. Some companies make no secret of this, and even integrate this human element into their sales pitch. For a number of years, the US firm Nuance Communications mentioned "human-assisted virtual agents" (HAVA) in its communication.[47] In the context of after-sales support, for instance, human workers and IT devices work in tandem to deal with customer inquiries. Between 2015 and 2018, Facebook's marketing strategy for its short-lived "M" chatbot was, in mimicking Nuance's HAVA, to tout the chatbot's "human-powered qualities." This experimental service sent unsolicited but personalized suggestions to subscribers, based on users' conversations and interactions. The service offered to book a cab, make an appointment, or even answer complex questions—through a combination of AI models and human support.[48]

This human involvement is required for both technical and commercial reasons. Artificial intelligence is largely based on automated learning that relies on human "supervision" and "reinforcement." That is, machines learn to interpret information and to perform actions by engaging with their users. The term "AI training" is often used in tech jargon to describe an adjustment period for software programs. In fact, the machine never stops learning; reinforcement is a process that involves extending training in order to help prevent economic and human catastrophes.[49] Sometimes, after machines have been trained, humans still contribute to perfecting and improving them. They do so by correcting errors and biases, thus creating a "reward system" for the AI model.

Of course, an artificial intelligence solution that tries to guess your musical tastes will be appraised less harshly than an artificial intelligence device used in legal decisions, medical diagnoses, million-dollar investments, or the delivery of raw materials to remote corners of the planet. The need for human input does not lessen as more and more areas of society come to rely on intelligent systems. On the contrary, the complementarities between humans and intelligent systems become even more essential.

The rhetoric of all-powerful intelligent systems is at odds with what tech companies actually produce and market. While transhumanist theorist and

Google Chief Engineer Ray Kurzweil waxed poetic on how "strong" artificial intelligence could outperform biological systems,[50] his employer focused on mass-producing "weak" or "narrow" AI. The fantasy of AGI is still present, but it conflicts with the only artificial intelligence that is currently possible, and it is limited and inefficient in the absence of human intervention.

So, again, society's focus should no longer be on the expert computer scientist who designs "general" AI systems or the computing engineers who set them up, but on the hundreds of millions of hired hands who, on a daily basis, pull the strings of puppet automatons. By performing this menial and imperceptible work, we humans become the trainers, stooges, and the maintenance staff of these "weakly intelligent" machines. Given the complexity, the range, and the diversity of the digital tasks required to train AI systems, digital labor can no longer be ignored as an object of study. Moreover, suggesting that artificial intelligence has not yet reached full automation increases the probability that it never will.

Artificial intelligence devices can be enabled, trained, and sometimes even impersonated by humans. In later chapters, you will meet the people working behind the scenes to help artificial intelligences. Where are they located? What is their employment history? Under what conditions do they work? How are they paid? Where are they recruited?

All these questions establish a connection between automation and digital labor, because they assume there are markets where that relationship can be negotiated. In fact, only occasionally does money play a role in this negotiation. The tasks that allow artificial intelligence to exist and function are posted and auctioned on dedicated subcontracting (or microsubcontracting) websites. Other times, there is no remuneration involved, as, for example, when humans participate in artificial intelligence in exchange for other economic goods (such as vouchers, an exchange of services) or for altogether noneconomic incentives, such as pleasure, recognition, and play.

Fully Automated "Little Hunchbacks"

If robots are operated by humans, if artificial intelligence devices are not so artificial after all, if machines are always driven by living labor, then the ontological status of these entities may arouse our radical suspicion. Might the automation that investors desire and that technophobes fear be essentially human labor made invisible? Research into digital labor shows the key role of human agents in software analysis, the individuals who produce and sort the data collected by and on platforms, and the massive presence of hidden "helpers" who work in tandem with the IT devices.

The figure that best illustrates how humans are placed at the center of technological devices is the Mechanical Turk, my focus in chapter 4. Half a century before Amazon named one of its best-known marketplaces for digital labor after it, Walter Benjamin described this invention in the first of his *Theses on the Philosophy of History*: "The story is told of an automaton constructed in such a way that it could play a winning game of chess, answering each move of an opponent with a countermove. A puppet in Turkish attire and with a hookah in its mouth sat before a chessboard placed on a large table. A system of mirrors created the illusion that this table was transparent from all sides. Actually, a little hunchback who was an expert chess player sat inside and guided the puppet's hand by means of strings."[51]

The rest of this text is well known: the Mechanical Turk is in fact a philosophical metaphor. Benjamin compares this fascinating but fraudulent device to historical materialism, a doctrine that always seems to win against any opponent, but which hides within it the "wizened dwarf" of theology—to him, a crude and unattractive form of metaphysics. To explain the immanent conditions of human societies, Benjamin seems to say, sooner or later one has to appeal to the transcendental.

But in the context of automation in the digital age, it is possible to turn the metaphor on its head: It is materialism, intended here as the attention given to the material conditions of existence of the producers of value, that is impaired, reduced to the role of a being too small to be seen or heard, locked away by an abstract belief in a truly artificial intelligence, in the theology of machine learning.

For a myriad of *bucklige Zwerge*, "humpbacked dwarfs" are hiding behind the ubiquitous bots, the infallible algorithms, and the all-powerful "neural" networks. The theoretical sleight of hand that consists in focusing on robotization, algorithms, and on "smart" solutions, perpetuates this technological imposture. Appropriating Benjamin's ableist analogy to better disarm it, we can use it to highlight how often AI models are operated by human labor.

When one looks at the digital through the lens of materialism, one is bound to see how work changes in its technical materiality—that it is underpinned by a computational infrastructure—in relation to economic incentives and adjustments in production that concentrate value in fewer hands. The fantasy of automation distracts public opinion from more direct threats, like corporate entities monopolizing assets and rare resources to protect their baselines.

Whether they present themselves as harbingers of the new disruptive economy as opposed to the comforts of old industrial capitalism, or as well-meaning multinationals, today's major digital platforms are involved in constituting an oligopolistic market. These productive oligopolies come into being when companies from different sectors converge. Mergers and acquisitions of independent corporate entities lead to economic concentration. In digital markets, rivalries alternate with cartel-like agreements, creating a situation of "coopetition."[52]

Today, internet multinationals are synonymous with economic hegemony on a global scale in the form of platforms. They are also emblematic of a style of workforce management characterized by the move to outsource, and to divide jobs up into tasks. To fully exploit the possibilities offered by globalization and production standardization, oligopolistic actors have set up long chains of subcontractors for both logistic and cognitive tasks. Outsourcing certain business processes related to how a social media application like YouTube is run and moderated is not essentially different from Apple's decision to offshore the manufacture of its electronic components to China. Likewise, secret agreements between big tech companies to cap salaries, or synchronized layoffs of white-collar employees across several companies,[53] are not just symptoms of collusion. Importantly, they are a mirror of the insecure and degraded employment status of the underpaid blue-collar workers or "volunteers" of their platforms.

Putting Off Automation

These are the strategies that digital oligopolies employ to capture the value produced by their users and by a multitude of nonmarket stakeholders. The inversion between formally recognized work and unpaid work done outside of the workplace is key to platform profitability. Digital platforms providing not jobs but tasks for workers, who get labeled as subcontractors and freelancers, or even consumer–producers, amateurs, enthusiasts, and simple users. Indeed, for platforms to thrive, the classic employer–employee relationship must be actively destroyed. This is, most importantly, how platforms come to embody a new paradigm of value creation.

Two principles are in play here. The first is that a platform is not reducible to a company. It is, above all, a coordination mechanism between certain actors (suppliers and customers, creators and audience, delivery workers and restaurants, etc.). Its model transcends commodity markets or the allocation of resources by a central authority. The second principle follows from the first: When platforms organize the supply and demand of workers, they propose all sorts of economic and symbolic incentives, such as salaries, fees, rewards, piecework payments, badges, points, stats, etc.

By doing so, they refashion certain institutions inherited from the modern industrial age, not just employment, subordination, and social security. These digital oligopolies are managing the labor force and the outside workers in such a way that automation—replacing human labor with software sequences—simply becomes a tool for disciplining work. It is for this reason that full automation is never completed, always delayed.

Humans and machines are highly interdependent, so the degree to which a process is automated isn't determined by the number of robots replacing people, but by the number of digital intermediaries or platforms that taskify and outsource work to people at each step. A maximum number of machines interact continuously with a maximum number of humans, who in turn communicate with each other. As sociologist François Vatin says, there is "some naivety in thinking that employers have as their main objective a wholly mechanized workforce. [. . .] Although always pursued, and always almost attained, this 'production without humans' nevertheless remains a fantasy, because humans always reappear somewhere."[54] Why is this accomplishment always deferred? In order to continue brandishing the threat of the "great technological replacement" before the workers. Automation is the stick used to discipline the labor force. And, ultimately, the carrot to attract investors.

From this point of view, debates have made little progress since 1970, when autonomist philosopher Franco "Bifo" Berardi argued that the major theoretical and empirical stumbling blocks in labor studies result from viewing technology as a way to increase productivity instead of as a symptom of political subjugation of labor. "Reducing the socially necessary labor," Berardi insisted, "intensifying productivity, automation [. . .]. Establishing control involves all of these factors."[55]

If these analyses still seem relevant now, it is because today's platform capitalism makes abundant use of the same old ruse used by factory owners in the past. Technology is stripped of its social dimension so that it appears to be a necessary step on an endless journey toward a vaguely defined progress. This is done to hide tensions and resistances that workers bring to production through their demands and aspirations. The technological discourse that accompanies the emergence of AI can thus be read as a method for preventing workers from organizing, and for reducing their bargaining power. Here robots are simply the convenient embodiment of the intention of platform owners to suppress opposition.

· 2 ·
WHAT'S IN A DIGITAL PLATFORM?

A Market–Company Hybrid

Although digital labor is on the rise, the terms associated with it obscure its nature. Fashionable expressions such as "disruptive technology"[1] or "uber-ization"[2] compete with "digital transition," "AI revolution," or "transformation."[3] Germany and the countries of Southern Europe are particularly fond of "Industry 4.0."[4] Those in urban planning and banking circles prefer to speak of "smart technologies" and "digital disruption."[5]

In this lexical profusion, the notion of "platformization" has an advantage. Nick Srnicek gives the following definition in his book *Platform Capitalism*: A platform is a digital infrastructure that connects at least two groups of individuals.[6] Srnicek differentiates advertiser platforms (such as Google Ads), which trade in the information provided by users; industrial platforms (such as Siemens Digital Industries Software), which specialize in life cycle management of electronic products; product platforms (such as Spotify), which commercialize access to goods or content; "lean" platforms (like Uber or Airbnb), which do not own their assets, such as cars or real estate; and "cloud" platforms (such as Amazon Web Services), which host third-party content and data.

Based on this initial classification, we can describe platforms as follows: They are multisided systems that use algorithmic coordination to connect different groups of users, who collectively create value. Even though they operate as businesses, they function like marketplaces in how they distribute the value that their users generate.[7]

Digital platforms are primarily characterized by being a specific type of multisided mechanism. Researchers view contemporary digital infrastructures as *multisided platforms*, following Nobel Prize in Economics winner Jean Tirole's pioneering work on intermediation services like credit cards and free media financed by advertising (called *two-sided* markets).[8] Free daily newspapers, for example, have two types of customers: advertisers who pay a fee to feature their ads in the newspaper and customers who enjoy these services without any cost. YouTube is an example of a platform that operates in the same way but has several categories of users: advertisers who pay, viewers who have free access, and creators who receive compensation for their platform use.

A second characteristic of platforms is that they rely heavily on users' personal data to support these coordination mechanisms. This is referred to as *algorithmic matching* between different categories of users. For example, Google Marketing Platform is an ad service that specializes in behavioral targeting (personalized advertising based on browsing behavior). It relies on collecting data from internet users on a vast scale, and transmitting it to real-time bidding platforms, which in turn sell each click to the advertisers who pay the highest price.

A third characteristic of platforms is that they capture the value created by their users. The vast amounts of data needed to run a search engine such as Bing, to price goods exchanged on the Etsy shopping platform, and to earn money on Instagram as an influencer, are gleaned from the ecosystem users built. Platforms thus blur the boundaries between inside and outside the firm. They juggle complex trade-offs between open access to information and proprietary enclosure.[9] As such, they act as new forms of organization, somewhere between markets and companies. Amazon is one example among many. It is largely a traditional company: in terms of work organization, it is particularly hierarchical, and reinvests profits rather than paying dividends to its shareholders. But, at the same time, it is a marketplace, where sellers (publishers, manufacturers of consumer goods) are matched with buyers.

The fact that platforms are market–company hybrids is particularly striking because, to return to Srnicek's definition, platforms have emerged, historically, to compensate for a twofold failure: that of traditional capitalist firms, which are relatively bad at extracting and using data, and that of traditional markets, which seem incapable of allocating resources efficiently without generating acute crises.

Markets tend to oscillate, and they are often affected by shocks and instability. The standard macroeconomic view holds that unregulated markets are unstable, fluctuating between expansion and unemployment, efficiency and poor resource allocation.

Other authors argue that platform capitalism is a new phase compared to market-based economic structures, caused not by the failures of the latter, but by the very obsolescence of markets. Anthropologist Jane Guyer argues that, as the importance of the market dwindles in the experience of individuals, it is perceived less as a physical place, making it an "abstraction" unable to give meaning to the real activities of economic actors. Platforms, by contrast, have come to represent coordination mechanisms that really influence economic lives. They're not just concrete solutions that consumers have in their own pockets and use every day. Platforms are also new ways of conceiving transactions and productive acts, in line with the current use of technology. Hence they appear as *exact* and *neutral*.

This much seems to hold true for the goods and services exchanged on platforms, but it is even truer for work. It is impossible to understand the real impact of platforms on the people who produce value (workers, free-lancers, or simple "amateurs") if we consider that platform workers are just another form of labor like that performed by traditional workers. In fact, these value producers are both users and workers in the digital economy. In this sense, they differ greatly from workers in traditional labor markets. "Whereas a market is depicted as place, people, and commodities," maintains Guyer, "a platform is made up of built components and applications, from which actions are performed outward into a world that is not itself depicted."[10]

So what is a platform? Its philosophical and political foundations need to be explored.

A Political Theology

From the mid-2000s, the term "platform" gained currency in IT milieus to describe the services that link up information, people, and goods.[11] Borrowed from the field of design, "platform" came to supplant the expression "computer architecture," being used to refer to technological structures.[12] However, "architecture" was used primarily for hardware, and "platform" for software.[13] While architecture always suggests the presence of a builder, platforms do not appear to have one. It is no longer implied that a designer has contributed to them.

As such, the platform is only a base, one on which other social actors (users, companies, institutions, etc.) can build. Sociologist Tarleton Gillespie suggests three reasons why the notion of platform lends itself to instrumentalization, especially in political contexts.[14] First, platforms are seen as mere intermediaries and not as instigators of social interaction and of strategic economic decisions. Yet, their supposed horizontality conceals hierarchical structures and relations of subordination that persist despite the consistent use of rhetoric describing them as "flat organizational structures." Second, the technologies themselves eclipse human contributions. Platforms' supposed neutrality absolves them of social responsibility and masks their impact on human communities. Third, and most importantly, the claim of these technical infrastructures to precision and autonomy underestimates the amount of work required to operate and maintain them.

Historically, architecture used the term "platform" to describe the foundation of a building, an elevated military fortification, or the podium where an orator stands. By metonymy, the term came to represent the words of an orator, especially one who gives political speeches. Subsequently, another usage developed in the English spoken in the United States in the period between 1648 (when the *Cambridge Platform* was written by the puritans who settled in the Massachusetts Bay Colony) and 1831 (when the International Platform Association, the speakers' organization founded by Daniel Webster, used the word to refer to the public discussion of a political program). In today's political world, a platform refers to a candidate's or a political party's positions on various issues. As Gillespie summarizes: "The US Democratic and Republican parties could support their respective presidential candidates by publishing their party platforms. [. . .] Curiously, a term that generally implied a kind of neutrality toward use—platforms are typically flat, featureless, and open to all—in this instance specifically carries a political valence, where a position must be taken."[15]

In fact, the concept of platform has been used with religious or political connotations for several centuries. In medieval France, for example, "*platte-fourme*" was in use from the fifteenth century onwards to describe a flat surface.[16] According to the 1582 English adaptation of Bartholomaeus Anglicus's work, the term refers to "fertile soil" and, more generally, to a productive resource.[17] A few decades later, when Sir Francis Bacon wrote his *Advertisement Touching a Holy War* (1622), he used the term platform figuratively, to indicate the ground on which he would build his "mix'd of Religious and Civil considerations."[18] In the seventeenth century, the Reformed Churches of England and the North American English colonies published a number of programmatic texts called platforms. Among them, the aforementioned *Cambridge Platform of the Puritan Congregationalist Churches of New England* (1648).[19] Similar documents exist in still other places. The *Savoy Declaration* of 1658, for example, established both articles of faith and rules for congregational governance, devoting a whole chapter to a "platform of discipline," which refers to the foundations of doctrine. The rules are seen as "the models and platforms of [a given] subject."[20] This latter sense, clearly in line with Bacon's usage, may have been the basis for the very specific political inflection the term acquired when it spilled over from its context of religious government in the years between the English Civil War of 1642–1651 and the Glorious Revolution of 1688–1689.

Thus the first clearly political use of the term "platform" to refer to a vision of society, and to the role of human beings in relation to the authorities and to each other, developed between the Establishment of the Commonwealth of England and Cromwell's Protectorate. It appeared in an essay written in 1652 by Gerrard Winstanley, called *The Law of Freedom* (originally *The Law of Freedom in a Platform*),[21] and was a founding text of the proto-communist movement of the Diggers. The political program (platform in the strict sense) drawn up in the Diggers' text, crafted to promote a society of free individuals, was based on principles that the Diggers developed in the course of their struggles: the *pooling of productive resources*, the *abolition of private property*, and the *abolition of wage labor*. Despite the appeal to divine authority, and the profoundly spiritual nature of these Christian anarchism movements, the term platform had already been cut loose from its religious origins and referred to an agreement made between a number of political actors with a view to the collective negotiation of a set of common resources and prerogatives.

The nature of this project did not escape the notice of Winstanley's contemporary, the Cavalier Colonel Winston Churchill, ancestor of his namesake. As he suggested in his 1675 book *Divi Britannici: Being a Remark Upon the Lives of All the Kings of this Isle*, the revolutionaries who put Charles I to death in 1649 wished to "erect a new Model of Polity by Commons only." To achieve this, they had "set up a new platform, that they call'd *The Agreement of The People*."[22] The concept, which initially described a covenant between religious bodies, evolved into an agreement between political actors.

Based on this etymology, the twenty-first-century use of the term "platform" in digital contexts extends and distorts that of the seventeenth century. In particular, the original political principles are preserved in the logic that underlies today's platforms: the *communal regime of economic and political organization*, or "polity by Commons," envisioned by Churchill, has become "sharing" on collaborative economy platforms. The *abolition of wage labor* (Winstanley's critique of the servitude of "work in hard drudgery for day wages") has mutated into insecure jobs and the exaltation of fake self-employment, as evidenced by the many freelancers now economically bound to gig platforms.

More interesting still, the *abolition of private property* (the agrarian communism advocated by the English Diggers) has been downgraded into the "opening up" of certain productive resources like data. One example can be seen in the term "platform-state," coined by futurist Tim O'Reilly to describe both a political and economic vision for platforms. In 2011 O'Reilly used this notion to describe a "Government 2.0," committed to radical transparency, innovation, competition, and a spirit of free enterprise. Beyond this profession of faith, however, lies the wish to impose everywhere the model of "bellwethers like Google, Amazon, eBay, Craigslist, Wikipedia, Facebook, and Twitter, [which] have learned to harness the power of their users to create added value."[23] Putting users to work and monetizing their "implicit participation" have been hallmarks of the digital economy since its inception: "First-generation web giants like Yahoo! got their start by building catalogs of the content assembled by the participatory multitudes of the net, catalogs that later grew into search engines. eBay aggregated millions of buyers and sellers into a global garage sale. Craigslist replaced newspaper classified advertising by turning it all into a self-service business, right down to the policing of inappropriate content, having users flag postings that they find offensive. Even Amazon.com, nominally an online retailer, gained competitive advantage by harnessing customers to provide reviews and ratings, as well as using their purchase patterns to make automated recommendations."[24]

Today, the concept of "digital platform" is more than just a metaphor. However, despite having deviated radically from its original meaning, the notion still carries philosophical and political weight. These implications can be better understood by comparing them with another economic paradigm—that of the firm.

A Borderless Structure

As we have seen, platforms can be understood as supplementing the (theoretical and practical) deficiencies of the market paradigm. The same applies to businesses and their units, namely firms. In the twentieth century, the paradigm of the firm dominated organizational models, but it became dysfunctional and dispersed. Like the market, it is going through a legitimacy crisis today. This is largely due to the unmitigated expansion of financial capital, which has disrupted the firm's method of value creation: the joint commitment of labor and capital to economic growth, primarily through investment and innovation. Even companies in traditionally risk-averse sectors like banking began trading (and often losing) on highly volatile financial markets in order to be able to promise huge and quick gains to their shareholders.

As a result, since the 1980s, companies have chosen to pay out dividends to shareholders instead of raising salaries or investing in production. In the United States, in the 2000s, big business spent 64 percent to 94 percent of its profits buying back its own shares to keep the stock market price artificially high and to skim off the profits, instead of investing in innovation and new technologies.[25] Over the three years leading up to the 2008 subprime crisis alone, the practice of stock buybacks increased from $31 billion to $144 billion for Exxon Mobil and from $9 billion to $50 billion for Pfizer.[26] Among companies in the S&P 500 index, buybacks have outpaced dividends since 2010. They have been the largest source of equities demand, reaching $923 billion in 2022.[27]

As a result of these practices, firms have lost sight of their historical mission of bringing together capital investors and workers who accept a certain "government of labor." This vision of value production, involving structured collective learning and rules of solidarity that transcend the annual distribution of financial results, has crumbled under financialization. Rather than retaining resources and investing them in new technologies, firms now opportunistically downsize and pay dividends to equity holders.[28]

This new philosophy of the firm has led to long cycles of mergers, acquisitions, and optimizations, inducing a decrease of the wage share (the part of income that goes to wages), compared to the profit share. Starting in the 2000s, companies have tended to keep only high-value-added services in-house, outsourcing the rest. It is a shift that has eroded collective labor structures within firms and forced them to reorganize. While, historically, the role of the salaried employee had emerged within the firms to limit bargaining, to insure a stable workforce, and to replace contracts for hire and for services, it appears to have lost its status as a model and given way to forms of nonstandard employment (freelancers, casual workers, subcontractors, etc.).

The logic of these transformations is epitomized in the digital labor of platforms, but the phenomenon is also a symptom of a broader trend, characteristic of the firm's decline as a paradigm: the presence of markets and contracts *within* the firm. This is a sure sign that the concept of the firm has outlived its usefulness as a way to regulate employment. Even though the firm was created to limit internal competition, the market sneaks in every time a freelancer or subcontractor is hired.

This change undermines the very "theory of the firm," as developed by the English economist Ronald Coase. In his pioneering article of 1937, he defined the *raison d'être* of the firm as a form of organization through which "one can eliminate certain costs and in fact reduce the cost of production."[29] Firms exist precisely to avoid having to resort to contracts and to the "price mechanism" (the market) every time a productive resource is required. They reduce *transaction costs* and in so doing increase productivity. There is no need to negotiate the price of raw materials, consulting costs, or bank account fees all the time. Firms replaced market coordination (through prices) with administrative coordination through defined hierarchy and managerial authority. Regarding labor, this hierarchy was expressed in workplace subordination: the relationship among employees, supervisors, owners, etc. It represented a curtailed freedom relative to that of the external service provider or freelancer.

In this way, companies were protected against price fluctuations, resource shortages, and information acquisition uncertainties. Moreover, they retained workers by making them salaried employees who were guaranteed an income every month.[30] But it would cost money to keep employees in a state of subordination.

The emergence of digital platforms today is in some sense the sign that the boundary between a place of hierarchy and security (the firm) and, at the opposite extreme, a place of free coordination by price (the market) has become obsolete. In this light, digital platforms have proliferated because the firm has betrayed its primary objective.

A Coordinated Ecosystem

As we have seen, platforms share certain features with markets and present themselves as an efficient alternative to them. They claim to be neutral and without hierarchy, given the alleged horizontality of the relations between their members. Customers and suppliers, artists and spectators, content moderators and creators are all—despite their diversity—just users of a platform. Interactions are thus reshaped on the basis of reciprocity.

The platform's success as a mechanism of intermediation is dependent on its technical performance, which is designed to minimize transaction costs (the costs of information searches, of contracts for each interaction, and so forth). Normally, a market operator would have to meet these costs, and historically that was the role of the firm. Platforms are gradually replacing these declining productive structures and, unlike them, they do not rely on defined hierarchies or on an instituted authority. Instead, they use the logic of algorithmic matching to record the declared or calculated preferences of each user as data, to assemble and process it, and thus to identify the other users or groups of users who would be able or willing to respond to those preferences. Personalized ads on Google pages are a classic example of this logic. When users enter terms into Google's free search engine, its algorithm sorts and indexes all the information that is available in its corpus, by relevance. An algorithm called PageRank determines a website's relevance based on its popularity. At the same time, another algorithm called Ad Rank uses these terms to select advertising content among another category of users, the advertisers, who pay for this service. The end result is a Google search page composed of a central column of "organic" results (websites or multimedia content that are not necessarily commercial), with blocks of advertising and sponsored links alongside.

Algorithms play a key role in this process. This fact should not, however, lead us into an unrealistic view of what platforms do. Due to commercial and political considerations, digital companies conceal how platform algorithms are designed and how user data is processed and matched.[31] This is not to say that algorithms are powerful in themselves. Far from it, for their effectiveness depends on very particular economic incentives.

Platforms, as I mentioned at the beginning of this chapter, act as multisided infrastructures.[32] Just like the two- or three-sided markets that existed before the digital economy, different economic incentives are proposed to the different categories of users with whom the platforms interact. In this, they resemble commercial television channels, where the public does not pay for entertainment or information, while advertisers pay a high price to reach their target audience. The same principle applies to certain food delivery platforms, but with a more complex system of rates and incentives: restaurants pay a commission in order to be listed and visible on the service, advertisers (supermarkets, tourism companies, etc.) pay subscriptions for their ads to appear among the offers, food customers are encouraged to sign up through discounts and vouchers, while the couriers receive piecework rates (or, more rarely, payment by the hour) for their delivery services.

One feature that digital platforms and the two-sided markets that preceded them have in common is that the price paid for using these platforms may be positive (like in the case of the restaurant owner, who pays a sum greater than zero), zero (like in the case of the consumer who receives a discount from 1% to 100%), or negative (like in the case of the courier, who is actually paid to use the platform). Unlike traditional markets, however, where it is generally easy to distinguish between supply and demand, digital multisided structures blur the distinctions between consumers and producers of services, content, and data. As a result, it is difficult to distinguish between transactions. Are platform payments intended to serve as an incentive mechanism, or are they intended to serve as remuneration?

Kevin Kelly, the founding editor of the magazine *Wired*, made this point in 2012, when he declared he would patent a system of compensation for reading e-books. The system was based on the premise that reading is work, and as such, it deserves reward. Each e-book sold would contain software that could track the reader's progress. The data collected automatically would trigger a reward for the e-book purchasers, such that they could gain more than the price paid for the book. If the e-book cost the readers $5, they would receive $6 for reading it, thus earning $1.[33] Consumers could also decide not to read the book, but then they would not be able to save or earn any money. There would therefore be a three-sided platform, with the following categories of users: the nonreading buyers who pay the market price for the book, the reading buyers who pay a "negative price" for the book (which means they are rewarded for reading it), and the publishers who use the revenue generated by sales to the former to pay the latter.

Kevin Kelly intended for publishers to use this model to respond to the perennial "crisis of the book" by rewriting economic incentives in a playful manner. It is in fact perfectly consistent with the new economy of electronic publishing, which, on the one hand, relies on large distribution platforms making content available immediately and almost free of charge, and, on the other hand, on "curator" services (selection, editorialization, and sharing) or on social archiving services (on LibraryThing or Goodreads), where buyer–critics add value through their reviews, their book summaries, and their recommendations. Since readers provide a service that is reflected in publishers' profits, we can understand better now why Amazon took over Goodreads. Media scholar Lisa Nakamura observes that, "Goodreads turns the reader into a worker, a content producer, and in this it extends the labor of reading and networking into the crowd. [. . .] Built on 'play labor'—the recreational activity of sharing our labor as readers, writers, and lovers of books and inviting our friends from the social graph to come, look, buy, and share—Goodreads efficiently captures the value of our recommendations, social ties, affective networks, and collections of friends and books."[34]

On digital platforms, *technical* coordination (through algorithmic matching) and *economic* coordination (through incentives) go together with *systemic* coordination. This can be described as the tendency of platforms to constitute ecosystems, that is, environments comprising users and companies, which reroute the production of value and the responsibility for innovation from within parent establishments (such as IAC/InterActiveCorp, which publishes the Tinder dating application, or Alphabet Inc., the conglomerate that oversees Google), to wide networks of subsidiaries, suppliers, public infrastructures, and even independent producers—as well as the consumers of information, goods, and services.

To function effectively, these ecosystems must be integrated, and all their component parts must be compatible. But they also essentially have to be divided up and transformed into independent units that can stand on their own.[35] Platform ecosystems can provide insight into outsourcing and taskification, two trends discussed in the first chapter. The platform tends not to produce goods or services internally, instead using flexible staffing arrangements to do the work. It is similar to what several companies did at the end of the twentieth century, except in a more extreme form, expelling their stable and committed workforce to then hire contractors or temp workers via employment intermediaries.[36] In today's world, this method of value production is ubiquitous. The new twist is that platforms do not delegate outsourced work to a person or a group, but to a network of production units.

A complex ecosystem made up of disparate individuals, communities, applications, databases, and businesses must be broken down into elements that can be translated and synchronized across them all.[37] Consequently, these elements should be standardized, normalized, and simplified. This is achieved by turning them into tasks or microtasks. The tasks typically involve producing data or content, providing services, or performing an action as simple as a click or data entry, to assist artificial intelligence in its operations. This standardization and the massive enrollment of users make platforms highly effective.

Despite a superficial similarity, the taskification demanded by platforms is different from the fragmentation of work typical of large twentieth-century companies.[38] In the Fordist view, work was broken down into units of time. The result was a reduction of heterogeneity that enabled individual contributions to be integrated smoothly into a collaborative effort to create value. Due to this, individual employees would become less skilled, since each of them only dealt with a fraction of the production process. Increased bureaucratization, however, would allow the development of productive know-how within the labor force as a whole. Platforms do not have the same problem.[39] Since they must constantly reinvent their business strategy, their identity, and their products, they are obliged to take on *task uncertainty*. This, in turn, is the path to task standardization and segmentation.

If platforms were judged by the same criteria as classical firms, their mode of organization would be considered confused and opportunistic. Underlying this impression is indeed a reliance on ad hoc solutions and improvisation. But their principle concern is to put interchangeable individuals to work on fragmented and standardized tasks. The taskification of work is necessary for platforms to develop and simultaneously to improve their services—a goal known as "scalability." Twitter, for instance, had only 7,500 employees before October 2022. By the time that Elon Musk had taken over the company and it morphed into X, the headcount had fallen to less than 1,000. The platform does not have an editorial committee per se, but instead promotes content based on an algorithm. Although there is a nominal effort to moderate content, users are in charge of creating, sharing, and selecting posts, images, and links. In essence, the platform simply delegates to users the fragmentary and simple task of posting messages. The posts of nonpremium users, who make up almost all of the user base, are limited to 280 characters. In this way, adding a new account to the existing user base is very cheap. As a result of hyperstandardization, the service has grown from zero to more than 500 million accounts since 2006.

A System for Capturing User-Produced Value

Unlike the modes of production that preceded them, platforms privilege value capture over value creation. Since the 1990s, the most significant successes in the digital sector have exploited user contributions, peer-to-peer collaboration, and above all, users' data. As a business model, it should be viewed less as an appropriation of untapped digital resources and more as an active extraction of value from users' work.[40] Using their network structure, platforms are able to take advantage of the activity of a variety of actors in every transaction. This step represents a major departure from the way firms have operated in the past. It was only through the subordination of workers that firms were able to extract value. Conversely, platforms not only utilize subordinates, but also search outside these relationships for those who are most qualified.[41]

The concern to capture value is a major strategic issue for platforms and has been since the beginnings of the World Wide Web.[42] Finding ways of increasing users' contributions, innovation and participation is vital, since these are the lifeblood of platforms. User-generated content has been one of the most visible forms of appropriation for commercial purposes. The power and success of the great social platforms of the 2000s (Friendster, Myspace, YouTube, Facebook in the Northern Hemisphere; and Orkut, Kakao, QQ, and VKontakte in Asia and South America) initially relied on the hundreds of millions of "volunteers" producing and sorting texts, images, videos, and sounds. Sometimes these users were regarded as amateurs (as in the case of Flickr's photographers), sometimes as pirates (as in the case of Napster in its early days), and sometimes as editors and authors (as in the case of large blogging platforms such as Overblog or LiveJournal). Beyond these specificities, the platforms, then as now, are fed by the information made available by users, who are the real bearers of their commercial value.

Capturing this value does not simply mean ensuring that products or services circulate on the platform. It also means finding a way to monetize them. *The Facebook Diaries*, for instance, a ten-episode reality tv series launched in 2007, consisted of video collages submitted by young users. Facebook engineered a lucrative partnership with Comcast to broadcast them on its own network as well as on Ziddio, another online video service.[43] The series was both a form of advertising for Facebook and an income source through the sale of distribution rights. In an economic context where the media industries had been suffering wave after wave of downsizing, this use of amateur content, Nicole Cohen argued, amounted to "outsourcing the work of production from the media companies to the producer-consumers."[44]

So platforms capture the value of their networks by marketing user-generated content, but that is only one of their methods, and not necessarily the most lucrative one. Since the 2010s, with the commercial success of platforms focused more on services for individuals or businesses than on the production of cultural goods, the profits of platforms have come for the most part from the monetization of their users' data and metadata. Personal information is regularly sold through commercial agreements to brands and advertising agencies for personalized advertising targeting; to insurance or credit rating companies; or to states for surveillance purposes.

More recently, a new source of revenue has emerged, with the use of personal data to calibrate artificial intelligence devices. Since 2015, X (then Twitter), like other platforms, has not only been selling its users' data to advertising agencies, but it has also been making this data available to companies that are developing powerful machine learning programs, such as IBM, Oracle, or Salesforce.[45]

I should stress that value is captured not only from cognitive tasks such as publishing messages or making personal information available. Platforms also extract value from users' behavior and their physical activities. This is what Google Trekker does in delegating to crowds of anonymous volunteers, through its "manual digital labor" program, the task of photographing the places that Google Street View's vehicles are unable to reach. Equipped by the company with a 42-pound backpack containing 15.5-megapixel cameras, which take a panoramic photo every 2.5 seconds, Google's trekkers set off to capture mountain summits and desert dunes. The images and geolocation data that they collect improve the Google Maps service. Apart from Google's assistance in the form of cameras and processing power for the information gathered, the service is provided wholly by users. They are the ones who go trekking, weighed down by equipment, to ensure that Google's service will continue to be extended and improved. Even if this data and multimedia content is not sold on directly, volunteers' contribution are anything but abstract. This contribution gives Google a competitive advantage, enabling it to generate more traffic on its platform, and thus providing further opportunities for extracting yet more geolocation data for use in targeted advertising.[46]

There are, therefore, several different ways for platforms to profit from user input. They can encourage the performance of tasks that keep the platform running and produce *content*. Platforms can then use this content to create metadata that they will monetize. Or, as we have just seen, the user's own behavior can generate resources that automation systems will use. The Airbnb platform exploits these methods of capture to the full. It encourages its users to contribute to the platform well beyond the paying hospitality service. Just like on a social media platform, Airbnb's hosts and renters alike must produce large quantities of content. Hosts provide photographs and assess the civility of guests; guests rate the accommodation and the professionalism of the hosts, including comments and stories about their stays. They must also provide a steady stream of geolocation data, personal information, and demographic information. To satisfy its thirst for data, Airbnb has also resorted to semilegal operations, such as when, in its early days, it took over the Craigslist ad site's directory so that it could identify apartment owners, to whom it would send targeted offers.[47] In a more conventional and less "disruptive" way, in 2015 it partnered with Foursquare to use a geolocated image base to create a collection of photographs of urban environments exploitable by the machine learning system of the platform Aerosolve.[48] On this platform, visitors were suggested nearby places to visit and hosts were suggested prices based on variations in demand.[49]

Niels van Doorn and Adam Badger developed the concept of "dual value production" in order to explain how platforms acquire surplus in multiple forms. What kinds of value are created through platform labor? Platforms capture two types of value from their workers and users: the ones associated with transactions (such as subscription fees, service charges, and purchase prices), "and the more speculative and volatile types of value associated with the data generated during service provision."[50] In today's platform organizations, data constitutes a particular class of asset.

An Inspiring Paradigm

Platforms that started online are not the only ones capturing value in this way, and the practices described above are spreading to brick-and-mortar companies as well. Platforms are not only an organizational model, but also a paradigm that has inspired a large number of players beyond the giants of the internet economy. Companies in the private sector and several state and para-state enterprises have adopted similar strategies to reproduce the platform model. The terminology used to describe this change has evolved over the years: the virtual corporation,[51] the cellular organization,[52] the hypertext organization,[53] and the N-form corporation.[54]

Today, automobile and energy companies, among others, can and want to be platforms, be supported by platforms, host a platform, and be an integral part of a platform's ecosystem.[55] The systematic exploitation of user-generated data is now widespread in the insurance and retail sectors. Large multinational companies like Admiral in the United Kingdom or BHV in France now ask their consumers to produce geolocation data or information about their habits and preferences.[56] There would be nothing new about this if this data's only purpose was to hone target markets. However, the constant tracking of consumers has another consequence: it stimulates and manages the contributions of users, from using self-checkouts to posting on social media.

In the transportation sector, companies exploit user data to generate additional profits from advertising income. In Germany, Düsseldorf's public transport company, for example, tested the WelectGo mobile application, which allowed the traveler to view commercials on a smartphone instead of paying directly for a bus or streetcar ticket. In the process, the company sold the user's data to advertising agencies. This was not the reason why the app was eventually discontinued. The system was a victim of its own success: far too many people used it, which led to a loss for the city transportation authority.[57]

Other companies have used similar strategies to launch new services. With the use of free Wi-Fi and cellular connections within trains and stations, French railway company SNCF has developed an ambitious data-driven innovation program,[58] Again, the data gathered is of course monetized.[59] But it is also used to improve the flexibility of the transport service. TGVPop, a short-lived experimental service, used demand-driven algorithms to streamline rail traffic. Users were asked to provide social data for a just-in-time reservation system. Trains would only depart if there were enough upvotes.[60] A logic of social participation involving contributions by, and coordination between, users has been driving this innovation. It builds on skills that users have already developed to promote their content on social news sites like Reddit, or to coordinate trips using collaborative platforms like BlaBlaCar. Users have to be connected, have to combine their efforts, participate, and make enough reservations for the platform to consider that the departure of a given train was economically viable.

The media industry has been particularly affected by platformization. The press faces a twofold value capture by large platforms such as Google and Facebook. The first is "infrastructural," with the online circulation of content. Platforms enable media outlets to measure the spread of their content (the number of times images are displayed, articles are read, or videos are viewed). In this manner, data is captured from the media and their readers. These analytics are expected to enable newspapers and streaming services to make better editorial decisions. Thus, platforms have become indispensable for both distributing press content and measuring its success.

The second capture is "institutional."[61] The media have traditionally acted as watchdogs of economic actors. But in their increasing dependence on digital platforms, they find themselves caught up in a classic situation of "regulatory capture."[62] The body responsible for establishing rules and limits for a given sector ends up—for reasons of social proximity and economic interests—adopting the sector's convictions and values.

Platforms Are Made of People

For digital platforms, the fundamental question is how value is created. As platforms transcend both companies and markets, they also mark the end of the two main instruments used to describe them: products and prices. Platforms must be flexible to combine the various elements of their ecosystem. As a result, they tend not to focus on a specific product—if the term is still relevant. A platform is too involved in the multisided coordination of user groups to rely solely on price to differentiate them. In a context whose historical hallmarks include markets that are liable to crash time and again, and firms that have lost sight of their ultimate rationale—to innovate, and to limit transaction costs—economic operators are faced with what some authors have called a "crisis in the representation of value."[63] Businesses no longer understand where value comes from. Platforms, on the other hand, allow value to be construed in an original manner. How they *represent* value is the central issue.

Some platforms mimic the communication patterns of traditional companies when they disclose their economic value, for instance when they go public or meet with potential investors. They also sometimes highlight their technology (the number of servers they have, the quality of their algorithmic solutions, or the computational power of their processors). But the source of their value remains the quality and quantity of the personal data they exploit, the dynamism of their communities, and the usefulness of the services they develop on this basis. Despite the rhetoric they produce for their financial market partners, their vision of wealth is not contained within a circumscribed company, but rather captured from an ecosystem of heterogeneous actors.

To understand this vision, it is important to let go of the traditional separation between value created during a manufacturing process and value exchanged during the circulation of goods and services on a market. In recent years, valuation studies have challenged this distinction, which is maintained in both conventional theory and Marxist approaches to value. Pioneers of science and technology studies like Michel Callon have maintained that the value of a product is inextricably linked to its exchange value, which governs how property rights are transferred. This axiom is all the more true for digital platforms, where users shape data, content, and services precisely for circulation. As platforms multiply ways to engage consumers, the latter get greater accessibility to products, contents, and services. The result is endless cycles of what Callon calls the "qualification and requalification" of goods. In the case of online content, for example, value is determined by the qualities of its creators. But the consumers are also responsible for requalifying it according to their own preferences and talents.[64] As users interact and exchange, they carry out the fundamental work of testing the tools the platforms make available. They also come up with use cases for the technical solutions.

This testing should not be seen as an activity reserved for experts: all Gmail users have improved the email service by using it (it remained in a beta version for five years); online games like Minecraft were based on players creating and testing new situations; the interfaces of mainstream social media apps such as Meta, X, and TikTok are constantly twitched.

Users are also given the crucial role of producing ratings (through comments, criticism, and reviews) and appreciation metrics (emoticons, likes, stars, etc.). This in turn generates information, an additional layer of data, which companies can extract. Users thus contribute to a cyclical process of qualification–requalification of any and every object, which gains value through this process.

Despite the emphasis on the production of data and analytics, this mode of value creation is not simply a process that transforms quality into quantity. It is not confined to putting figures on different production factors. On the contrary, it is a "work of qualification" carried out by the user, and valorized by the platform. Taking metrics into account does not run counter to judging content and services. Callon, for instance, uses the term "qualculation" to describe this mix of the qualitative and the quantitative.[65] As part of qualification, the customer also classifies, ranks, and evaluates the product or service based on comparisons and connections.

On platforms, qualification can never be divorced from work. Users' data, behavior, preferences and inventions, which produce and measure value at the same time, can be viewed as a series of productive acts aided by digital technologies. Their qualification work is carried out via algorithms whose purpose is to provide and link together information that will allow users to create value.

The analysis of a platform requires analyzing the digital labor it captures, since *users' actions can be assimilated to forms of labor embedded in social relations.* Users interact with each other, and the value they produce cannot be considered anything other than a collective effort.[66]

This collective endeavor is not, however, characterized by solidarity between the users and the owners of the platform, between the engineers and the hired hands of digital labor. Far from it. Despite marketing professionals' efforts to promote a narrative of technology fostering networking and collaboration, platforms are still marred by conflicts over value. Users can engage in conflict with other users, support staff, vendors, etc. in an ecosystem through attacks on their discussion forums, apps, or marketplaces. In other cases, conflict can arise from the mechanisms of differential appropriation of what users produce, like the opposition between monetization of content and collective ownership on a video-streaming service.

Value capture by platforms is an agonistic process, in which conflicting definitions of an object's value compete, because conflict is inherent in the work of qualification and value creation by users. As users create content, metadata, and information, the output of their digital labor becomes the object of a struggle over what is valuable, who owns it, and how it can be disposed of. Traditional markets were arenas where supply and demand met, where confrontations occurred. In the course of these confrontations (again, this is Michel Callon's argument) "goods emerge, fragile and unstable, their qualities having been fixed at least temporarily."[67] Disputes over value used to be resolved through transactions that would ultimately lead to monetary payments. In the hybrid firms-markets that are platforms, digital labor can expose problematic situations where antagonisms lead to micro- or nonpayment, predation, or even extortion.

*

To analyze the impact of technology on work only by focusing on the supposed displacement of humans by more or less intelligent machines would be reductive. Technological devices are just tools, and tools have never been absent from human labor and production. Working today means figuring out how production works in light of a double collapse: the decline of the market as an effective coordination mechanism and the departure of the firm from its historic purpose of promoting innovation. A new economic and technical paradigm has arisen, and henceforth it affects work. Its transformations must be analyzed in terms of the notion of the *platform*.

A platform is built by the people who dedicate time and effort to ensuring that its technical structure functions in a meaningful manner. These people are mainly its users. Today, platforms are multisided coordination mechanisms that capture the value generated by the three types of digital labor performed by their user: *on-demand work*, *microwork*, and *social media work*. Different types of work are associated with different techniques, tasks, and degrees of conflict over the recognition of the work itself.

· PART 2 ·

THREE TYPES OF DIGITAL LABOR

· 3 ·
ON-DEMAND DIGITAL LABOR

Let's consider the first of three types of platforms for digital labor. "On-demand" platforms connect those seeking and providing work in a specific location. The ride-sharing app Uber and the food delivery service Deliveroo are good examples. Their apps offer many small jobs to workers, who provide transport, accommodation, delivery, personal assistance, and repair or maintenance services. The scale of these projects is usually local: a city, neighborhood, or region. Behind the scenes, as with all types of digital labor, workers and users produce data. Although the workers appear to enjoy a certain level of sociability, driving, hosting, and managing a delivery are arduous tasks and far from being leisurely activities.

To a platform owner or a third party, this type of digital labor seems to require only a moderate level of skill, and the work is remunerated by contract or by payment per hour or per task. On-demand labor is sometimes considered part of the "gig economy"; it emphasizes flexibility and freelance short-term positions. Sometimes, in what's been called the "collaborative economy," individual workers are required to liaise and coordinate with each other. We shall see that platform capitalism is about anything but "sharing." If a sharing economy exists, it results from cooperation between workers *despite* the platform, not because of it. Workers' jobs are insecure, and they're often underemployed. Despite this, workers have cooperated to improve working conditions and pay, and to obtain the right to organize.

Nonstandard Work: The New Norm?

On-demand digital labor covers a multitude of activities. Some platforms propose services related to the home (shopping, cleaning, plumbing, furniture assembly, or babysitting), while others compete with sectors of the traditional economy (transport, housing, delivery, customer service, or IT assistance) at preferential rates.[1] The ways that workers get paid also vary greatly. Applications like Deliveroo can pay some of their workers on a minimum hourly basis, even if the norm is to pay per delivery. Tips make up a big part of workers' earnings at Instacart (even if those are standardized and compulsory). Before switching to hourly rates, platforms such as Task-Rabbit encouraged workers to bid for work and rewarded those offering to do the task for the lowest wage.

These differences between apps lead to confusion about the nature of the platforms. Pundits initially equated on-demand labor with disparate phenomena such as the "collaborative," "sharing,"[2] "circular," or "functional" economy, and even the production of commons or of peer-to-peer transactions.[3] These ecosystems have managed to conceal the work necessary for their existence by calling their users' work a "ride share," "a favor," or "a helping hand." Even the term "gig" implies a one-off performance rather than workers investing their time on a regular basis. However, the true nature of these services becomes apparent when, from time to time, conflicts emerge between users and platforms. Then the terms "work" and "workers" resurface, such as during media debates, wage disputes, or court cases.

Users of an on-demand platform also perform work. A matching algorithm connects a certain category of user that consumes a service with another that produces or supplies it. On-demand digital labor has a distinctive feature: the platforms function both online and offline. The matching of customers and service providers takes place on an application or website, but the tasks take place in the real world.[4]

Labor law experts have debated the consequences of on-demand labor for nearly a decade.[5] Generally, these platforms have high turnover, and the workforce has grown dramatically in recent years, especially since the COVID-19 pandemic. The Pew Research Center revealed that the share of Americans who have engaged in online gig work by participating in activities such as ride hailing, grocery shopping, household tasks, or package deliveries jumped from 8 percent to 16 percent in 2021.[6] The years of the global health crisis witnessed a substantial surge in this form of platform work, both in terms of its share in the total workforce and within the broader context of the gig economy. The number of people doing app-based, on-demand work like Uber and DoorDash more than doubled during the pandemic. The size of the workforce does not correlate with the earnings of these workers. There were two million workers by 2018, but three million new workers joined after COVID-19, most of whom worked for transportation or delivery platforms. By 2021, platform gig work accounted for 3.15 percent of the overall US workforce.[7] This estimate is conservative. Other studies suggest that even before the pandemic, on-demand platform workers made up to 7.6 percent of the US workforce.[8]

Platforms' marketing departments will typically stress the platform's ability to give workers access to employment, flexible working hours, and additional income. Platforms' detractors, instead, highlight worker exploitation and precarity. From the start of the gig economy, the size of the on-demand workforce has not correlated with income, especially for drivers, whose earnings fell 53 percent in the lead-up to the pandemic.[9] The situation in many parts of the world only got worse thereafter. In Kenya, for example, ride-hailing platforms didn't lower their commissions in 2020–2021.[10] This fact, combined with greater competition among drivers, resulted in lower fares. In this business, it is the workers that shoulder the economic risk: if the platform can't provide work, they don't get paid.

The employment status of on-demand platform workers is a recurrent source of conflict around how working hours are defined, pay systems, health and safety, social security, rights to representation, and training. Many companies, from the household chores app Handy[11] to the urban transport giant Uber,[12] have been taken to court more than once for misclassifying their workers as self-employed. Emergent and traditional trade unions have successfully negotiated collective agreements on on-demand labor in Europe. Most of these initiatives are for food delivery workers, but they also target ride hailing and other services.[13] Laws have been changed as a result, as have platform working arrangements. Cleaning services platform Helpling, for example, has adapted its workers' legal status to the local laws of each country where it operates—hiring through temp agencies in some countries, offering long-term contracts in others.[14] After several legal disputes, Uber began to propose an array of employment statuses, spanning from independent contractors in the US to permanent employees of partner car rental companies in Germany. As part of the 2021 "riders' law" in Spain, on-demand platforms are required to employ all delivery workers as dependent employees. In reality, this is just a legal presumption, and platforms have adapted differently to it. Some followed the letter of the law, and others kept hiring self-employed independent delivery riders.

The varying nature of these arrangements complicates things. The fact remains that, despite platforms' best efforts, these jobs don't really qualify as self-employment in the classic sense. In many countries, self-employment is a long-standing tradition. From graphic designers to gardeners, to plumbers, it has to do with freelancers renting their time and skills. "Liberal occupations," like lawyers, accountants, agents, etc., also provide highly valued intellectual services in the interest of their clients, but under their own responsibility.

The "self-employed" individuals who swell the ranks of on-demand digital platform workers have nothing in common with these white-collar professionals—whether in terms of income, training, or social protection. The more than 70,000 Uber drivers who are classified as "workers" in the

UK serve as an example.[15] Under British law, "worker" is a unique employ-ment category and, although "not employees," they receive minimum, holiday pay, and pensions, none of which are rights for self-employed in-dividuals. According to some observers, a gray area is emerging between salaried and self-employed jobs.[16]

The on-demand economy alone doesn't explain this trend, but it's an-other sign of burgeoning job insecurity. Platform workers operate like ca-sual workers, who do on-call work, voucher-based jobs, and zero-hour con-tracts. In these flexible arrangements, workers are only paid for the hours they work, or only for tasks completed. This is why official statistics tend to lump precarious and platform workers in the same category of "nonem-ployees," who total, in Europe, from 14.4 percent in the United Kingdom to 24.9 percent in Italy, as being in nonstandard work arrangements in the US.[17] So these figures should be correlated with other indicators, such as job stability. In the European Union, since 2009 a steady percentage of the workforce has been on fixed-term contracts. In 2020, they accounted for 11.9 percent of the total employed people aged fifteen through sixty-four; after the COVID-19 pandemic, the share increased to 12.1 percent.[18] Other estimates put the figures even higher. According to a European Commis-sion study, Europe had twenty-eight million on-demand workers in 2021 and is projected to have forty-three million in 2025.[19]

Platforms like Etsy, Airbnb, and Lyft emphasize flexible hours in non-standard occupations. On-demand platforms seem, at first glance, to be a good alternative to full-time employment as a primary source of income. That's not the whole story, though. In some cases, platform work is a non-standard worker's full-time job, while in others an individual works full- or part-time along with their on-demand job.

Authors Caroline Bruckner and Jonathan Forman wonder whether it is easier to define these workers "by what they are not," i.e., "workers in tra-ditional employer–employee relationships earning salaries or wages and eligible for employer-provided fringe benefits."[20] Since nonstandard work-ers come from a wide range of circumstances, they are difficult to measure in terms of their income. Whether they work on digital platforms or not, it appears that workers in alternative employment relationships consistently earn less than workers in traditional employment relationships.[21]

In the terms of the International Labor Organization, there are four nonstandard employment regimes: temporary jobs (fixed-term contracts; seasonal, casual, or day labor; project-based work; or, as in the case of digital platforms, task-based work); part-time work (including zero-hour contracts and on-call work, or on-alert work for mobile applications); multi-employer work (temporary work, employer groups, wage portage, brokerage, or jobbing on the internet); and concealed work (disguised em-ployment relationships, dependent self-employment, undeclared work, and misclassified work on platforms). Companies use these nonstandard types of work to optimize their labor force, to minimize wage costs, or to attract individuals less interested in permanent or full-time jobs. Already in 2010, 23 percent of European companies made frequent or intensive use of nonstandard labor. This percentage rose to 40 percent of employers, in low- and middle-income countries.[22] In a 2020 US Census survey of businesses, 33 percent of companies reported using contractors and 50 percent reported using part-time workers. In 2021, 89 percent of the S&P 500 companies (the largest companies in terms of market capitalization traded on American stock exchanges) mentioned their use of some type of contractor arrangement in their annual financial reports.[23] At a global scale, it seems that a transition is in place from a homogeneous staff composition to a "blended workforce" where full-time employees work in tandem with on-demand workers connected via digital platforms.[24]

The growth in nonstandard employment worldwide has several causes: the emergence of the service economy, the cascades of outsourcing in many industry sectors, and labor law reforms in several countries. But it also reflects changes in the demographics and composition of the labor force. In particular, the growing presence of women on the global job market is also the result of a market doubling of domestic tasks. Platforms that provide care, personal assistance, cleaning and home cooking services represent this trend, as women are overrepresented in these professions.

Another significant trend is the persistence of informal labor, particularly in the Global South. In North America and Western Europe, the informal sector only makes up 9.8 to 13.1 percent of jobs, but in Latin America it reaches 52.6 percent and in Africa it comprises 84.3 percent of jobs.[25] Informality and platform work have much in common: working conditions are bad, health insurance isn't available, and earnings are uncertain. Platform workers and workers in informal economies also share the ambivalent feeling of being both exploited and in control. "Hustling" is a commonly used term in the US. It can have positive connotations, insofar as it represents the ability to make deals and to get by that allows people to overcome insecurities. Marginalized communities in industrialized countries and the masses in low-income countries see informality differently. It can certainly be a way for underpaid domestic workers, artisans, and street vendors to feel self-sufficient and self-reliant. However, there is a general awareness that workers are mostly running a survival economy without public protection or private support.[26]

The Inflexible Flexibility of On-Demand Work

It's common for platform owners to downplay the risks of on-demand labor. A widely cited study underlines how the insecurity of Uber workers is offset by their self-employed status, with its flexibility and autonomy, allowing them to choose their working hours and sell their labor to several applications if they wish.[27] In theory, they can switch freely from one platform to another. In practice, this proves extremely difficult. While possible, multiactivity is restricted by several factors, some cognitive, others technical. Among the cognitive constraints are the time required to learn how to use an interface or the rules of interaction on a particular platform. Airbnb or Etsy, for instance, have their own etiquette that takes time to master. Another example is the psychological effort ride-hailing drivers have to put in to figure out the parameters of pricing algorithms on different apps. This makes them less likely to switch from their default platform to a competing one.[28]

On the technical side, as in the case of delivery services, couriers must equip themselves with tools such as smartphones, external batteries, charging cables, and bicycle accessories. Standards of quality can vary too, and limit multiactivity. Drivers for certain ride-hailing services may have to meet requirements, like free Wi-Fi in their vehicles, which other services might not have. Another obstacle is how the service itself is provided, such as the requirement that some drivers wear suits and ties, the app logo on the equipment and uniforms, or the fact that they speak a certain language. In some companies, this is both a way to build brand recognition and to reduce the likelihood of workers switching platforms.

The flexibility of work schedules is also limited. Although officially workers may choose to put in time only when it is profitable or when their schedules allow, many platforms reward those who make themselves available at off-peak hours, through symbolic (badges, points, or stars) or economic (performance-related prizes) incentives. On the other hand, they sanction workers who do not comply. Already in 2015, a group of Californian drivers took Uber to court; in company materials revealed during the case, it became clear that Uber had tools to "deactivate accounts" (a euphemism for "fire drivers") if the journey acceptance rate was too low.[29] Instacart and DoorDash have been accused of suspending, shadow banning, or deactivating accounts if their completion rate (the number of deliveries made during a particular period) is too low.[30]

To be able to generate just-in-time incentives, the apps must acquire personal data related to both their drivers and their passengers. This is how another risk, specifically linked to real-time production of information, emerges: users are tracked all the time. By taking liberties with the legislation on the consent of individuals to data use (the GDPR, for instance, the European Data Protection Regulation), platforms acquire large amounts of information about their users. Platforms accumulate information without explaining where it's stored, how it's processed, how it's protected, and, most importantly, how it's shared with other companies.

This is nothing new, of course. As part of their regular human resources management, traditional companies collect a lot of data about their staff, too, like their identity, background, and performance. But digital platforms go a step further, by collecting geolocation data and browsing histories and claiming the right to access the images, calendars, and contacts on their users' smartphones.[31]

The surveillance associated with these mobile applications is, however, almost never an issue in the negotiations and struggles between platform users and owners. In recent years, disputes between these digital service providers and their subscribers and collaborators have largely been work-related. The most important causes of platform worker protests are pay, employment status, and health and safety.[32]

It is easy to recognize driving, delivering, and cleaning services as "work" because they are already listed in descriptions and contracts in traditional jobs. Those jobs are thus more likely to follow more typical work arrangements. Since the beginning of the gig economy, on-demand workers have taken industrial action and filed lawsuits to get the same salary and employment protections as traditional workers.

Back in 2017, the showdown between Etsy and its users focused on classic occupational claims: its "independent sellers" wanted rights to health insurance, six weeks' parental leave, and standardized severance pay.[33] On Etsy, many buyers were businesses that used the platform to find reliable artisans and subcontractors at affordable rates. So Etsy makers and shop owners demanded that the platform be recognized as an intermediary. Again, in 2022, sellers went on strike against the platform, shutting down their stores for a week in protest against fee increases.[34]

Similarly, beauticians and spa attendants working on the Indian platform Urban Company went on strike in 2021. The protest was sparked by a lack of insurance, low wages, and high commissions. There was collective dissatisfaction with the platform's policies. The workers of the platform, mostly women, are often assigned night shifts, creating unsafe conditions. The company required workers to purchase beauty products from them at higher prices than those available on the market. And if workers failed to complete a minimum of thirty jobs per month, they were subjected to charge of 2,000 rupees.[35]

Other claims involved work accidents and occupational diseases, especially from couriers. After an Uber Eats delivery worker died in 2018, the #Niunrepartidormenos ("Not one courier less") initiative spread from Mexico to several other countries. Since then, at least 285 couriers have died at work throughout Latin America in traffic accidents and criminal assaults since the campaign started.[36] Even in countries with universal healthcare like France or Spain, self-employed workers struggle to get full accident coverage. In case of sickness, workers get a daily allowance; platform couriers have to be sick for at least four days (in some cases, eight) to have the same benefits.[37] Regarding family allowances and replacement income (unemployment benefits and retirement pensions), on-demand platforms do not always contribute to social insurance programs and national contribution funds. They tend to pass on to workers the administrative burden of checking with the official bodies of each state that using the app gives them such entitlements. Digital platforms use all legal loopholes possible to corner a market. Then, taking advantage of their dominant position, they circumvent labor regulations.[38]

Platforms also use the promise of giving higher wages than their offline competitors as an excuse for flouting the rules on payment of user-workers' social security contributions. One example is the Belgian platform Ring Twice, which pays between 0.88 percent and 15.71 percent more than traditional companies for housework, babysitting, gardening, and transportation.[39] Frequently, a platform will pledge to pay more than the minimum hourly wage but will make no guarantee of the actual number of hours a subscriber will work.

Uber and Lyft promised their workers they would make more than California's minimum wage of $15 per hour if a bill, known as Prop 22, passed into legislation in 2020. The proposal, which prevents state policymakers from reclassifying drivers as employees, was approved, but since then the average driver only makes $6.20 an hour.[40] The amount they earn actually depends not on the time they spend on the app but just on the "engaged time" (the time between accepting a trip and finishing it) measured by the app's algorithm.

Uber regularly commissions studies to show how much better it is to be a "partner" than a salaried employee. One of these, carried out by economists from the elite French business school HEC, maintained that on-demand platform drivers earned up to twice France's minimum wage.[41] It is not all rose-colored, however. When car maintenance, taxes, and service fees are taken into account, Uber's drivers actually make a much lower income. They must work up to sixty hours per week, which is almost twice the country's thirty-five-hour legal working week.[42] In the US, 33 percent of Uber drivers were caught in the "debt trap": they borrowed too much money, hoping to earn a living on the platform.[43]

Delivery platforms have also been criticized because their flexible schedules translate into unpredictable working hours and demanding conditions for couriers. Those who work on Postmates had to put collective pressure on the app designers to add a "stop" button, to ensure that refusing a job at the end of a long day did not affect their score on the platform.[44] Clearly platforms interpret "flexible hours" as they see fit.

There is little truth to the claims that "you can choose when to work" and "make more money" on on-demand platforms. So it is not surprising that many independent contractors want to be reclassified as employees to protect their rights. An unmistakable connection exists between the Seattle City Council's 2015 decision protecting Uber drivers[45] and the Dutch Supreme Court's 2023 ruling that Deliveroo must pay pension contributions to riders[46], and this is a reminder for on-demand platforms that labor law cannot be ignored.

Tracking and Datafication

So far, organized workers and civil society groups have succeeded in challenging platform labor law violations, but their victories are largely limited to the visible part of the work of drivers and couriers: transporting people and delivering goods. Wins are fewer when it comes to the inconspicuous parts of these jobs, such as producing data and feeding algorithms. In 2021, the Italian data protection authority fined the app Foodinho almost $3 million for collecting data from couriers in a way that did not comply with European privacy regulations.[47] British Uber and Ola drivers successfully challenged the platforms in a Dutch court in 2023, seeking access to their own data.[48] The following year, French Uber drivers managed to have the Dutch data protection authority fine the company $11 million for mishandling their personal information.[49]

On-demand applications are large databases that are not always handled with the same standards of ethics and transparency as the data stored by large public statistics institutions. Some platforms select users by gender or ethnicity, whereas discrimination based on gender or skin color is normally prohibited. Platforms typically recruit women and people from immigrant backgrounds to perform lower-paid tasks, and so perpetuate labor market inequalities.[50] There is also discrimination against other categories of users, such as passengers in ride-hailing or apartment-renting apps. A study of transportation platforms in the United States found that African American passengers were subject to longer waiting times and higher cancellation rates.[51] Guests with non-English names were 16 percent less likely to be accepted by Airbnb hosts,[52] and this situation worsened in 2020, when anti-Asian sentiment in the US rose sharply at the beginning of the COVID-19 pandemic, causing Asian American Airbnb users to face greater discrimination.[53]

When exposed, the platforms concerned were, of course, quick to draw up policies to limit these abuses by prohibiting what, on the surface, appeared an outcome of the racist behavior of certain individuals. Some experts, though, argue that systemic bias has developed. Legal scholar Veena

Dubal argues that a systemic new form of discrimination is put in place by platforms, dubbed *algorithmic wage discrimination*.[54] A growing number of low-income and marginalized racial minorities are negatively affected by algorithmic management in the United States. On-demand platforms collect their workers' data and change fares accordingly. It's what platforms call "personalized wages."[55] This makes hourly pay unpredictable and variable for workers. As a matter of fact, this logic mirrors what's called "price discrimination" in consumer economics, where the same product or service is sold at different prices depending on who's buying it. As well as harming workers economically, discrimination damages them morally. It disrupts their long-standing expectations of wage equality, a minimum wage, and inclusive workplaces.

The extraction and linking of information are integral to these applications, and the data concerns not only the products sold, but also the people who sell them. Users of an app, whether they are workers or customers, are encouraged to fill out their personal profiles and upload photos. Such profiles can facilitate discrimination by perceived ethnicity, gender, age, or other aspects of the users' appearance.[56]

Other experts have argued that the data collected by digital platforms would, on the contrary, make it possible to detect any form of systematic discrimination—and this more surely and more rapidly than in traditional companies.[57] In other words, they maintain, the discrimination problems that the on-demand economy seems to suffer from are not the result of a real increase in discriminatory behavior, but simply the consequence of the wider availability of information about forms of racism that would exist regardless of the platforms.

Whichever argument is the more convincing, this situation highlights the risks incurred by extracting, circulating, and processing large volumes of data, with all the potential for surveillance and algorithmic control over the value producers that this implies. As I hope is starting to become clear, tracking and extracting data are not secondary activities for these platforms. Data is at the very heart of value capture, which platforms achieve by monetizing the personal information of users and workers through targeted advertising. But data is also what makes platforms function in the first place. Evaluation data—by both drivers and passengers, buyers and sellers, etc.—generate the information required for the service to run at all.

Several studies have examined on-demand platforms' methods in the transportation industry. The ubiquitous Uber, a true social landmark of the 2010s and 2020s, is emblematic of an algorithmic and data-driven management style.[58] One of Uber's most important activities is tracking its drivers: calculating the speed and performance of their vehicles, locating them, and determining whether they are working for a competing service. This surveillance allows routes to be assigned according to criteria that are well defined but not made explicit to the drivers. Sometimes, simply being close to the customer is enough to be assigned a trip; sometimes, impeccable driving is necessary; sometimes, loyalty to the platform is also a factor.

Drivers, for their part, try to obtain more detailed knowledge of these criteria and of the algorithms that, from their point of view, are opaque. They also must overcome the shortcomings of the Uber application interface, in order to try and bypass the automated dispatch system. But to do this, they must perform a true reverse engineering of the platform. In addition to consulting two smartphones at the same time to compare different applications, they must learn to manage their own data and interactions with the app. This is how they can eventually control a large number of parameters and other information.

According to several online forums devoted to sharing tips and advice, Uber drivers appear to spend most of their time on their smartphone screens performing computer tasks.[59] Sometimes their digital labor consists of making sure their schedule is filled for a specific period of time (usually a week). At other times, they must evaluate their performance against the average of others, or against their own scores. The application regularly issues alerts to encourage them to go to locations with low vehicle density. As a result, drivers have to sort out which notifications are important for their business and which are simply ruses of the interface to get them to accept rides.

Drivers' discussion forums can also help us understand another important contribution to their heavy digital workload, namely, avoiding wage theft. Due to the complexity of the app interface, they need to check it constantly. The reason for this is not a design flaw, but a business model based on maximizing the difference between what passengers are willing to pay and what drivers actually earn. Driver attention is required at specific times. For instance, a driver is entitled to compensation in the event of cancellation of their journey by passengers if they cancel it more than five minutes after the reservation. But the Uber application does not display the elapsed time, so drivers are required to undertake complex procedures to claim their compensation. Additionally, the driver's fare is based on the distance measured by the interface. Drivers must then use one or more applications to check that the length of the route has not been miscalculated. It is the same for incentives to work on the weekend, to travel a certain number of miles, or to attend certain events.[60]

But digital labor on Uber is not restricted to the activity of deciphering the opaque procedures that govern its interface. The "social" aspect plays an equally critical role to the technical work. Drivers must make an excellent impression on customers. This implies spending time filling in or updating their profile on the application, choosing a good photo, and managing the relationship with passengers using text messages or calls. In this sense, Uber can be viewed as a social network or a dating site for drivers and riders. Managing reputation scores is a considerable effort for drivers, a real threat hanging over them. In the case of Uber, the driver's e-reputation is not simply a question of popularity or a way of boosting social capital. It has, rather, wholly material consequences: if the score drops below a certain level, the driver's account may be deactivated. To avoid being excluded from the service, the driver must pay attention to communication with the passengers, who are the ones to evaluate the driver and award a score, from zero to five stars, based on service, punctuality, responsiveness, and driving abilities, as well as the vehicle's cleanliness.

Working for Uber involves, first and foremost, clicking on a mobile application; driving only comes second. The time spent prospecting on the app, looking for new passengers, or getting closer to areas of high service demand, is not remunerated by the platform. Drivers call this unpaid time "dead miles," and they naturally seek to minimize it. It can nevertheless occupy up to two-thirds of their day.[61] Research indicates that workers spend an average of twenty-five hours each week on the app, though this number can rise to thirty-eight hours in some cases.[62] Half of those hours can be spent waiting. Although traditional taxi drivers also deal with long waits and empty rides, the on-demand ride-hailing economy has a unique asymmetry between drivers and platforms. Uber drivers receive payment for only a portion of their working time, while the platform also profits from downtime.

Even when not driving passengers, Uber drivers continue to generate data, which the platform exploits. Obviously, it can monetize it and share it with other companies, applications, and public authorities. The information can also be used to improve its products, such as calculating travel times more accurately, assessing routes, or recalibrating suggested routes.[63] Improvements in these areas contribute to its automation projects, enabling it to develop new functions and enhance the performance of autonomous vehicles. Another use for this data—and the one most closely related to Uber—is to align supply and demand. A key feature of this platform is its variable pricing algorithm, also known as "surge pricing."

"Milan like Stalingrad"

Just a few miles from Milan, coronavirus makes its first landing in Europe. We are in April 2020, and Milan, Italy, is the epicenter of the European pandemic. "Milan like Stalingrad," states some graffiti on a wall on Visconti di Modrone Street. Everybody is in lockdown. This city is under siege. There are long lines outside the few supermarkets that are still open. Security agents with thermometer guns screen people for fever and make sure they have written authorization to be outside their homes.

At night, those who are unable to fall asleep step out onto their balconies, which look out onto empty avenues. All the curtains are drawn in the buildings opposite, out-of-service traffic lights flicker orange in the streets. Only the bicycle couriers zoom through the darkness, sometimes solo, sometimes in fleets, to deliver to those craving the small solace of sushi or ice cream, to the fearful who don't dare leave their homes, to the "symptomatics" who are no longer allowed to leave their rooms.

Around midnight, scattered, frenetic, the couriers converge at a metro station to catch the last suburban train home. Destination: the old working-class suburbs in the north of the city. Seveso, Comasina, Nova Milanese. There's an air of Sovietness about their names, too. As the metro train pulls into the station, a hissing sound escapes from the conductor, as if he were wrestling with a canker sore inside his mouth. "There are dozens of bicycles waiting to board. There's no one else but them couriers." No couples, no urban dwellers out for a stroll. Nothing but multicolored insulated bags with Uber Eats, Glovo, and Deliveroo logos.

For people who already had precarious and contingent jobs before COVID-19, the lockdown means being out of work. Globally, unemployment is on the rise. In the first quarter of 2020, twenty million Americans are jobless, the largest increase in unemployment since the Great Depression.[64] Labor markets in Europe are not doing better.

The longer the lockdowns last, the more deserted the city becomes, and the easier it is to spot couriers. They have become unexpectedly visible, and their apps increasingly popular among the newly unemployed as a way to earn a few bucks and escape their tiny apartments. People who formerly drove Ubers now deliver food as well, to make up for the decline in passenger transportation earnings. The number of workers on last-mile platforms has increased in almost all countries affected by the outbreak.

These days, it seems that more couriers are arriving by bike, scooter, and even by car. The more they sign up, the more competition there is on the apps. Competition means more people are ready to work for less, especially if they are new to the platform and want to increase their score. This drags down both prices and earnings on the apps. This is great news for customers, but bad news for couriers, whose pay decreases day by day.

Their jobs are also coming with more risks. Many international delivery platforms have started offering "contactless delivery." The courier leaves the meals in the hallway, not in the hands of the customer, and takes only prepaid orders. Even though the delivery is contactless and relatively safe for the customer, it is still a potentially dangerous task for the courier, who still has to interact with restaurant staff, move through public spaces, and touch potentially contami-

nated surfaces. For delivery workers, the concept of "contactless" is entirely relative.

Sure, union lawsuits force on-demand platforms to acknowledge some health risks and to provide workers with protective gear (like gloves, masks, and sanitizer gel). Around this time, over half of 120 platforms in twenty-three countries have also started offering bonuses and sick days for COVID-19 sufferers.[65] But for the unions these measures are just "fair-washing" and don't really reduce the health risks. In the early stages of the pandemic, without vaccines and widespread testing, couriers struggle to avoid getting infected, and they also struggle to show that they have actually contracted the virus to claim benefits.

Italy ravaged by the coronavirus needs app-based delivery workers, not only healthcare workers and supermarket cashiers. Couriers have indeed morphed into essential workers who provide a vital service. Almost everywhere in the world, they are now included on the lists of key jobs needed for the ongoing operation of the economy.[66]

Platforms, however, continue to value them not for their services, but for the data they produce. Besides delivering, couriers generate information and train GPS systems, targeted advertisements, and dynamic pricing. Grassroots groups have started demanding that delivery workers be compensated for their data as well. A local union called Deliverance Milano has devised "the ten commandments for couriers." As is stated in one of them, "whenever delivery workers are online, they are working and must be paid for the time they are engaged, as if the app is active and the courier is waiting."[67]

Activists, technologists, and researchers are paying more attention to data, especially the data generated by delivery workers. Suddenly, labor laws no longer suffice to protect workers' rights; data protection laws now have to be observed as well. Somewhere in Milan around this time, algorithm analyst Claudio Agosti has decided to work on artificial intelligence. His curiosity culminates in a technical report published by the Brussels-based ETUI Foresight Unit; it explains meticulously how platforms manipulate riders and harvest their personal data.

The report, titled "Exercising Workers' Rights in Algorithmic Management Systems," details the complexity of the machine learning models that govern couriers' work on Foodinho, an Italian subsidiary of the Spanish food delivery giant Glovo.[68] The algorithmic management systems of Foodinho track riders' locations outside of work hours, raising questions about the line between work and personal life. Aside from that, location data and other personal information are disseminated to undisclosed third parties, a notable deviation from the app's promise of transparency. Likewise, Foodinho generates a "hidden rating" based on the rider's profile, which has nothing to do with the official "excellence score" that is supposed to represent each worker's performance.

Then as now, these revelations show how the regulatory framework that governs labor rights and data protection needs to be reevaluated in this AI- and data-driven world. It seems that there is a class conflict over data at the moment. Those who own the platform want to concentrate them in their own hands, while the workers increasingly start wondering how to redistribute them.

Managing Users

"We don't set the price," Uber's founder Travis Kalanick said in an interview dating back to the platform's inception, "The market sets the price [. . .]. We have algorithms to determine what the market is."[69] Like other platforms, Uber strongly defends its claim to neutrality. In its official communication, in front of investors or in court, the company presents itself as a "technology company." Uber spokespersons describe the platform's activity as "real-time data analysis," in sharp contrast with official rulings, such as the one of the Court of Justice of the European Union, which found that the American company is part of a global service whose main element is transportation.[70] As Uber uses this argument every time it is required to comply with transport regulations and labor laws, it may legitimately raise suspicion. The platform's founders are correct, however, in stating that automation, not just transportation, is a major goal. This automation has taken two forms over the years: improving pricing algorithms and pioneering self-driving cars.

To understand Uber's pricing algorithm, I will ask whether the platform, which repeatedly denies being a transportation company, has any managerial authority over drivers and passengers. Uber, according to my own definition of a platform, is both a company with the power to manage labor, and a market that regulates different groups of users (riders, drivers, partner companies, etc.). To grasp what Uber thinks of these groups, it's helpful to consider the patent for its algorithm. There is no mention of passengers or drivers, but instead "a plurality of requesters for an on-demand service" and "a plurality of service providers."[71] As much as the platform claims to provide flexibility and autonomy to both drivers and riders, it is the Uber algorithm that determines prices, standardizes behavior, and assigns rides. The way this algorithm operates reveals how Uber captures value, as well as how it views passengers as producers and not merely as consumers.

An article by the economists Jonathan Hall, Cory Kendrick, and Chris Nosko illustrates this point with a concrete example.[72] On March 21, 2015, Ariana Grande's concert was ending in Madison Square Garden, New York. Thousands prepared to return home. Some were planning on public transport; others had their own car. But a large number were going to open their Uber app to book a ride. Potential customers must do this to flag their presence to the service. Uber used the number of app activations on the smartphones in the Madison Square Garden area over a fifteen-minute period to estimate the number of people who were "on the market" and to gauge potential demand. Uber's problem was to ensure that the supply met the demand. For obvious logistical reasons, it was impossible for thousands of vehicles to wait outside the venue. The platform operates a just-in-time system and, unlike taxi companies, does not manage a stable fleet of vehicles or have access to reserved parking spaces. How, in such a situation, can suppliers and requesters of transport be connected? In other words, how did Uber ensure that the right number of drivers arrived for those seeking rides from the concert hall?

This is where the pricing algorithm comes into play. The algorithm does not calculate the price of a trip from a high-demand zone based solely on journey distance, time of day, or type of vehicle. The algorithm estimates a rough cost, and then multiplies it by a coefficient between 1 to 50. Importantly, the app displays this coefficient on both drivers' and passengers' smartphone screens. For the riders, this means the price is higher than usual; for the drivers, this means that they have the chance to earn more from the ride. The increase, however, is temporary. If a passenger does not finalize their purchase within a certain timeframe, the coefficient will decrease. Passengers develop strategies according to whether they are prepared to pay the higher fare (because they are in a hurry, for example, or because they can afford it), whether they can get home some other way, or whether they decide to try their luck again in a few minutes. In the latter case, surge pricing encourages the customer to log on several times, thus increasing the time spent on the app, the quantity of data produced, and the quantity of data captured by the platform. As they watch the price rise or fall, users have to focus their attention and energy on making the algorithm work for them. For example, they can move to a nearby location to test whether the price drops. The application will of course record the corresponding geolocation data. Or they can open another transport app and compare prices. Uber will be able to deduce this behavior and may then adjust its price.

In a way, Uber acts as a market for services and as software for matching supply and demand. But Uber is most importantly a real-time data analytics company; the algorithm functions as a tool for inciting passengers to produce data. Potential passengers must use the app sufficiently for the pricing algorithm to allocate each of them to a vehicle successfully. The main point here is that when Uber describes the surge pricing algorithm as simply a way of coordinating supply and demand—and not *also* as a data production tool—the app conceals the digital labor of one category of users: the riders.

The contribution made by passengers to Uber's services is particularly clear in this case of variable pricing. However, the production of data by passengers does not stop there. At any given moment, app users who do not opt-out produce data which is sold by Uber to third-party companies. Some of these have direct access to real-time data through an application programming interface (API),[73] which allows them to develop secondary software and services. For example, Uber Trip Experiences API enables passengers to read the news, listen to music, book a restaurant or hotel, or connect to a home automation system during a ride—all of which are activities producing, you guessed it, high volumes of data.[74]

Uber riders, just like their drivers, are rated. The rating system allows drivers (or other passengers, in the case of a shared ride) to assess passenger punctuality, their responsiveness, their good manners, and other criteria that are sometimes arbitrary, but that show that any user (a passenger or a driver) faces the same injunction: to rate in order to be rated, to enter data into the app, and to make sure one's rating does not fall below a certain threshold, after which one is de facto banned from the service. Even though passengers cannot be "fired" by Uber except in particularly extreme cases of abusive behavior, it may become increasingly difficult for them to find a driver willing to take them if their reputation score is too low.

Last, in the words of Carolyn Elerding and Roopika Risam, passengers perform large quantities of "affective digital labor."[75] It involves producing certain moods and behavioral tendencies in other users, through face-to-face interactions and through mobile apps. The experience of the journalist Maureen Dowd highlights this aspect. Upon discovering the rating system for Uber users, she is immediately reminded of how much effort must be put into managing her reputation: "Revealing that I had only 4.2 stars, my driver continued to school me. 'You don't always come out right away,' he said sternly, adding that I would have to work hard to be more appealing if I wanted to get drivers to pick me up. [. . .] Except then I learned that sitting in an Uber car was pretty much like sitting in my office: How much have you developed your audience? How much have you been shared? How much have you engaged your reader? Are you trending?"[76] As with drivers, riders are subjected to an algorithmic management system that influences their behavior, thereby turning them from consumers into workers.

Who Drives the Autonomous Vehicles?

So far, I have reviewed the main methods that on-demand platforms like Uber employ to extract value from the digital labor of its users: work from its drivers, fares from its riders, and data from both.

Some of these methods involve a *labor of qualification*: both drivers and riders rate each other, fill out their profiles, choose routes, and decide how much they want to pay or receive as a payment. Both workers and passengers trigger the Uber algorithm, which organizes rides. For most rides, the company charges a 25 percent commission, but this can be lowered for some routes (such as those that take longer than expected) or in certain countries (like Ghana, where drivers' protests led the platform to lower its service fees to 20 percent).[77]

Other methods produce value through the *labor of monetization* performed by drivers and passengers. They both generate a constant flow of data, which the platform monetizes by making it available to third parties for a fee. This second revenue source is common to business models in the tech sector.

However, Uber's value capture is also focused on new technologies. Its users must provide a *labor of automation* to ensure that the platform's innovation efforts remain productive. As part of this labor, workers and passengers participate, by working or consuming, in the development of automated systems.

"Self-driving cars" are a good example of this type of automation labor. Both large digital platforms and major automakers have been competing to develop autonomous vehicles (AVs). Similar efforts are underway at Google, Apple, and China's Baidu and BYD, as well as Tesla, General Motors, Volvo and Ford. This is a field full of promises, hype, and outright lies. Even though platform owners are overoptimistic, the road from manually controlled cars (also known as "level 0 automation") to vehicles that do not require human intervention ("level 5") has been long and bumpy.[78] Human drivers tend to have accidents every million miles of driving, whereas AVs have accidents every 300,000 miles.[79]

Nevertheless, big tech companies have been massively investing in autonomous vehicles since the late 2000s. Along with Lyft and Google, Uber is part of a lobbying group, "Self-Driving Coalition." In 2015, Uber opened the Advanced Technologies Group in Pittsburgh, Pennsylvania, a research center devoted to mapping, vehicle safety and autonomous transport. The following year, the Coalition launched its first experimental program of self-driving cars.[80]

In subsequent years, various nations have developed regulatory frameworks for automated vehicles, and fully autonomous taxis have been authorized in several countries, from the US to China.[81] In the meantime, the self-driving car unit of Uber has merged into Aurora Innovation, a startup backed by Amazon.[82]

There is a fundamental misunderstanding concerning this technology. It is assumed that autonomous vehicles are "driverless." AVs are marketed under the assumption that robots may replace humans as drivers, one day. Sadly, this day just never comes. In 2016, Elon Musk announced Tesla's Full Self-Driving solution. Since then, he has repeated that autonomous vehicles would be available within "a year"[83] but never delivered on that promise. To date, "fully autonomous" car experiments have failed miserably. The deployment of GM's Cruise robotaxi in San Francisco has been an unmitigated disaster. Launched to great fanfare in February 2022, Cruise ended its operation the following year, after a particularly gruesome accident where one of its cars dragged an unconscious victim several feet before stopping.[84]

The fantasy of a completely autonomous vehicle ignores the reality that these cars are operated and overseen by humans. Human input is required to "teach" vehicles how to drive, and on the other hand to continue driving them despite their supposed full autonomy. The vehicle is continuously supervised, either by people inside or by remote workers. "Live persons monitor the vehicles and communicate via the vehicle," according to San Francisco law enforcement guidelines.[85] This leads to the central question: at what point does "live control" become "remote driving"? Do self-driving cars really operate like remote-controlled toys at 1:1 scale?

I should stress that there are no truly autonomous vehicles today. At the inauguration of the Uber self-driving unit, Travis Kalanick was forced to admit that "no one has developed software that can drive a car safely without a human."[86] Years later, the situation remains unchanged. This is why the experimental vehicles that the platform has launched since then include one or even two "vehicle operators," who sit at the wheel to see to the car's maintenance and deal with unforeseen circumstances—to avoid hitting vehicles, killing cyclists, and other accidents.[87] Operators are described as "safety drivers," and are front-line workers forced to take risks inherent in the design of the car. Are the brakes malfunctioning? Does the software fail to recognize traffic lights? The safety driver is there to handle the situation.

Unfortunately, when unreliable technologies are combined with the burden of responsibility placed on a single worker, tragedy can strike. This was the case in the fatal accident on March 19, 2018, involving an Uber AV, which ran over a woman crossing a dark road in Tempe, Arizona.[88] The onboard camera of the autonomous vehicle shows a person seated behind the wheel—the operator. The prosecutor determined that the platform was not criminally liable, and thus that this operator remained the only person liable in court,[89] After a smear campaign in the press, in which the operator was accused of streaming *The Voice* on their phone instead of checking the road (in fact, they were distracted when monitoring the Uber tablet displaying the car data),[90] they were charged with negligent homicide and took a plea deal by agreeing to three years of supervised probation.[91]

The need for in-vehicle operators is regarded as temporary by AV manufacturers. Despite its frequent debacles, they assure the public, fully automated cars are just around the corner (given their danger, thankfully, not literally). However, vehicle engineers' ambitions meet limits independent of technological innovation. Even if they managed to digitalize all the processes needed to drive a vehicle, they would have to overcome the hurdle of passengers' psychology. Riders need to feel at ease and in control. Autonomous vehicle manufacturers have accordingly planned to intersperse the journey with voice messages, with alerts on applications, with videos—and, of course, with advertisements. Passengers must therefore always carry out the work of paying attention to and managing these signals. Another limit, physiological this time, is the discomfort passengers feel at the slightest acceleration. The proposed solution is turning passengers into, well, drivers, entrusting them with the responsibility of either handling the vehicle itself or the other interfaces present inside.[92]

A document from Waymo relating to the security of transport on-demand services gives us a glimpse of how all this functions. Unlike with Uber, there is no "operator" inside a "fully automated" car. So, as part of their experience, passengers have to perform a series of actions on control screens:[93] hail the ride through a mobile app, sit inside the vehicle, press the "Start the journey" button. On board, they must pay attention to the automated driving by checking the route and assessing how well the vehicle understands its environment. Seeing "via the vehicle," or rather via the AI that drives it, thus becomes an activity from which it is possible to anticipate a danger. In the case of an unforeseen event, the occupant presses a "stop" button, which triggers a procedure to identify the safest place to get off the road. In the event of a collision, or of poor weather conditions, the passenger must press another button to call the technical support team and report the situation using the audio and visual information provided by the interface of the onboard system.

For the supposedly self-driving car to function, passengers must perform all these tasks. A "driverless vehicle" means that the passenger has to become the driver. Or rather, AI-based transportation solutions combine the roles of passenger and driver into one category—the operator.

The passenger not only performs the not-so-hidden work of driving, but also produces valuable data. First, a lidar (Light Detection and Ranging) is used to capture and transmit to a server the visual recordings made by a driverless car. A lidar is a rotating laser coupled with a 360° camera and a radar that scans the vehicle's environment from its roof. Additionally, it is equipped with a GPS, a gyroscope, and an accelerometer. The purpose of all these sensors is to detect other cars, pedestrians, and nearby objects. Since anyone traveling on the same route as an intelligent vehicle may feature in these recordings, this technology poses legal questions regarding privacy and ownership of personal data.[94] These questions are just as relevant for the occupants of the vehicles themselves, whose information feeds the platforms' databases.

All of this information is crucial for the automation process. Machine learning requires examples, concrete cases. There are billions of examples to be found in autonomous vehicles equipped with lidars and cameras mounted on their roofs. Ultimately, the performance of a self-driving car depends on how much data it collects: Waymo has created a public dataset of data accumulated by its cars over twenty billion miles,[95] Tesla hoards petabytes of data produced by its own customers,[96] and Uber's continued involvement in Aurora's self-driving car startup can be attributed to Amazon's data storage and computing capabilities.[97]

Automated cars are data collection devices on wheels that may or may not deliver passengers to their desired destination. They certainly deliver valuable information to other cars and to the central servers of the platforms that develop them. This does not mean that they are able to interpret and give meaning to the data. For that to happen, it is necessary to emphasize again how much human labor is required. Harvested data must be annotated, and this work is carried out by users of other platforms.

Armies of *microworkers* scattered all over the world, "teach" automated systems to distinguish between the different elements that make up the ever-changing environment of the vehicles. In Uber's promotional videos, these workers are described as highly specialized "cartographers" who label each frame of every video sequence recorded by the lidar of autonomous cars.[98] The reality is very different.

As Anthony Levandowski, the former project manager of Google Street View and disgraced chief engineer of the autonomous vehicle program at Uber, puts it, the individuals who tag images of roads, of traffic signs, of the weather conditions, are nothing but "human robots."[99] He is not referring to half-humans/half-machines hybrids, but instead to workers who perform tasks that robots cannot. Sitting behind their screens, they draw red squares around trees, add a filter to adjust the lighting of an image of a tunnel, highlight road signs, and distinguish a child from a lawn ornament. The safety of pedestrians and passengers depends precisely on the accuracy of this human-performed work. In the next chapter, we will meet these hired hands.

· 4 ·
MICROWORK

If we follow the data extracted from Uber, Airbnb, and the other on-demand platforms described in the last chapter, a likely next step is "microwork" platforms, where workers refine that data, further increasing its value. In microwork, tasks are fragmented and performed by hired workers on websites, apps, and other services. Amazon Mechanical Turk and Clickworker are examples of these platforms.

Various terms may be used to describe this occupation. It's sometimes called "crowdwork" because of the huge workforce platforms require,[1] but also called "cloudwork" due to its remote nature via cloud computing infrastructures,[2] or "data work" simply to emphasize how important this kind of digital labor is for data labeling and processing.[3]

The term microwork refers to the execution of "microtasks." It involves standardized, routine tasks that are considered low-skilled, but actually require certain competencies. Microtasks are performed for clients, sometimes called "requesters." Among them are companies, public bodies (especially research centers), and individual citizens. This type of digital labor is inseparable from "human-based computation," which historically consists of tasks that machines cannot handle and are delegated to humans.[4] Microwork thus involves performing small chores such as annotating videos, sorting messages, transcribing scanned documents, answering online questionnaires, correcting values in a database, and linking similar products in an online sales catalog. Platforms pay workers for these activities, offering them sums ranging from a few dollars to less than one cent per task.

As in the on-demand economy, platforms charge a commission on any transaction (usually to requesters, not to workers), although unpaid forms of microwork are also common. Struggles and conflicts related to workers' rights have erupted here too, but workers are less organized, and protest is less frequent, than in the on-demand economy. Often, conflicts arise when investors and designers attempt to place these tasks outside of the context of "real work." Microwork platforms' interfaces can be designed to be playful and appealing, so that workers will forget the drudgery of the tasks they perform. Compared to the jobs in the "on-demand" economy, it is relatively easy to deny that this is standard work, in part, because the pay levels are very low: users and platform owners agree that microwork can at most provide a supplementary level of income, especially in countries with high cost of living. Microtasks typically have a very low entry level requirement, which explains why they are often the first steps into the labor market for people with extremely varied educational backgrounds, language skills, and work experience.

The Mechanical Turk, or the Ruse of Artificial Intelligence

In 1836, Edgar Allan Poe devoted a long essay to "analytical machines," considered today to be precursors of computers: "What shall we think of the calculating machine of Mr. Babbage? What shall we think of an engine of wood and metal which can not only compute astronomical and navigation tables to any given extent, but render the exactitude of its operations mathematically certain through its power of correcting its possible errors? What shall we think of a machine which can not only accomplish all this, but actually print off its elaborate results, when obtained, without the slightest intervention of the intellect of man?"[5]

Despite his string of grandiloquent questions, Poe seems unimpressed by Babbage's achievement. Poe cites the father of modern computing as a foil for "the most marvelous invention of mankind," the real protagonist of the text, namely the chess-playing automaton decked out in Ottoman costume, whom the Baron Wolfgang von Kempelen presented to the Viennese imperial court in 1769.

Babbage's calculating engine, Poe argues, is a simple machine "modeled by the data" it receives from its inventor. Kempelen's mechanical chess player, by contrast, generates its own data, and thus reproduces the functioning of the mind. The chess player is, in a way, a first example of artificial intelligence. Like many of its modern counterparts, this technology also requires a "human agency within" to function.

Poe reviews various speculations on the "mystery of the automaton," among them the hypothesis of a "well-taught boy, very thin ... so that he could be concealed in a drawer almost immediately under the chessboard"; or that of "an Italian in the suite of the Baron" who was of "medium size," with "a remarkable stoop in the shoulders" (a century later, Walter Benjamin would make him into a "little hunchback.")[6]

The "Mechanical" Turk is a perfect illustration of how human labor is still crucial when machines must do more than simply carry out the instructions given to them, but also to issue instructions themselves. The crowning achievement of the scientific and industrial program of machine learning is when machines manage to reproduce human behavior; but to do so, they must first learn from humans. It is hardly surprising, in this light, that the inspiration behind Amazon's microwork platform was the eighteenth-century Mechanical Turk.

Sometimes, even the simplest activities are too complex for a machine. For a long time, Amazon's problem was eliminating duplicates from its vast catalog. In the early 2000s, when software solutions to the problem proved ineffective, Amazon's computer engineers had the idea of recruiting large numbers of people to detect duplicates in a few pages each, paid on per piece rate. From there, it was a short step to letting other companies benefit from this system and taking a commission for brokering the connections: in 2005, Amazon Mechanical Turk was born.[7]

When Jeff Bezos wants to seduce new investors, he shows them drones delivering goods or algorithms recommending the perfect product. But, behind the scenes, hundreds of thousands of "Turkers" are sorting addresses by hand or classifying every product in the catalog by relevance. This upstream work (providing examples to train the software) and downstream work (checking that the results are correct) is vital for the business successes of Amazon and the companies in its ecosystem. This, in the jargon of the platform, is an example of "artificial artificial intelligence."

In a 2006 talk given at MIT, the CEO of Amazon presented Mechanical Turk as one of three pillars—with cloud storage and data security services—of Amazon's big data strategy. The fact that Mechanical Turk was placed on par with these billion-dollar businesses suggests that the microwork platform was a major strategic choice. Mechanical Turk, Bezos maintained, switches the respective roles of computers and human beings. It can code and integrate "human intelligence into software." Amazon may be best known for online shopping, but at the time of this talk its business model was transitioning to selling IT solutions in the form of "software-as-a-service": software is not installed on the computers of those who purchase it, but rather on a remote platform owned by a tech company. The same applies to microworkers. They are not located on the premises of the companies that hire them. They exist only as a remote workforce available through a platform.

"This is basically," Bezos concluded, "people-as-a-service."[8]

The Reserve Army of Artificial Intelligence

The idea behind Amazon Mechanical Turk, or MTurk, is pretty simple. The platform is a web portal that can only be accessed by choosing one of two roles, "requester" or "Turker." A requester is a client who needs to set up a workflow that can't be handled by a single person and for which a machine would be insufficient. For example, a company that has scanned its accounting records for the past fifty years would be faced with a mass of images of

handwritten pages that text recognition software can only partially interpret. This work would take twenty years for an employee equipped with a
computer, a whole year for twenty employees on fixed-term contracts, six
months for forty interns, and so on. If the company posts an ad on Amazon
Mechanical Turk, asking 200,000 people to transcribe two lines each, it
will cost infinitely less than twenty years of salary.

The majority of the platform's workers, the "Turkers," are US citizens.
The percentage was 72 percent in 2017,[9] but it rose to 90 percent before
the pandemic and has remained stable ever since.[10] The other workers are
mainly Indian.[11] Turkers located in Europe, South America, Africa, and
Oceania who remained active on the platform make up a tiny percentage.[12]
Due to US tax laws concerning platforms,[13] and to limit the coordinated
actions of microworkers, who in the past have tried to take advantage of
the service,[14] Amazon confined new user registrations to US workers alone.
However, there is substantial evidence to suggest that Turkers use virtual
personal servers and other tools to mask their locations, giving the impression that they are based in the US.[15] The platform is primarily used by
microworkers between the ages of twenty-five and thirty-five. In terms of
gender distribution, MTurk showcases almost perfect gender equality.[16]
Different countries have different gender ratios; women account for only
33 percent of microworkers in India, compared with 50 percent in South
Africa and 48 percent in the United States.[17] Earlier studies linked gender
with income inequality. In the US, Amazon's Turkers come from house-
holds with a median annual income of around $50,000, so the platform is
primarily used for earning extra income. In India, where men are the majority of workers, microwork is sometimes the primary source of income for
families with a median annual income of around $10,000.[18]

The different rates of pay for the tasks negotiated on the platform are
thus dependent on the demographics of the users. The price of each task can
theoretically be a few dozen dollars, but before the 2010s, 90 percent of the
tasks paid fewer than 10 cents, and the vast majority paid barely 1 cent.[19]
These rates have been subsequently increased, but even if some workers
reached an hourly wage of up to $8, for more than half of them, it never went
higher than $5.[20] In fact, ever since a study claimed that the median hourly
wage on Amazon Mechanical Turk was $1.38,[21] requesters have tried to im-
pose this rate on taskers as a kind of "maximum hourly wage." As a result,
median wages have hardly increased at all; according to the most recent estimate based on almost four million tasks, the median hourly rate was not
higher than $2, and only 4 percent of the workers earned more than $7.25
per hour.[22] There is also a significant wage gap between women and men:
women, especially if they take care of children, earn 20 percent less per hour
on average, despite no gender differences in task selection or experience.[23]

These discrepancies in pay levels provoked angry protests among Turkers. But there is a further problem because hourly rates are difficult to calculate for jobs paid by the piece, and the rate is subject to many variables: How long can workers stick at the job, or how much time are they willing to devote to this activity? With better skills, can they access better-paid tasks? How quickly can they accept the tasks and complete them? The possibility for Turkers to earn a sum equivalent to a minimum wage monthly is in fact jeopardized by the limits placed on the number of tasks that can be performed in a day (between 100 and 3,800, depending on the status of their account, but the number is actually way lower, since workers are sanctioned for performing tasks too quickly).[24] The use of bots, scripts, and generative AI to speed up certain tasks is frowned upon, and workers avoid it for fear of being banned.[25] Mechanical Turk is a valuable service because each microtask must be performed by a human.

However, Turkers experience systematic misrecognition of the value of their contribution. In the platform's jargon, every microactivity performed by a Turker is a "human intelligence task" (HIT). This does not mean that workers perceive it as such. Apart from the rare HITs when workers need to write structured product reviews or complete lengthy questionnaires, microtasks almost always appear to be mundane operations that require little skill. Here are some examples: "read a web page and whenever you come across an email address, write it down in a .txt file," "put together a reggae playlist," "watch a fifteen-second video and choose three words to describe it;" "take a look at the scanned image of a receipt, then transcribe it," "spot all the hooded people in a sequence shot by a surveillance camera," "select all the images of hot dogs in a series of ten images of food products," or "watch a movie and take a screenshot of a Hollywood actor expressing fear or disgust." These tasks require little expert knowledge, provided one is tech-savvy enough to operate a computer or a smartphone. The main contribution they make is to inject some human common sense in a digital operation. In fact, they are not always easy to perform, and may require studying lengthy instructions or downloading unsafe software. Microworkers are typically viewed as unskilled by requesters and platform owners in order to justify low wages. Across microwork platforms, however, workers display a high level of education and possess a wide array of skills.[26]

Microtasks often call on thinking that requires a nuanced and subjective analysis. This is why a large part of the platform specializes in developing large language models, in complex image recognition,[27] and in "sentiment analysis."[28] Polling companies can use this type of analysis to assess the general outlook of politicians based on the social media messages about them during a televised debate, or to rate the impressions a product generates by the comments posted on a consumers' forum. In the US presidential campaigns of 2016 and 2020, Pete Buttigieg, Ted Cruz, and Donald Trump all used Amazon Mechanical Turk during their campaigns.[29]

Gamification and Qualification

As with all forms of digital labor, debate surrounds the nature of work on Amazon Mechanical Turk. Can a HIT be qualified as work? According to researchers who study microwork, Amazon invented Mechanical Turk so that it could outsource certain processes and reduce its costs.[30] There is no doubt that Mechanical Turk is a commercial operation, built around client demand for digital labor. The workload on the platform is based on the working week. The weekly distribution of the HITs follows the calendar of the requesters, which runs from Monday, when the HITs are put online, to Friday. With tiny differences due to geographical time zones, workers complete these tasks between Tuesday and Saturday.[31] Weekends are relatively quieter, and on Sunday, HITs seem to be accepted by older, less experienced Turkers with a full-time job the rest of the week.[32] Turkers have access to a micropayment sheet that summarizes all the tasks they have completed and all the payments they are entitled to.

However, the platform always tries to position these activities outside of work, and to characterize them instead as sociability and play. The interface gives Turkers the impression that they are on social media. The workers access Mechanical Turk directly through their Amazon account, where they can still do what anyone using Amazon does: shop and leave comments on products. The fact that many microtasks involve managing multimedia content (watching videos or reading reviews, for example) adds a fun element to the work and allows the platform to claim that there are few tasks of drudgery, and that its workers are also consumers. Additionally, the platform's designers "gamify" the work: to assess the skills of the Turkers, many of the HITs are actually questionnaires resembling personality tests. The most skilled and zealous workers receive badges that they can add to their profiles. Profiles also display their performance scores, including, for example, the percentage of tasks completed and positive comments. The service emphasizes emulation, social interaction, and a sense of community in combination with visible control mechanisms.

This "community" is not necessarily based on affinity and cooperation, and the playfulness often takes a competitive turn. Amazon Mechanical Turk makes users compete to access tasks. In most cases, these are only advertised for a short time, ranging from a few hours to a few days, and then vanish. Some workers install "HIT scrapers," small software programs or web browser extensions that help them to select the microtasks that will make them earn more, and more quickly. After each task is completed, the requester evaluates the work and can validate the task or refuse to pay for it. Turkers receive payment in the first case; otherwise, they get nothing or even lose accuracy points. The fact that the requester can refuse to pay the Turker at any time is another aspect of gamification. It suggests that, for the platform owners, the monetary exchange between the two parties is only of minor significance. And indeed, Amazon tends to construe microtasks as symbolic incentives only, called "rewards,"[33] and not as wages. In contrast, Turkers cannot be expected to work for free, because they tend to have lower incomes and are more likely to be unemployed than the general population.[34]

To add to their economic difficulties, there are a large number of Mechanical Turk activities that are unpaid. They consist of tests that allow Turkers to gain skill certifications. According to a survey conducted by the International Labour Office on 1,100 microworkers, a typical week involved an average of 28.4 hours' work, of which 6.6 hours were unpaid. Nearly a quarter (23.2 percent) of working time was therefore spent on unpaid tasks, which amounts to saying that "for every hour of paid work, workers spend eighteen minutes on unpaid prospection or preparation activities."[35]

In the event of a conflict between worker and requester, Amazon refrains from intervening. This "neutrality" exposes the Turkers to exploitative practices,[36] to which there are two possible solutions: to negotiate directly with the requester, or to retaliate by evaluating the requester negatively. The ratings of predatory requesters are often displayed in "halls of shame" in dedicated forums. An advocacy group called Turkopticon developed a homonymous application that places a button next to each requester and highlights those whose reviews are available. At one point, the app became the standard reference for measuring the honesty of requesters because of the relative subtlety of its criteria ("communicativeness," "generosity," "fairness," or "responsiveness").[37]

Gamification thus introduces competition, and mutual surveillance between users (both requesters and workers), as ways of disciplining work. It also enables Amazon to capture the value of the mutual evaluations data. As with Uber, where drivers and riders rate each other, requesters and Turkers of Amazon Mechanical Turk assess each other as well as the goods, content, and data that make up the microtasks. This labor of qualification is necessary for Amazon to be able to select the most motivated workers, and the most scrupulous requesters. And this, in turn, ensures that the marketplace for microwork continues to maximize the company's profits.

Monetizing Microtasks

The contracts proposed by the platform to the microtaskers likewise attempt to skirt round the issue of employment. Using a "participation agreement" instead of work contracts, the platform avoids any suggestion of subordination between the Turkers and the platform or the requesters.[38] As with on-demand applications, the platform depicts workers as independent contractors who "perform tasks for requesters in their personal capacity." The platform systematically blurs its role as an intermediary and presents itself simply as an online resource in which requesters and workers spontaneously connect. Yet Amazon serves as an intermediary, and its monetization is another source of value that the platform captures. Three tools are used to this end: commissions on microtasks, the resale of personal data, and the Amazon payment service itself.

For commissions, whenever a requester publishes a task, the price they pay to Amazon consists of the "reward" for the microworkers, and a commission for the platform. Officially, this can range from 20 percent to 40 percent of the amount paid to the Turkers. But there are added costs depending on the type of work required. If the requester wants their microtasks to be carried out only by expert workers ("Master Turkers"), they pay an extra 5 percent. If they want them to be carried out by specific segments of the population, they will have to choose between 132 criteria ("premium qualifications": age, gender, education, sports activity, digital literacy, language, etc.), and pay an additional sum.[39]

Amazon thus manages to earn the highest commission for every task posted by a requester. But there is another aspect to this higher price. The premium qualifications actually include personal information like education, location, and income that Amazon requires from its Turkers. This data can be monetized by charging an extra commission on microtasks. In other words, the microwork platform, as with other types of digital labor, engages in a systematic commodification of its users' data from which it gains a regular income that is not redistributed to its workers.

MTurk thus prospers through renting out the Turkers' labor to clients and capturing their personal data in the process. This labor of *monetization* is, in addition to the labor of qualification routinely, carried out to keep the platform running. Another income source derives from the specific payment methods demanded by Amazon for posting HITs. The platform acts as a bank. It requires clients to purchase prepaid credits through Amazon payments before they can post a task. These credits can only be spent on Amazon services and the amount purchased must be greater than that promised to the workers. Amazon payment is offered both to the requesters and to the workers, who receive their "rewards" in the form of dollars on Amazon gift cards. They can leave them in their account, use them to buy products from Amazon, or transfer them to a bank account (although this last option is limited to the US and a few other countries). This further monetization allows Amazon to create cash reserves that it can invest elsewhere.

The Third-Party Beneficiary

Amazon has created an equilibrium in which requesters and microworkers constantly compete, while the platform extracts data from one group of users to sell to the other. Despite the tensions between workers and clients, by portraying the Turkers' productive acts as play, the platform evades questions of fair pay and labor law. Logically, competition and play, which are inseparable from the methods of extracting value through qualification and monetization, should balance each other out. They allow Amazon Mechanical Turk to create a perpetual state of mistrust between microtaskers and requesters, without leading to a major conflict.

With Uber, Deliveroo, and other on-demand platforms, labor disputes target the platforms themselves, their algorithmic management rules, and their terms of service. Due to their local nature, tensions tend to concentrate in one place, as anti-Uber rallies, customer boycotts, and courier strikes in cities all over the world have demonstrated over the last decade. Microwork platforms differ in that the work itself is remote and workers are dispersed geographically.

Amazon has managed to present itself as completely neutral, a technical tool that disintermediates work, and fades into the background except when enabling communication between users. Thus, for the Turker, it is the requester more than the platform that resembles an employer. Conflict manifests itself as a dispute between two categories of users rather than as an employee lawsuit against the platform.

I have already mentioned Turkopticon, which allows microtaskers to monitor requesters and comment on their behavior. Similar tools exist (MTurk Crowd, Turker Nation, TurkerView) that not only allow workers to review other users and HITs, but also to coordinate and document abuses. Amazon does not feel threatened by these online resources, and it doesn't interfere with Turkers and requesters evaluating each other. On the contrary, by creating a competitive environment for users, it lessens their chances of organizing collectively. Moreover, as a *tertius gaudens*—a third party that benefits from requester and Turker disputes—Amazon captures the value of the ratings provided by microworkers and requesters, and thus makes the most of its role of multisided coordinator.

Despite its efforts, Amazon has not completely avoided the crossfire. In 2014, Amazon Executive Chairman Jeff Bezos received "Christmas cards" directly into his email box from Turkers expressing their demands and grievances, especially regarding the platform's design and functionality.[40] The microworkers tried to describe the features that were directly responsible for earnings loss or for needlessly arduous work. Some asked for the creation of an independent arbitration panel to resolve disputes; others wanted to know the precise criteria for becoming a Master Turker; some challenged the negative reviews by their clients, resulting in an overall score decrease. These demands, which seemed to focus on technicalities, in fact showed how greatly users' activities corresponded to what is known as "work" from a legal standpoint. According to Turkers, Amazon should assume responsibility for managing conflicts and promoting careers, just like an employer would. Bezos, who is known for his cavalier replies to emails, did not respond directly to the Turkers' messages.

In any event, the users' protest was quickly stifled. It was mainly the "guild" of microworkers, WeAreDynamo, that suffered. This quasi union was created by American and Canadian researchers and activists who had connections with the creators of Turkopticon. Initially, its activity confined itself to exhorting requesters to maintain a level of ethical conduct in their recruitment and remuneration of microtaskers. Their campaigns did not seem to bother Amazon. However, when Jeff Bezos was addressed personally in the letter-writing campaign, the platform got wind of the fact that WeAreDynamo aspired to become a channel for union organization rather than just a tool for the Turkers. Amazon nipped this in the bud. As a requirement for becoming Dynamo members, workers had to verify they were Turkers by completing a short task. Amazon shut down the account Dynamo used to post those tasks, so the guild was unable to recruit new members and eventually fizzled out.[41]

In the history of MTurk, this guild's vicissitudes mark a crucial milestone and a first significant effort to organize. The way this initiative was stifled angered the Turker community and led to attempts to circumvent Amazon by inventing an ethical alternative to the Mechanical Turk. In the wake of the Dynamo debacle, researchers and workers created Daemo, an experimental service presenting itself as a "self-governed" microwork platform where power is distributed in a balanced way between requesters and microtaskers. Tasks, objectives, and pay were co-defined. A distinguishing feature of Daemo is that it was born out of a labor dispute and aimed to continue the tradition of cooperative self-management and control that existed well before the tech sector.[42]

Ultimately, Daemo failed to compete with profit-oriented platforms such as MTurk and fell into oblivion, but it paved the way for subsequent platforms to comply with labor laws. Since 2019, LeadGenius has been positioning itself as "a software company with the heart of a social enterprise," by offering mentoring between expert and novice microworkers, as well as pay in line with the living wage in the users' countries.[43] In 2022, Appen, an Australian platform with over a million microworkers globally, implemented a crowd code of ethics that emphasizes minimum wage payment, inclusion, privacy, and worker feedback.[44] The following year, a consortium of German crowdwork platforms, which claim several million microtaskers, adopted an ethics code that outlines legal compliance, minimum pay, training, respectful interactions, job clarity, workers' rights to refuse dangerous or questionable tasks, and data privacy.[45] Even Google Deepmind, whose ethical standards are not legendary, incorporated into their internal operations the guidelines for responsible data supply chain management developed by the NGO Partnership on AI in 2022.[46]

Although Daemo is an influential example, it doesn't lessen the difficulty of collective mobilization in a platform economy. If Amazon refuses to assume its status as an employer, dissatisfied workers can of course go over to a competitor, but they will encounter everywhere the same modes of value production. While several platforms today emphasize the importance of organizational transparency and open governance, they practice the same type of microwork that enables requesters to outsource huge quantities of business operations at prices that are systematically lower than their profit margins. WeAreDynamo and Daemo's experiences prove something else: microwork platforms are becoming hubs for protest movements. These movements have the potential to reform microwork by showing that the labor performed on them is "real" work, and that the platforms are employers with real responsibilities toward their workers.

In the Wings and Behind the Scenes

Like Uber, Amazon has been experimenting with autonomous technologies. Jeff Bezos's company has launched a number of drone and robot delivery programs between 2016 and 2019, but they failed to meet expectations.[47] However, Amazon's innovation efforts mainly focus on capturing third-party value from its users' productive acts, the *labor of automation* to feed its machine learning solutions. In addition to the labor of qualification (when users evaluate products and rate each other), and monetization (when Amazon resells data, takes commissions on HITs, and centralizes payment methods), humans also step in for machines in the "human-based computation" approach already discussed. That is, average humans do the work that intelligent systems and software are not capable of doing.[48] In traditional machine learning, domain experts acted as a "teachers" and "supervisors" of the algorithms by annotating data and evaluating the model's performance. Unfortunately, these specialists were expensive and ill-suited to working on large-scale projects. Moreover, science and technology projects today are often beyond the analytical capabilities of local research teams or of a single expertise field.[49] This led to the adoption of a new method that relied on untrained individuals. Progressively, machine learning started using large masses of interchangeable nonspecialists, who could be paid less and hired by the task.[50] Amazon clearly pioneered the second method, as the company discovered that crowds of microworkers were just as competent as a few hand-picked experts when it came to disambiguation in automatic text interpretation,[51] creating sound archives of conversations to calibrate voice interfaces,[52] and annotating images for the visual recognition of forms.[53]

Microtaskers are thus almost indistinguishable from software units, as suggested by the design of the Mechanical Turk's API, which makes it possible to publish a HIT in a single line of code. Imagine a company that wants to devise an application that can suggest places to visit in New York in real time. It can first make a file of tourist sites in the city, and then program some commands that will select a random entry from it when a user opens the application. But these random suggestions may quickly become irrelevant. To resolve this issue, the company can sign up as a requester on Mechanical Turk and replace the initial lines of code with others that will automatically launch two HITs: "Suggest an interesting place to visit in New York today," and "Vote for the best place among those suggested." These lines would look roughly like this:

```
ideas = []
for (var i = 0; i < 5; i++) {
idea = mturk.prompt(
"What's fun to see in New York City?

    Ideas so far: " + ideas.join(", "))
  ideas.push(idea)
}

ideas.sort(function (a, b) {
v = mturk.vote("Which is better?", [a, b]) return v == a ?
  -1 : 1
})54
```

Recruiting human labor can thus be as simple as pressing a button on an interface—quite literally. Other platforms have followed in Amazon's footsteps: Scale AI for example, promises to have a task carried out on demand by human beings, with a simple line of code.[55] Developers of mobile apps can call on microwork platforms to run their image recognition software (for example, a user takes a picture of a mushroom, microworkers adjudicate that photo against a database of mushrooms, and the app tells the user whether the mushroom is poisonous), audio transcriptions (a user records a conversation, microworkers transcribe small segments to create the full text), or information systems (users enter the words "lawyer + German commercial law," microworkers suggest three profiles, chosen from the professional network LinkedIn).

In these cases, humans play the part that platform designers conceived for machines alone. This blurs the line between automated work and work delegated to microtaskers. Automation boils down to a simple formula: a facade with engineers who boast of the spectacular performance of their machines, and a backroom where workers toil away at microtasks.

But humans are not only substitutes for machines. An important element of Amazon's business model and of other microwork platforms is teaching machines. Whether teaching a piece of software to recognize patterns in moving images, or to read text or interpret voice commands, machine learning requires large numbers of examples drawn from everyday experience, organized in pre-processed databases. Crowds of human users must produce the examples, perfect the databases, and check and sort the results produced by the machines. AI solutions often incorporate this process directly into their name. The "P" in GPT, for example, stands for "pre-trained": the AI model has previously been trained by human annotators.[56]

These microworkers produce the sounds, images, and texts that feed the internet giants' databases. Their tasks may include recording brief conversations, taking screenshots, or even writing short texts. But once these examples have been collected, they comprise just a mass of "raw" data that is often unlabeled, nonstandard, and contains duplicate or incorrect information. The work of streamlining and refining the data can be the most difficult and labor-intensive phase of machine learning.

Developing a translation software, for instance, requires a large number of examples of conversations in several languages, as well as accurate annotations of each word to identify polysemy and idioms. Microwork naturally seems to be an excellent way of reducing costs and time. Microtasks are put in place to transform raw data into "annotated" data that can be used to train algorithms that perform various operations systematically and precisely. Here's a simple translation algorithm: "Select a term 't' in German, choose another language 'a,' and produce the equivalent term 't(a)' that has the same meaning according to the examples contained in the United Nations terminology database." But to ensure that it always selects the best possible pair of terms 't' and 't(a),' it needs to be repeated over hundreds of thousands of cycles for several values of 'a.' It is useful to compare it to other algorithms as well. Some of those are based on criteria that are more suitable ("Choose a term 't' and a language 'a,' and select the entry 't(a)^{b}' from a bilingual dictionary"), while others are less relevant (e.g., "Choose a term 't' and a language 'a,' and produce the term 't(a)^{s}' that begins with the same letter"). The machine "learns" by comparing different models and different ways of functioning. But then, as in school, it has to pass a test. It is at this point that humans come back into the picture, to evaluate the results by voting, measuring, and selecting the best outcome. What, for instance, is the best English translation of the German word *Bild*? Is it a "picture," a "metaphor," or the title of a Berlin newspaper? The context will tell us. Unlike machines, humans are excellent at assessing context. It is humans who will ultimately decide which algorithm has given the best answer.

Obviously, these examples downplay other aspects of machine learning, namely the parameters used by the algorithms (the "software layer" in engineering sciences) as well as the computing power (the "hardware layer"). On microwork platforms, the "human layer" trumps them both. But its victory is bitter-sweet: the loss of dignity of work in the age of machine learning can be attributed to the informal nature of microwork's contracts, the lack of visibility of human contribution to AI, the competition among workers, the low remunerations, and the considerable latitude given to platforms to make arbitrary managerial decisions.

It is because humans' work is so devalued that artificial intelligence is an ethical and social issue. As a result of platform architecture, workers are unable to see what the ultimate goal of their activity is, since their labor is divided into a variety of microtasks.

Still, there is a deeper anthropological issue. Algorithms are artificial objects that must produce meaningful results in a human world, which they cannot experience. They do not inhabit a cultural and social environment, so they have to rely on humans to make sense of it. This need for pragmatic understanding has even been incorporated into the commercial names of Amazon's services. Clients can now access MTurk workforce using an interface called "GroundTruth"—an allusion to the commonsense knowledge of the world that microworkers possess.[57]

One of many paradoxes of digital labor is that the productive activity that allows Amazon to capture the *value of automation* is a dehumanizing process that reduces human creativity to small, data-driven tasks. And yet these tasks remain the most properly and irreducibly human facet of work in the age of platforms.

The Trainer, the Verifier, the Imitator

Is microwork going to stop someday? Will we still need it after we have trained all the AI models and annotated all the data? Isn't this 'human in the loop' labor likely to be a temporary thing? Unfortunately, no. Here are a few examples from my own fieldwork in three countries.[58] A microworker's contribution to machine learning solutions is not limited to the preparation phase. I would summarize these three contributions as follows: artificial intelligence needs trainers, verifiers, and imitators.[59]

The easiest contribution to illustrate is that provided by an AI trainer. An AI trainer is someone who prepares data for AI. A trainer operates before the algorithm is ready. Adriana is an excellent example. She lives in Belo Horizonte, Brazil. Today she'll be taking pictures. It's not her real job. She's not a professional photographer. To make money, she just logs into a platform and performs microtasks: snapping pictures, annotating videos, transcribing conversations, and so on. Adriana's tasks vary daily or weekly. Today, she has to take pictures of objects in her apartment. For this microtask, a particular object is specified: taking photos of her dog's poop. For what purpose? Because these images will help a US company improve a robot vacuum cleaner image recognition algorithm. The small vacuum cleaner is still learning how to recognize objects. However, only if thousands of workers like Adriana feed them with enough examples of objects in an average apartment, like shoes, bags—and dog excrement, will they work. Adriana, the trainer, generates data for a technology that isn't on the market yet.

A verifier is, on the other hand, a microworker who improves existing technologies. Like Justine, who lives in Saint-Etienne, France. A few years ago, she joined a Chinese platform that recruits microworkers for a major US tech company. She says that doing microtasks is better than working as a waitress or secretary for a boss. The title she holds officially is "transcriber," but she has been acting more as a "listener" over the past six months. She listens to conversations recorded by smart speakers. The people in the recordings speak French, sometimes with an accent from her region. The microtasks assigned to her involve listening to three to fifteen-second audio files that the smart speaker records. The number of files she receives in an hour may reach 170. It's up to her to judge whether the smart speaker understood what the user said. A new file, for example, has just arrived. This is a very short audio sample that the smart speaker's AI has tentatively transcribed. There's something odd about the transcription: "Bin Laden the word then." What does it mean? Is it dangerous? Is someone asking their smart speaker to look up the dead terrorist Osama bin Laden? That's not what Justine thinks. After listening, she corrects: "Well, give it to me then!" In French, the two sentences sound quite similar.

Justine's case shows that verification microtasks are not about improving prototypes that are still in the development stage. Her smart speaker is already available on the market, sold as a fully functional device. This raises a disturbing question: how do you distinguish AI that works automatically from AI that works with human intervention to fix its bugs, sometimes in real time? AI imitators provide a clear example of the latter case.

Miandrisoa works for European supermarkets but is based in Africa. As a matter of fact, he works for a company that makes intelligent cameras that are sold to supermarkets in Europe. It is even more accurate to say that he works for a small informal company in Antananarivo, Madagascar, which subcontracts workers to a company that sells intelligent cameras to supermarkets all over Europe. He commutes to his "office" every day. It's a nondescript small house. Each room hosts 20–30 microworkers connected to a platform. In the garage, Miandrisoa works as part of a special team. It is the team's job to "run" the smart cameras. His team monitors the stores remotely to prevent theft. It's like being a security guard who keeps shoplifters at bay. Miandrisoa bookmarks the live stream when someone steals chocolate or wine. He then copies the bookmark link, pastes it into a text message, and sends it to a cashier, who intercepts the shoplifter. Nobody knows about Miandrisoa and his colleagues in the garage. Everybody thinks intelligent cameras send the text messages automatically. Some European supermarket chains purchased the intelligent cameras five years ago. Would anybody suspect that an AI solution hasn't been developed yet? In five years, no algorithm has been implemented. It's been five years of fake AI. Five years of manually watching, copying, and pasting to compensate for the lack of automation.

Getting rid of "humans in the loop" is not going to be as easy as one might think.

The Other Mechanical Turks

Mechanical Turk was first patented in 2001.[60] At the time, Amazon had only been around for seven years. During this early stage of the company's history, Jeff Bezos already saw humans as the driving force behind future AI. Microwork powerhouses geared toward improving machine learning algorithms have burgeoned ever since: Apple, Facebook, Google, IBM, and Microsoft have created similar services, often in the form of internal platforms that manage sensitive industrial projects that are only accessible to workers after a more extensive selection and training period. Whereas access to Mechanical Turk requires only an Amazon account, these other platforms require workers to know manuals of several dozen pages by heart and to pass difficult tests. In 2004, Microsoft created the Universal Human Relevance System (UHRS) and, in 2008, Google developed EWOQ (which later became Raterhub), platforms where human microwork is used, among other things, to make a particular type of algorithm function correctly, the one used in search engines.

The search engine Bing, for instance, is constantly scanned by "quality raters," also called "judges," who assess whether the results best answer the users' queries. Judges have evaluation grids to assess search results. The most relevant websites end up appearing at the top of the results page, while malicious and lower quality sites are relegated to the bottom of the ranking. To find the "perfect" search results, human digital labor is essential. According to Microsoft's fifty-two-page guide for its microworkers, if search engines were already reliable, they wouldn't need human judges.[61]

Google places ads on its results pages, to cash in on the popularity of its search engine. It thus uses the work of microtaskers not only to rank the results of a search, but also to make sure the search terms and the ads match each other. During their training, microworkers at Google Raterhub are instructed that their reviews are fed to automated decision-making algorithms and models.[62]

Unlike Amazon Mechanical Turk, these services are not made available to third parties. The workers themselves can only access them through other microwork platforms (such as Clickworker), or through outsourcing platforms that specialize in information gathering, content filtering, or database management (Lionbridge, Leapforce, Appen, and OneForma, among others). These companies will sometimes pay higher rates to their microworkers, on average $13.50 per hour, which is almost double the US federal minimum wage. Some workers are initially recruited as self-employed or freelancers, others on part-time contracts, and still others on full-time contracts.

According to my own research, piecework and more stable forms of remuneration may coexist from time to time. Usually they can be found in intricate networks of subcontracting platforms, often used to avoid employer responsibilities and protect proprietary information about AI. One platform might recruit workers, another might develop an online app to perform microtasks, another one might process payments, and so on. This multi-layered structure, dubbed "deep labor," is a web of intermediaries, contractors, brokers, and suppliers that is reminiscent of the complexity of deep learning neural networks.[63] All this makes it hard for workers to know who they are working for, what they are doing precisely, and how much they are earning.

Other big tech conglomerates have adapted their methods, depending on the nature and sensitivity of the work. In the past, Apple has used piecework apps like TryRating to improve its Apple Maps, but more recently it has entrusted "fixed-purpose workers" hired by IT firm Globetech to grade Siri's recordings of users in private conversations.[64]

Mergers and acquisitions further complicate deep labor chains. One microwork platform originally known as Spare5 changed its name to MightyAI, was acquired by Uber, and was eventually sold to Aurora Innovation. To transcribe texts, label images, and assess natural language queries, its 200,000 workers use their own smartphones. Their tasks also included the training of automated vehicles and data preparation for IBM Watson, which won the TV game show Jeopardy in 2011.[65]

You Say "Freelancers," I Say "Microworkers"

Amazon Mechanical Turk pioneered the global microwork market. While contributing greatly to the development of machine learning solutions, the platform has also revealed that there is no such thing as a truly "artificial" intelligence.

But let's unpack other ways that digital giants delegate their microwork to companies that are part of their own ecosystem. These use more varied contractual frameworks than piecework alone. Microtasking sits alongside other forms of outsourced work, for example, crowdsourcing, "working from home," and freelancing intermediated by web services, which traditional companies have practiced since the 2000s. These recruitment methods have won the day particularly in the creative professions such as graphic design, accounting, specialized writing, and IT. In particular, COVID-19 has resulted in the adoption of remote work in the short term. The health crisis has been described as a successful "work-from-home experiment."[66] Prior to 2020, telework was only available to employees under specific contracts.

Despite its popularity, telework isn't replacing traditional work. Work at home doesn't cover the full range of jobs. Even in high-income economies, only one-third of existing jobs can be performed entirely at home, with significant variations by industry and location. In lower-income countries the rate drops to as low as 10 percent because of economic, occupational, and infrastructure constraints.[67] Even when telework is possible in principle, not all workers have the right equipment (computer, broadband, printer, headphones, etc.) and a home workspace.

The pandemic hasn't generalized preexisting telework, but it has opened up a new type of technology-mediated contingent work, which is focused on digital platforms. Both microwork and freelancing are available on these platforms.[68]

Freelancing in principle draws on a more qualified and better-paid workforce, which carries out complex assignments or projects. Microwork, on the contrary, represents a general trend toward insecure and poorly paid work for "intelligent crowds." Is the latter simply a byproduct of the former?

There is evidence to support this conclusion in two World Bank reports, published in 2015 and 2023. Titles emphasize the "global opportunity" and "promise" of online work to set the mood.[69] The World Bank researchers take a resolutely international approach to exploring online platforms that promote online work, which is sometimes characterized as microwork (Amazon Mechanical Turk, Figure Eight, etc.), and sometimes as freelancing (Upwork, Freelancer, etc.). The analysis also pulls data from five major platforms, tracked by the Online Labour Index of the Oxford Internet Institute.[70]

The 2015 report states that "often, microwork and freelancing overlap, the main difference between them being the size and complexity of the tasks, as well as the payment proposed." The focus in 2023 is on the "lower barriers to entry" of microwork over online freelancing. So microtasks are seen as a way to generate income for unemployed and underemployed people "with few or no specialized skills."

Despite being characterized as less skilled and unemployed than freelancers at the beginning of the reports, the data show that microworkers tend to be younger, already employed, and have a good level of education. In contrast, freelancing platforms are mostly for people looking for work, and they're usually in the thirty-five to fifty age range.

There is also another relevant phenomenon between 2015 and 2023: the growth of the estimated population of online workers from a few dozen to hundreds of millions. Even though platform workers' numbers should be treated with caution,[71] the figures can help clarify the relationship between microtasks and freelance work. According to the 2015 report, 47.8 million workers were registered on platforms, of which 42 million were on freelancing sites and 5.8 million on microwork service platforms. The situation had radically changed by 2023. The more recent report estimates that there are between 154 and 435 million online workers worldwide. The lower estimate is the number of online workers, mainly freelancers, who use platforms for their primary source of income. Microworkers are more likely to be found among the rest, who use online labor platforms as a secondary or marginal source of income. The estimates show that online workers, both freelancers and microworkers, make up between 4.4 percent and 12.5 percent of the global labor force.

These numbers may seem overinflated, but they read as conservative when compared with the platforms' own user statistics. In the mid-2010s, platform self-reported figures were already staggering: Freelancer, ZBJ, and Upwork alone claimed to have 51 million users, plus 41 million workers counted by Taskcn, Crowdsource, Microworkers, Epweike, and Fiverr. Pandemic shock and lockdowns led to a greater influx of new online workers reported by platforms. The Australian giant Appen declared "a record number of new contractor applications," which "has further added to the diversity of [their] crowd, and to the depth and breadth of [their] contractor skill base."[72] The German platform Clickworker proudly announced climbing from 800,000 to 2.8 million registered workers in 2021.[73] According to a 2023 report from Upwork, 64 million Americans, or 38 percent of the US workforce, worked as freelancers.[74]

As compared to these figures, microworkers are on a completely different scale, far outnumbered by the freelancing giants. Amazon has barely 250,000 Turkers.[75] According to the data published by Oxford's Online Labour Index, a total of sixteen million microworkers can be accounted for across fifty-five platforms worldwide. This represents 10 percent of all online workers. I believe this estimation to be very low. A number of local platforms, all proprietary platforms like Google's Raterhub and Microsoft's UHRS, as well as several freelancing platforms offering microtasks do not figure into these calculations.[76]

Even if we attempt to be optimistic about the "global opportunities" offered by online labor, the World Bank's theoretical framework is highly questionable, and indeed seems to have a structural flaw. The logic underlying the two reports is that microwork platforms generate a global labor market in which younger and more skilled workers have higher productivity, receive less pay, and perform less skilled tasks. In contrast, freelancers are mostly older and less skilled. But, contrary to expectations, they get higher pay and work on more high-value projects. If this were true, the standard economics descriptions of labor markets, in which older and less skilled workers are offered lower-wage jobs, would be completely incorrect.

What this contradiction suggests is that microwork cannot be restricted to specific platforms. It is an inherent part of the platform-based remote work, sometimes indistinguishable from freelancing. After all, the World Bank researchers themselves admit that users of freelancing platforms also carry out a lot of less skilled work. They participate in online video games (in order to train bots to play, or else to obtain credits, a practice known as "gold farming"), or they share and like specific brands, to optimize their online presence in line with the indexing algorithms (a practice known as "cherry blossoming").[77]

By overrepresenting high-skilled tasks (like software programming, legal advice, accounting, and IT systems management), institutions like the World Bank promote online labor. But microwork is undeniably on the rise. The 2023 report identified 545 online work platforms across the globe. Nearly three-quarters of these platforms are local platforms. 40 percent of their traffic comes from low- and middle-income countries like India, the Philippines, Indonesia, Pakistan, Nigeria, Brazil, and Mexico. These are countries where microwork is more common than freelancing. In the Global South, issues relating to connection speeds, equipment and international payments can be obstacles to pursuing real freelance careers. As the authors of the first World Bank report conceded, almost reluctantly, that the only form of work that can be proposed by *all* these platforms is microwork: "Online outsourcing is most effective and most likely to function like a truly global market when the task is less complex and involves fewer local institutions and less communications. It is easier to outsource online sign-up to websites, search and click, and vote, than medium to high complexity tasks such as web and software development, and customer service."[78]

Undocumented Microwork

If we look closely at how the platforms that present themselves as free-lancing services operate, the paradoxes I have just pointed out seem to disappear. After all, how can these platforms base their business model on the completion of complex tasks while attracting a workforce willing to perform simpler ones? How can they at once promise requesters a high-quality, cheap service, and promise workers attractive pay?

The answer lies in the process of reintermediation by which complex tasks are fragmented and allocated to microtaskers. According to researchers from the Connectivity, Inclusion, and Inequality group of the Oxford Internet Institute, who have studied this aspect, reintermediation is in reality concealed: only the faces of users are visible on the platform, "but behind a face there is often a small network of organized people."[79]

Reintermediation happens when some workers acquire a direct connection of trust with a client. As a result, these workers turn into intermediaries themselves, taking on more work than they can handle personally, hiring other workers on the platform to help.[80] This is visible in the distribution of earnings on these platforms. Their workers can be divided into two distinct groups. On one of the largest platforms, Upwork, most users earn only a few dollars a month or do not even manage to get tasks assigned to them.[81] On the Chinese giant ZBJ, the situation is even more extreme. Data analysis shows that almost 80 percent of workers have earned less than $10 in total, over six years of activity.[82] Of course, that could mean that most users on these platforms are only (very) occasional workers, and that there is a very high turnover rate. A user signs up, creates an account, performs a task, and then forgets about the profile. This is also a reflection of the fact that there are many unpaid tasks on these platforms. Yet another important insight is that, compared to those who earn barely any money, a very small percentage of workers earn large amounts on these platforms, sometimes several thousand dollars per month. This "elite," which on ZBJ amounts to barely 0.4 percent of users, consists of subscribers who, because of their seniority or their reputation, can convince clients to give them more complex and better-paid projects—that only experienced freelancers can handle. They then fragment the projects into microtasks and delegate them to other workers, often off the platform to avoid the commission fees.[83] The other workers carry out these tasks for less pay. Above all, these tasks are not displayed on their profiles. Consequently, their experience and reputation ratings do not improve. The reintermediated microtasks are farmed out through informal channels such as email and private messaging, so they do not feature in platform figures; they can be considered unreported (micro) work. This widens the gap between the elite platform workers, often posing as freelancers, and the microtaskers even further. The former receive more and more work, and can even start bargaining or offer discounts to some requesters.[84] Meanwhile, microworkers lacking the qualifications and the direct access to clients, fall into a "poverty trap" on the platform.

However, intermediaries aren't just profiteers who rely on their reputation and seniority to earn a steady income. They, too, must prospect for work, make contacts, manage their relationship with the requesters, and also train and supervise the other microworkers. It's not uncommon for them to translate requests into their local language, write guidebooks, and oversee the quality of results. Above all, they must reduce, standardize, and fragment complex projects into microtasks, which they then have to reintegrate into a final deliverable.

One of these elite workers says: "I wear the proofreader's hat, editor's hat, whatever else that's left and they [local subcontractors] do just the raw transcription. [. . .] I look at myself as someone who is employing other people."[85] Behind the smokescreen of freelancing, which is paraded as being enriching both financially and in terms of professional experience, the countless invisible ramifications of platform work prove that microtasking is indeed its basic and constant form.

One-Click Offshoring

Even in the absence of the reintermediation specific to some platforms, the separation between an elite workforce and a mass of less visible workers with no bargaining power reappears on microtasking sites. On MTurk, this division is expressed between different populations of users whose activity follows, roughly, the Earth's rotation. When the clock strikes 6 p.m. in San Francisco, American Turkers usually finish their day. That's when Indian microtaskers in Bengaluru start getting down to work. As for the North/South distinctions, in the North, relatively specialized American women go online to supplement their income, while in the South, it is less specialized Indian men, who are often entirely dependent on the platform for their livelihood.

The political economies of online microwork thus vary from country to country.[86] The availability of tasks at certain times, differences in internet connectivity, levels of language proficiency, computer security issues, and stipulated payment methods all generate major inequalities in the incomes to which microworkers can aspire. Non-US users denounce these global asymmetries[87] and compare their devalued condition to that of sex workers, fast-food staff, or agricultural day laborers.[88] Since no national borders determine the allocation of tasks on the internet, the traditional geographies of work are disrupted. Before the 1900s, labor was localized and rooted in a specific territory. Today, the platform economy recruiters come mainly from the richest countries, and the labor supply is dispersed across the globe.

Data flows between the Global South and the North suggest that since the early 2010s the majority of those requesting digital labor on English-speaking services are concentrated in the United States, Canada, Australia, and the United Kingdom, while the bulk of microtaskers live in the Philippines, Pakistan, India, Indonesia, and Bangladesh,[89] as well as in the Middle East.[90] The clients of microworkers in French-speaking Africa, and in Portuguese- and Spanish-speaking Latin American countries, are located both in North America and Europe.[91] Digital labor configurations thus reflect significant inequalities across the globe—as well as cultural, linguistic, and economic dependency.

For example, microtasks involving managing and sorting information are delegated to inhabitants of middle- and low-income countries in line with recognizable North–South relations of economic dependence (offshoring, foreign direct investment, and so forth). As with traditional outsourcing practices, investors are attracted by the availability of labor and its low costs. Another major advantage is the modular and flexible nature of this type of "micro-offshoring": gone are the days when only large corporations would open offices abroad. Platforms enable small and medium enterprises to access the global workforce. It's not necessary to set up shop in a foreign country to hire those workers. The smallest startup can outsource a few tasks for a few hours.

The Mirror of Unemployment

The geography of microtasking mirrors the globalized labor market and its inequalities. The ILO's *World Employment and Social Outlook 2023* indicates that worldwide there are 473 million workers who are without a job and seek one, resulting in a global "jobs gap rate" (a new statistical measure of unmet employment needs around the world) of 8 percent in high-income countries, while in low-income countries the rate goes up to 20 percent.[92] Additionally, there are two billion people who work in informal arrangements. Actually, 46.6 percent of the workforce worldwide consists of self-employed individuals, both workers who work for themselves, and unpaid family workers. Again, there are notable differences between countries, with Norway having 4.7 percent of self-employed workers and Colombia having 53.1 percent.[93] Across the Global South, where working-age populations are growing steadily, individuals compete for both formal and informal jobs—and for microtasks. Workers from low-income provide the bulk of the workforce for microtask platforms across the world.

Taken globally, high unemployment rates and informality go hand in hand with increased occupational risks and inadequate social protection. Microwork flourishes particularly in countries without sufficient employment protection (no agreed pay standards, no guarantees against dismissal, and no rules regarding temporary work).[94] Even in the United States, unemployment and use of microworking platforms are strongly correlated, especially among those with lower education levels.[95]

Microwork draws on large quantities of unpaid and casual work, without social protection. Only 30 percent of the working-age population is covered by comprehensive social security systems that cover everything from child benefits to old-age pensions. 4 billion people are only partially protected or completely unprotected.[96] Especially in Africa, the Arab states, and Asia, social protection isn't comprehensive or adequate. It is often restricted to small sections of the population, primarily standard public- and private-sector workers, with the costs falling (in part or entirely) on the employer. Since state social protection programs applicable to all are often absent in these countries, rural workers, the self-employed, part-time employees, very small businesses, and the entire informal economy have no safety nets. In addition, the coverage of social risk is not uniform, with some areas often omitted, such as maternity and disability.[97]

The new generations entering the globalized labor markets are thus in the confusing situation of both working, and being excluded from the social protection and stability that should be associated with employment. They are vulnerable to the promises of on-demand work and microworking made by the platforms. Platforms entice the candidate with the prospect of abolishing geographical boundaries, getting into "direct" contact with potential employers located in richer countries, and maybe even engaging in exciting IT careers.

Microwork is marketed as an alternative to standard employment. In reality, it is a reflection of unemployment, with standard employment being unattainable. In low-income countries, even more so than in Europe or North America, businesses and policymakers praise microwork as the best and "only possible future of work."[98] Their giddying rhetoric is full of clichés about personal autonomy and saying goodbye to long working days, expensive transportation, and annoying workplace hierarchies. The idea of flexibility is particularly seductive, conveyed through the buzz words of "working at your own pace," "having no boss," and "becoming an entrepreneur." The low wages, piecework, and precarity inherent in this kind of work are swept under the carpet.

In this context, some platforms advertise conditions that are more respectful of the expectations and needs of the populations that make up the recruitment pool for microwork. Since the Rockefeller Foundation introduced digital jobs initiatives in Africa aimed at extending economic and social benefits to non-employed populations, "social leverage" has become increasingly important to big tech companies investing in third countries.[99] According to its inventors, "impact sourcing (IS) employs people at the base of the pyramid, with limited opportunity for sustainable employment, as principal workers in business process outsourcing centers to provide high-quality, information-based services to domestic and international clients."[100]

An early example of a platform designed to combine microwork and impact outsourcing is Sama. Founded in 2008, it started as a nonprofit, then transitioned to a hybrid for-profit model in 2019. The social mission focuses on providing earning opportunities; microworkers are from populations living below the poverty line and who suffer long-term unemployment. Workers are allegedly trained in computer skills and paid a living wage that is indexed to the cost of living in their country of residence.[101] Due to the lack of connectivity and equipment in Kenya and in Uganda, Sama has built centers where workers can perform microtasks. This can happen when the projects are highly sensitive (as when moderating violent content for Meta) or secretive (as when working for OpenAI to train data prior to ChatGPT's release). Workers can perform microtasks from home for low-sensitivity projects. Workers utilize SamaHub, a proprietary microwork platform that is adapted to the requirements of the clients.

Since the discovery of the terrible toll taken on Facebook content moderators' mental and physical health,[102] the low salaries for ChatGPT annotators,[103] repression of the unionization drive and mass layoffs in their Nairobi office,[104] Sama's reputation as a "responsible employer" has been badly damaged.[105]

Other impact sourcing platforms use entirely remote workforces. The French company Isahit, with services including content moderation, video filtering, photo tagging, and expense report entry, has its microtasks carried out by women living in mainly in African countries such as Cameroon, Côte d'Ivoire, Nigeria, and Senegal.[106] According to the platform, the daily rate is equivalent to $22, which is ten times higher than the minimum rate needed to get above the extreme poverty line. While there is no guarantee that enough microtasks will be available to obtain the proposed monthly compensation (between $54 and $136), the program is particularly attractive to those who would otherwise be unable to receive such wages.

However, the employability of microworkers is not these platforms' priority. Although Isahit encourages women to open bank accounts and exit the informal economy, it does not create formal employment. In order to do microtasks, workers have to declare themselves self-employed.[107]

The purpose of these platforms, despite their social impact facade, is mainly to convey entrepreneurial values to the women who microwork for them and select them on the basis of their ability to develop a business idea, which must be approved before they can use the platform. Prospective workers, or "hitters," present slogans in short videos, to show that they can create and manage a company. One aspiring hitter, a bracelet maker from Côte d'Ivoire, describes the platform as a "gas pedal" that will give her the means to "scale up" her business;[108] a student from the Republic of the Congo hopes her clickwork will allow her to "showcase [her] marketing skills";[109] another woman from Côte d'Ivoire is awarded the title of "hitter of the month" and thanks the platform's "financial support" that helped her open a small accounting consultancy firm and "start prospecting for clients."[110] Behind these enthusiastic testimonials, the reality is dim. In several African countries, Facebook groups and discussion groups have been created by "hitters" who complain about low wages.[111]

The concept of impact sourcing sits at the crossroads of microwork and microfinance—a humanitarian movement that aims to give very small loans to impoverished borrowers who lack collateral, steady jobs, or a credit history. Some have branded impact sourcing as "philanthrocapitalism," which combines a spirit of charity with today's massive trend toward digital labor platforms. The analysts most critical of this phenomenon maintain that the discourse of economic empowerment of the most disadvantaged citizens in low-income countries in fact conceals a logic of domination and spoliation masquerading as social responsibility.[112] However ethical these services try to be, they do nothing to help establish public welfare services or collective systems of mutual solidarity. Even if microworkers receive adequate pay, they encounter the same old problems: The hourly rate does not automatically mean an improvement in living conditions due to variable workloads; they have no power to choose the tasks they perform; they do not benefit from the social protection of standard employment; and their rights and skills are not transferable from one platform to another. Moreover, impact sourcing platforms work with corporations that are only really interested in cheap labor for machine learning, content moderation, and transcription. Thus, however genuine their social and ethical goals, these platforms end up being part of the sweeping tide of outsourcing and international social dumping that is ultimately a global race to the bottom in terms of wages and working conditions.

The competition introduced between workers from industrialized countries and workers on microwork platforms in the South is particularly problematic. So, although impact sourcing seeks to reduce inequalities through fair employment, it potentially fails on two counts: it does not really encourage formal employment in the Global South, and it increases inequality and insecure work in the North.

Voluntarily Chained: Embracing Microwork

There is a close connection between offshoring and platformization. Companies in high-income countries assign computation and data management tasks to microworkers in low-income countries, forming "global value chains,"[113] which can continually add or subtract new workers to adapt quickly to changes in both markets and management decisions. Offshoring decisions are based on efficiency and transaction costs, even when they involve microwork platforms. In richer countries, companies intentionally maintain market power asymmetries to increase their surplus and keep contractors from upgrading.[114] Meanwhile, they keep the cost of resources imported from low-income countries low—and this includes labor costs. They often create excess capacity by fostering competition among their suppliers. Digital labor platforms exhibit this same trend toward labor surpluses and stiff competition between microworkers.

Adaptability is undoubtedly an important asset for microwork platforms, giving them an advantage over traditional forms of outsourcing—which are progressively turning into platforms, too. The digitization of logistics and production chains is increasingly being observed in certain services, such as call centers and remote support teams, and in specific sectors like fashion.[115] Actually, there's no clear line between traditional offshoring and platforms. Both derive from earlier organizations that disaggregated information-intensive functions into smaller units.[116] Both are technology-driven and aim to operate as virtual organizations using remote service providers.[117] And both pursue business process outsourcing and reduce their costs through global labor arbitrage.

It is little wonder, then, that global value chains are not composed solely of digital pure players, like Amazon MTurk, but also of private and public sector companies, bigger and smaller, both formal and informal. Each of them is engaged in the process of platformization. For these firms, microwork transforms the very nature of the tasks that are being outsourced. No longer is it activities with low added value (accounting, IT support, customer services), but the very core of their new business models, namely the training of AI and algorithms. Since data management and automation are essential to these companies, digital labor platforms are providing services that are absolutely central to all companies undergoing a digital transformation.

However, global outsourcing chains are not the only way that platforms allocate information processing tasks to human beings. The outsourced microwork discussed so far effectively expands a company's pool of available workers geographically. The relationship between workers and platforms, even if the activity is generally piecework and workloads fluctuate, can still be defined as a *relation of production (Produktionsverhältnis)*. This Marxist concept describes the social relationships essential to the production of economic resources. There are certain production relations that lead to the formation of slavery, serfdom, and wage labor. Microwork, and digital labor in general, can be viewed as a result of other relations of production associated with the emergence of platformization.

Another approach adopted by certain large platforms is to *outsource work directly to the consumer*. This, too, is a relation of production. For example, companies can ask consumers to label images or transcribe short messages in the course of their daily browsing, or as part of an online game, using applications specifically designed to extract free work.

Alphabet has been able to develop unpaid microwork better than other platforms because of its size, the critical mass of its users, and the quasi monopoly enjoyed by its services of search and app market. Crowdsource, an app by Google, was launched in 2016. It openly requests its users to contribute to the improvement of artificial intelligence systems. Users choose a category of tasks to perform, such as translating sentences, describing images, or confirming whether an urban landmark is visible in a photo. The application is doubly "free": Crowdsourcers do not pay to use the app, and clients (which in this case are all Alphabet subsidiaries) do not reward the work done.[118] This work is clearly framed as a joint effort by a community of enthusiastic Alphabet users in the interest of humankind as a whole. The FAQ page reads: "Will I get paid for my answers? No. Crowdsource is a community effort—we rely on the goodwill of community members to help improve the quality of services such as Google Maps, Google Translate, and others, so that everybody in the world can benefit."[119]

Instead of micropayment, users earn badges that allow them to level up like in video games. Moreover, users are shown "agreements" that indicate how many times others have given the same response. So, by "agreeing," the microtaskers themselves are the ones who endorse each other's work. The app Crowdsource extends the logic of reciprocal grading and gamification of microwork described previously. The few cents that the Turkers earn for each task are replaced here by purely symbolic rewards: having fun, while "improv[ing] Google both locally and globally."[120]

There are more than one million users on Crowdsource, but this is a relatively small number for Alphabet. Nevertheless, it reflects a general trend across Google's services, namely *making users work for free*. It combines several features present in other platforms. Google Translate, for example, a much more common tool, is based on this same principle. Users enter sentences or text samples in one language, and these are automatically translated into another language. At the same time, the application encourages users to suggest their own alternative versions, thus turning them into unpaid translators and simultaneously into trainers of artificial intelligence, since the texts also feed into a "neural machine" called Google Neural Machine Translation. This is not the first time Google has called upon myriads of amateur translators, since from the outset it has relied, as the linguist Ignacio Garcia explains, "on crowdsourcing to translate its interface into many 'minority' languages."[121]

The same process also optimizes the predictive text algorithm that suggests search results in real time on the Google search engine. According to communication and linguistics scholars Amanda Potts and Paul Baker, "each query to Google has two effects: the first, visible result is that the user is shown a series of 'responses' to his or her query, ordered by 'relevancy'; the second, more inconspicuous effect is that the entry of any query essentially casts a vote vouching for the popularity of the search string."[122] The users of these services, though unpaid and unacknowledged, are conscripted into microtasks just like the "quality raters" of Raterhub or the users of Sama, to whom the same Google offers remunerated computing work.

Among the wide range of Google services that require active participation from users, the "Quick, Draw!" open online "AI experiment" pushed the logic of gamification to its extreme: "Can a neural network learn to recognize doodling? Help teach it by adding your drawings to the world's largest doodling data set, shared publicly to help with machine learning research. Let's *Draw!*"[123] The app challenges users to draw a random object on the screen of their smartphone: "Draw me a horse, or a swing, or a tree, in less than twenty seconds," the app exhorts. While users try to demonstrate their artistic skills, a neural network starts guessing what the object is. The network is called a "generative adversarial network" because it learns by playing against an opponent[124]—in this case a human being. The machine thus tests itself in shape recognition, and eventually "learns how to learn." Between 2012 and 2019, *Quick, Draw!* collected a massive dataset of fifty million handmade drawings, divided into 345 categories,[125] from which the neural machine can identify hundreds of common objects, each backed up by hundreds of thousands of images used as models for recognition and for checking the accuracy of the results.[126]

In promoting this app, Alphabet has never mentioned making profits for a private company. Users are told they're expanding the frontiers of science and working selflessly. They are not wasting time doodling; they harbor a noble passion that should be fostered by joining a community devoted to furthering automation for Google.

This is where algorithms become an ideological narrative, which the platform leverages to conceal the work required to update and develop its services. Although Alphabet is a company saturated with a technological language of computing and computer engineering, as the sociologist Paško Bilić explains,[127] its algorithmic fantasies are also infused with a missionary spirit, which assigns a quite different value to the generous participation of unpaid users and to the paid microtaskers located in third-party countries.

It is American or British users who, "for fun," devote their time and data to artificial intelligence, contributing nobly to the cause of complete automation, while microworkers who perform the same tasks from Venezuela, Kenya, or the Philippines, are hidden as an embarrassing reminder that the prophecy of automation has yet to be fulfilled.

This vision of AI as a journey of the human spirit is also characterized by other tensions. With the popular reCAPTCHA system, it's precisely the opposition between the "humans" who solve it and the "robots" they prove not to be that makes the free microwork of reading short sentences or recognizing traffic lights more valuable. ReCAPTCHA is activated whenever it is necessary to authenticate users as, for example, when trying to retrieve a forgotten password or post a comment on a private forum. Initially, all users had to do to demonstrate that they were "not a robot" was to transcribe two distorted words from texts scanned in the Google Books catalog. The platform already knew the solution for one of these words; the other word was illegible to the "bots," the text recognition software. The new version of this system, with the unlikely name of "No CAPTCHA reCAPTCHA," follows the same principle, but presents microtasks of a different kind, such as matching ten images of animals, objects, road signs, etc. The first image is known (for example, it already contains the description "cat" or "storefront") and it can be used to compare the other nine, which may or may not contain an object of the same type. As they click, users help detect locations and reconstruct scenes, to enhance Google Images, or perfect Waymo's self-driving cars. The irony here is that a service that is supposed to distinguish humans from robots is actually making humans work to produce robots.

Although I have argued elsewhere that reCAPTCHA provides an example of "concealed digital labor,"[128] Alphabet clearly does nothing to hide it. On the contrary, the reCAPTCHA home page champions its ability to turn humans them into "human bandwidth" for its information flows, and claims to "benefit people everywhere."[129]

This has been the company's position from the start. Luis von Ahn, the system's designer and a recognized specialist in human-assisted computing, said as much in a 2008 article published in the journal *Science*. He even estimated how much human labor the first iteration of this technology would displace, namely "more than 1,500 people deciphering words forty hours a week (assuming an average rate of sixty words per minute)."[130] Clearly, his idea wasn't to replace the human workforce with a machine. Instead, it was that technology could offer a cheaper outsourced alternative to that, arguably remunerated, work. This was the same kind of calculation as the one made by the designers of Amazon Mechanical Turk in 2005, when they speculated on whether it would cost them more to pay thousands of people full-time for manual data deduplication in their catalog, or whether they should outsource this work. The answer is now known, and in their case the solution was to pay their Turkers on a piecework basis. Google reasoned in the same way, but the incentives provided to its microtaskers were different. For some, who have no choice, reCAPTCHA is a chore they must get through to continue with their navigation; for others, namely the myriads of volunteers who further machine learning, it is a matter of supporting the *grand oeuvre* of artificial intelligence, and contributing, in the words of von Ahn, "to the digitization of human knowledge."[131]

What's more, as I have intimated throughout, there is a continuity among paid outsourced and atomized labor, and another form of labor, consisting of the same tasks and acts, but presented as free and enthusiastic contributions. The same "human computation" services that microworkers perform—in exchange for very little pay and sometimes under duress—is possibly performed by billions of workers of another type, without any pay at all. This is where microwork morphs into the third type of digital labor: the labor performed on social media.

· 5 ·
SOCIAL MEDIA LABOR

Few people using Facebook, Instagram, or other social media platforms see sharing content with friends as "work" in any traditional sense. That applies more to individuals who appear in the media as content creators, influencers, digital marketers, and authors. They are often portrayed as successful businesspeople. But as you may suspect, having read this far, even for average people, publishing photographs, articles, and videos and commenting on others' posts are forms of networked labor. In fact, those who perform it have more in common with blue-collar workers than with creatives at the highest level. Especially when they are just regular everyday users.

The two types of digital labor discussed in the previous chapters give us the perspective necessary to see how social media platforms create value through qualification, monetization, and automation. Social media users' work is less conspicuous than that of on-demand workers like Uber drivers or microworkers like those on Amazon Mechanical Turk, since the allocation of tasks is less systematic. As with these other digital workers, virtually all social media users do not have formal employment contracts or contractual affiliation to the platform. Yet, just like in the other cases, these workers receive incentives for completing their tasks. These incentives to perform certain actions on social media may be symbolic (reputational rewards) or economic (in the form of money or commercial bonuses).

I am certainly not the first to argue that the demands that social media platforms place upon their users (to engage with content, to connect with others, and to contribute time and attention) amount to "free work." These demands have led to controversies over the past two decades, between those supporting and opposing the idea that being online could be considered as work. And yet these discussions have not generally captured the diversity and complexity of remuneration and labor—whether one considers "organic" social media users (consumers, fans, readers, and spectators) or professional participants (influencers, commentators, and public figures). On social media, "organic" users' work is concealed and unpaid. It may range from investment in creating or polishing a profile to monitoring invitations to visit a website or to open an app. In all cases, using social media relies on others' paid labor, in particular the work performed by moderators, influencers, and the countless hired hands, who produce vast volumes of ratings, shares, and "likes" of all kinds, including from "click farms" in China or Thailand. The status of "content producer" thus encompasses a wide range of roles and a range of contractual arrangements, well beyond the notion of voluntary work.

Because platforms make great efforts to disguise this work, its drudgery, its lack of independence, and its cost, it's not surprising that few workers interrogate this kind of work directly, and fewer still organize and engage in industrial action. Most often, though, labor disputes appear under the guise of questions about the principles of platform governance, the definitions of the boundary between private and public, the regulation of intellectual property, online civility and social relations, and, more recently, payment for user-generated data and content.

Work in the Age of the "Produser"

Every year, *TIME* magazine's December issue features a "person of the year." This isn't always an honorific, and the distinction has gone to both Vladimir Putin (2007) and Elon Musk (2021). In fact, personhood isn't strictly a requirement. In 1982, the magazine celebrated a "machine of the year," the computer. In 2006, the cover again showed a computer screen and three letters: YOU. That year, the magazine championed a "revolution" represented by "the small contributions of millions of people" on social media platforms. The editorialist Lev Grossman wrote the history of digital media not in the usual way, as one of consumption of information and entertainment services, but of work. For years, he pointed out, we internet users have not only viewed online content, "we have also worked. Like crazy. We made Facebook profiles and Second Life avatars and reviewed books at Amazon and recorded podcasts. We blogged about our candidates losing and wrote songs about getting dumped."[1]

From blogging platforms to multiplayer online role-playing games, to friend- and job-finding sites, to video-, image- and text-sharing pages, social media platforms seemed at first to be, as billed, a means of producing one's own media and sharing it with one's chosen community. But they also rapidly became content archives, inspiring traditional media to use these sites to cement audiences and recruit talent. Brands encouraged consumers to contribute to social media sites, and "markets" themselves were replaced by "conversations" about products.[2] Social platforms tempted those without any previous fame to become small-scale media stars and brands in their own right. But the appeal of personal branding and the promotion of "microcelebrities"—whether on the earliest social media such as Orkut or Myspace, or later on YouTube, Instagram and TikTok—concealed the real requirement of these platforms in return for these opportunities: constant content creation.[3]

In the mid-2000s, "Web 2.0," first championed (largely by the popular press) as a place where everyone could express themselves, revealed itself as a place governed by algorithms and advertising agreements. On the largest platforms like Facebook, China's Tencent WeChat, Korea's Kakao, and Russia's VKontakte, the algorithmic recommendation, selection, and promotion of contents and celebrities is more visible.

Mainstream platforms today resemble vast conglomerates. Users can publish texts, stories, videos, and live streams, and additionally manage a discussion group on a mailing list, find contacts in a directory, and so on. The convergence of different functions in one "place" is a way of netting users; once registered, users are unlikely to delete their accounts, block browsing history trackers, and even find the log out button (buried as it is within layers of the interface). These super-apps and "world platforms" lock their subscribers into their structures so that every post, every picture, every keyboard entry can be turned into value for the platform's owners. Social media captures value by encouraging users to add and edit content, and to abandon their roles as passive spectators to become site ambassadors by sharing and promoting that content.

In some ways, the triumph of the social web seems to confirm the visions of futurist Alvin Toffler, who predicted in the early 1990s that the consumer and the producer, separated during the first Industrial Revolution, would fuse into a "prosumer."[4] In the new cycle of wealth creation, he argued, the customer's role would include paying for a good, launching it on the market, and producing the information necessary to promote it. When services such as Flickr and Wikipedia first appeared, with their active "new digital media" publics, Toffler's hypotheses seemed to be bearing out: the "prosumers" (now also called "produsers," a mix of producers and users) were an emerging workforce reshaped by media and entertainment industries.[5]

The multiplication of websites and services based on user-generated content over the last decades then came to be interpreted as the sign of a new "prosumer capitalism."[6] Its rules are radically different from those of the older, production-based, and the newer, consumption-based, capitalism.[7] Unlike production-based capitalism, the extraction of value and the circulation of products is now performed by subjects who are not paid, and who access goods and services for free. Unlike consumption-based capitalism, the abundance of content and information takes precedence over the scarcity of goods and services.

The produser is not a "craft consumer" who uses their own production in a self-fulfilling and self-regulating way.[8] Whatever form the contribution takes (uploading video montages, translating texts produced by others, etc.), it is eminently social, that is, based on the circulation of content between individuals, and on reciprocal evaluations of that content. Even the simple and isolated act of sharing or evaluating content on social media counts as a social act. If a simple "upvote" or a "like" is recognized as an active contribution, then, with the standardization and generalization of these technologies, nonspecialized contributors have every chance of becoming producers. Unlike the craft consumers, or producers, of pre-internet capitalism, this user population may have little social and cultural capital. In fact, if they turn to social media, it is to create both the product and the community that will appreciate it.

Pain or Pleasure?

But are social media users "workers" in the ways we have discussed previously? Both academics and public opinion are divided on this issue. One view is that large platforms extract value from social media users' labor in an exploitative relationship. This perspective becomes clearer as both amateur and professional artists inadvertently fuel generative AI systems with their work, collected without consent from the internet[9]. The other view is that users' content production gives them pleasure and they freely contribute to a new culture of amateurism devoid of labor.

This divergence has often been treated hastily as a conflict of schools, with each camp accusing the other of being a hive of Marxist academic activists, or of industrial researchers paid by the tech sector.[10] In truth, each side harbors as many Marxists, liberals, and even libertarians as the other. Nor does the dividing line run between the social sciences and engineering, or between those close to, or distant from industry. Within each camp, computer scientists and sociologists rub shoulders with marketing and media experts, and public sector employees talk to private-sector researchers. Yet, theorists often find themselves caricatured by their opponents. People who emphasize that there is a labor component to social media content production are accused of technophobia and social pessimism, as well as of political victimization.[11] People who see it as voluntary and "empowering" participation by the audiences in the cultural industries are accused of idolizing technology and of political naivety.[12]

Taking the debate to a higher level by giving a fair and equal hearing to both parties should not prevent us coming down on one side or the other. As you may suspect by this point, I mostly argue in favor of the "work" perspective as opposed to the "pleasure" interpretation. But with some nuance.

Contrary to the mistaken idea that calling social media activities "work" is simply a reaction to the idealization of the social web of the twenty-first century,[13] the "work" approach emerged at the same time as this euphoria, and even precedes it. This viewpoint has been discussed in the context of legacy media for several decades. "As a matter of general theory," Raymond Williams pointed out in 1978, "it is useful to recognize that means of communication are themselves means of production."[14] The concept of television audiences as media workers "for the media, producing both value and surplus value," was also influential.[15] Scholars have focused on three issues specifically: unremunerated online contributions, exploitation of users, and the integration of playful elements into productive acts.

In a pioneering article published in 2000, Tiziana Terranova used the term "digital labor" to describe "networked, immaterial work." It included "the activity of building Web sites, modifying software packages, reading and participating in mailing lists, and building virtual spaces on MUDs [Multi-User Dungeons] and MOOs [MUDs Object-Oriented]."[16] This list of outdated online services shows how clearly far ahead Terranova was of the advent of social media. Social platforms had not yet acquired the phenomenal success they would achieve a few years later when she identified the two main characteristics of the new digital work processes: that labor was performed for free and that it had a social dimension. Digital labor was "free" because it was unpaid. It was "social" because it was performed as part of daily life and not in the places usually allocated for value production (factories and offices). For Terranova, "free labor is a desire of labor immanent to late capitalism." Network economies promote this occupation, while confining it to "under-retribution"—if there is any retribution at all.

The authors who support the "work" approach maintain that the gap between the value generated by users' contributions and the income received by them is a sure sign of exploitation.[17] Even if we take into account the costs incurred by platform owners and the services that social media provides for users, the gap remains.[18]

But can we speak of exploitation without considering what pay the users expect? Without this parameter, it is impossible to distinguish between the worker who performs some of his job "for free" and the volunteer devoted to a cause. If platform users have no expectation of reward, despite possibly knowing that their contribution creates value for the platforms, they can hardly complain of injustice.[19] In fact, those who are unsure about remuneration need only to read the legalese of the platform's fine print, through which platform owners protect themselves from any possible claims by users.[20] Facebook's terms of use, for example, despite years of disputes and emendations, still give the platform extensive rights over any content provided by its subscribers. The platform has a "transferable, sublicensable, royalty-free, and worldwide license to host, use, distribute,

modify, run, copy, publicly perform or display, translate, and create deriv-
ative works" of the content. If the content and profile are deleted, the platform can make back-up copies and continue to profit from user-generated
content if other members have shared that content. The terms do not mention any obligation for Facebook to provide direct compensation to users
for the content they produce.[21]

Christian Fuchs and Sebastian Sevignani, taking an orthodox Marxist
perspective, interpret the exploitation of social media labor rather differently.[22] For them, it is not simply working "for free" that enables the digital platforms' accumulation of capital. Content producers lose "control of
their labor power" and the sense of their activity as well. Once they no longer are aware that they use the means of production in order to generate
wealth, they can no longer recognize the "exchange value" of a post, picture, or emoji monetized by the platform. The "use value" of each thought,
image, or emotion out of which a post, picture, or emoji sprang is also lost
to them. Exploitation and alienation reside in this total penetration of capital, which reifies and expropriates not only users' social relations, but their
cognitive capacities as well.

But if all digital activity is alienation, what do we make of the fact that
these platforms offer opportunities for play, relationship-building, and
personal fulfillment?[23] Without abandoning the "work" perspective, these
benefits of social media use can be reexamined in the light of the influential collective book edited by Trebor Scholz on the erosion of the distinction
between productive acts and leisure in the age of online social media.[24]
The fact that content producers may themselves blend the two in order to
optimize their social capital does not exonerate the platform economy for
the way it exerts cultural and political pressure on users in order to exploit
the tenuous border separating the private space of leisure from the public
space of labor. When private forms of human sociability are transformed
into profit, Scholz maintains, "the entire fabric of our everyday lives, rather
than merely our workplace toil, becomes the raw material for capital accumulation."[25] The convergence of play and work does not herald a bright
future where there will be leisure for everyone. Rather, it signals a conflictual dialectic between users' desire to participate, fueled by the promise of
freedom through sharing, and the platforms' logic of appropriation, which
devours collaborative networks by means of the private, proprietary infrastructures built around them.[26]

The "pleasure" approach stands in opposition to the "work" perspective. In this camp some, like Adam Arvidsson and Elanor Colleoni, concede
that the activities of content producers can be compared to work, but the
comparison is figurative. They argue that a platform's value is not a function of the time each user spends contributing online. Instead, wealth is

created in the development of what Marx calls the general intellect—the surplus of knowledge, feeling, and sociability that results from human-machine collaboration.[27] Arvidsson and Colleoni give an original interpretation of an extract from "The Fragment on Machines" (a text published posthumously in the *Grundrisse*),[28] where Marx states that the increasingly important role of technology in production will eclipse the value generated by time devoted to human labor. By participating in a network, users really *want* to contribute their knowledge, creativity, relationships, and skills. Contrary to what theorists of the "work" approach believe, value is not created by humans but by algorithms and automated processes.

Furthermore, in digital economies the convergence of the machine and the human (which in Marxist terms corresponds to the rise of the "technical composition of capital"—a measure of the amount of means of production divided by how much labor power is used to produce something) will generate such powerful "network effects" that we will witness the emergence of a "social brain" whose neurons are the connected individuals. This analysis has similarities with Howard Rheingold's theory of "smart mobs"[29] and Yochai Benkler's "wealth of networks."[30] According to Clay Shirky, millions of people have turned away from the passive consumption of traditional media and joined large online projects involving collective sharing and peer production, resulting in a "cognitive surplus."[31] The positive externalities of these online networks will benefit all the parties involved, and the notion of exploitation will therefore become irrelevant.

Who is a "produser" in this context? Those who argue that social media activities are purely for pleasure define produsers as a "connected amateurs" engaging in the "creative reception" of media and cultural content, out of which they make secondary works, by remixing this content, commenting on it, and entering public space through it. According to sociologist Patrice Flichy, we are witnessing the emergence of a "society of amateurs that is more democratic, less elitist, and open to all knowledge," which bespeaks the rise of a subjectivity based on "ordinary passions."[32] In this emerging world, passionate laypersons are not the opposite of experts, professionals, authors, or artists. The two categories merge into a new group known as "pro-am."[33] Professional-amateur contributions, however, do not qualify as work, according to these authors, because they are "free."[34]

This argument overlooks the fact that payment is not sufficient for an activity to be qualified as "work": from forced labor to slavery, human history abounds in examples of unpaid activities that are clearly types of work. The proponents of the "pleasure" approach argue that the fundamental difference between the social relationship of labor and the activities of networked amateurs is, most decisively, that the latter *enjoy* a service that they use without being constrained to do so. In other words, if online partic-ipation is enjoyable, it cannot be considered work, and even if it were, it would be rewarded with symbolic and psychological compensations. The exploitation evoked by the other camp—which considers working for free as a form of wage theft from an economic viewpoint, and as the domina-tion and dispossession of one human group by another from a political viewpoint—would simply be a matter of how one feels about it. And, im-portantly, the positive feeling is conditioned by the fact that the work is free. As Clay Shirky clearly states, "the pleasure of competence, once paid for, stop[s] being a pleasure."[35] But is this not a tautological argument? An activity cannot be exploitative because it is free and enjoyable, yet it stops being enjoyable and becomes potentially exploitative once it is paid for.

Both the "work" and the "pleasure" approaches are seriously flawed. Both camps analyze the content posted by subscribers on social media, but minimize the real issue, which is the production of metadata and personal information. The pleasure argument overlooks the considerable dual ef-forts deployed by the platforms: to prevent subscribers organizing into a workforce capable of fighting its corner, while also encouraging them to stay online as long as possible. The work argument overemphasizes "free labor" and the disappearance of the divide between work and everyday activities in the digital economy. In turn, users' ambivalence toward plat-forms and the drudgery of social media usage are underestimated by the proponents of this approach. In the following pages, I will discuss another crucial element that they fail to recognize: the continuity between social media labor and microwork.

Between YouTube and OnlyFans, the Walled Garden of Amateurism

How can the "pleasure camp" explain that platforms are always delegating productive tasks to social media users? The "work" approach refers to us-ers' false consciousness and alienation. It goes so far as to blame users for indulging in these activities.[36] In so doing, it minimizes the contradictory messages received from the platforms, which both encourage regular con-tributions and hinder users from taking the leap to making a proper job out of these activities.

Let us consider, then, another approach, a more historical one this time. We should bear in mind that an internet "made by users" was never a spon-taneous process. When *TIME* magazine celebrated the social media "rev-olution" in 2006, the platforms were already facing a problem that still ex-ists today, namely subscribers' resistance to taking an active role. It was only with the introduction of social media that the famous empirical rule of "90–9–1" emerged. It divides internet users into three groups: the vast majority of readers and lurkers, who are considered passive (90 percent); a minority, who share posts and write comments (9 percent); and an elite (1 percent) of "publishing members" who produce content, messages, and other material.[37] From the studies made on the first social platforms in the mid-2000s[38] to those on Facebook,[39] and then to those on online influencers and creators in the late 2010s and early 2020s[40] these proportions have changed, but content creation on platforms is still carried out by a tiny minority. The platforms consequently treat only the most active 1 percent as worthy of embodying their ideal produser, and they strongly restrict the agency of the others. In its early days, YouTube (a case studied by José Van Dijck) used to privilege its most active users through programs to support communities of content creators, and competitions to choose the most popular videos, whose authors received tens of thousands of dollars as rewards.[41] The situation did not change much when Google bought YouTube, except users were invited to join the Partner Program, which allowed amateur content creators to monetize their videos by adding advertisements. Platform owners, however, were quick to see the risks. The monetary rewards made the most committed and influential users aware of the value of their contributions, as well as their economic dependence on social media. There is a general reluctance among YouTube creators to share their earnings with platforms that act as intermediaries. At the beginning, this was largely due to the fact that creators came to YouTube from online communities that the creators themselves ran. This is the case even now, primarily because of the popularity of creator-owned streaming services that contributors use to release videos not governed by advertiser-friendly content rules. To prevent online activities from appearing too explicitly as work, YouTube tried to restrict the user to the walled garden of amateurism. It has consistently adopted a vocabulary in which contributors are referred to "as 'hobbyists,' 'amateurs,' 'unpaid laborers,' and 'volunteers' [...], [rather than] with the words 'professionals,' 'stars,' 'paid experts,' and 'employees' commonly attributed to people producing traditional television content."[42] Over time, the platform has blurred amateurism and professionalism. In YouTube's neoliberal corporate growth model, creators are forced to play the role of professional amateurs. The platform's frameworks reward performativity and individualism and insist on authenticity and fun.[43]

With this strategy, YouTube defines the "actual user" as the constantly active and inventive member whose contributions could not in any way be equated with a job. In other words, everyone is encouraged to attain the giddy heights of celebrity on social media, but still to behave as passionate enthusiasts, and never hope to have their endeavors recognized as work. This perspective creates an artificial divide between professionals and amateurs, with the former incarnating mercenary goals and the latter, high moral qualities such as disinterest, dedication to the common good, altruism, and generosity. As a result, even though professionalism has increased in recent years, amateurs outnumber professionals overwhelmingly on platforms such as Instagram, YouTube, and Twitch. Social platforms have constructed the ideal picture of harmonious, collaborative amateurism *to counter any understanding of labor*, particularly that of media producers, who are forced to become content creators and professional amateurs in a participatory medium.

Another example of this divide between "genuine" and "mercenary" content creators is the phenomenon of amateur adult content, which has flourished on the internet and changed the relationship between professionals and casual users (the sector had particularly low "barriers to entry" anyway). On adult video platforms like PornHub, it's rare to see homemade videos trying to mimic professional productions, especially since unverified profiles were banned to prevent abuse.[44] Instead, professional sex workers attempt to imitate amateurs by conveying a sense of authenticity and intimacy. Before the early 2000s, the "big hits" in porn tended to be distributed in cinema film formats; today successful porn videos last a few minutes and might suggest that the shooting was unplanned, the subjects were unaware that they were being recorded, and that the video's circulation is unofficial. Amateurs are encouraged to post their erotic romps and adventures online, while the porn industry itself has adopted the formats, styles and codes of DIY porn—"amateur" being now the highest compliment. Professional porn actors take to pretending to be ordinary couples or friends out for a good time, a subculture attesting to common practices, or sensation-seeking partners. Similarly, "amateur" performers rely on other services for visibility and income. Patreon, OnlyFans, and Instagram, are all places where sex workers present themselves as typical platform users, who just happen to be revealing their life behind the scenes.

In particular, OnlyFans leverages the rhetoric of amateurism and authenticity, both in straight and LGBTQ productions. Beyond producing adult content, creators are encouraged to engage with their audiences by sharing personal stories, staging real desires, connecting across platforms, and even expressing political and ethical views.[45] This results in a new type of monetized social interaction described as "subscription intimacy."[46]

The intimacy between OnlyFans adult content creators and their digital patrons is necessary for social support dynamics to develop. Specifically, the platform requires its users to conform to the rhetoric of "fandom" (and not that of purchasing sexual services). Content creators use concepts like community-building, collaboration, and emotional support to describe the connection to their subscribers.[47]

Even though OnlyFans creators play down the work-related aspects of their online presence, several authors compare them to on-demand digital workers. The two groups are promised freedom, supplemental incomes, and autonomy. In reality, they face heightened economic precarity, volatile demand, and algorithmic management.[48]

The adult entertainment sector more generally projects the image of an ideal contributor who has no commercial relationship with the platform. In porn today—as also in generalist content and in the collaborative gig economy, the "real people" are the amateurs. When they go online in search of authentic lived experience, the motivations of amateurs are described as sincere. Professionals, on the other hand, are delegitimated and portrayed as venal individuals who threaten the collective spirit of participation and communion.[49] By creating a dualism between professional sex workers and amateur adult community-builders, platforms try to conceal the fact that, for these individuals, content creation *is* digital labor—and a form of monetized labor, too.

No matter how theorists describe it, whether as "immaterial affective labor"[50] or as "assetized work" (we'll talk about these in chapter 6), all agree that adult creators of OnlyFans represent the "extreme end of work that incorporates embodiment, subjectivity, and other inalienable aspects of self in commercial practices."[51]

Opposing professionalism and amateurism is thus a way of imposing a "governmentality of work" that defines the conduct of content producers.[52] The platforms' framing of contributors as "professional volunteers" proscribes the language of work, while demanding high-quality investment, and denying any structured relation of dependence between service owners and content producers. Although creators wish to portray their occupation as "real work" and gainful occupation, not all succeed in monetization, and their labor often remains uncompensated value that platforms appropriate. Besides, many content producers fall victim to the capitalist narrative that they are "entrepreneurs,"[53] or rather "fantrepreneurs,"[54] and argue that they are their own bosses and their work is driven by fun.[55] For social platforms to work effectively, creators have to be both free and docile, creative and disciplined, enthusiastic and discreet. This applies to adult performers on OnlyFans or other online content producers, influencers, and creative workers who are constrained to adopt the status of amateurs.

When Users Organize to Get Paid

When social media outlets box their users into the role of contented and disinterested amateurs, they are also attempting to avoid one of the constitutive elements of the dialectic between labor and capital, namely conflict. Yet conflict, in the real form of lawsuits and protests, has been present from the start.

The first and perhaps best-known legal action concerned America Online (AOL), which was the target of a class action suit by its user communities in 1999. Without compensation, almost 14,000 users, including 350 minors, moderated chats and discussion lists and generally encouraged people to use AOL.[56] These "volunteers," whose workload could be as much as sixty hours a week, were required to fill out time sheets, submit detailed reports describing their activities, and undertake several months of training in the Community Leader Program in exchange for free access to a premium (paying) version. The class action lawsuit to recover back pay resulted in a $15 million settlement in 2010.

Other US cases were not so successful. In 2015, reviewers on Yelp sought to be reclassified as employees; their plea was rejected.[57] In 2016, in response to the rules governing the identification of paid subcontractors on the Steam gaming platform, members of the volunteer translation communities have attempted a virtual strike; the most militant users were banned and ultimately Steam shut down its volunteer translation program and relied on Crowdin, a new service to access professional translation agencies.[58] Reddit moderators' strikes to express solidarity with fired employees[59] or to protest high commissions[60] achieved little, apart from forcing an already embattled CEO to resign.[61]

This services' reliance on the work of unpaid users is rendered apparent once they withdraw their labor. Attempts at remuneration also struggled in Europe. In 2014, the Austrian class action "Europe vs. Facebook" mobilized 25,000 citizens. Given differences between the US and European legal systems, this case did not focus on reclassifying users as workers, but on acknowledging that the personal data extracted from the platform had a value. It led to the demand that each plaintiff be paid a symbolic sum of 500 euros. The 12.5 million euros that Facebook would have to have paid in case of a favorable decision would have created a precedent for similar legal actions. Unfortunately, however, the Court of Justice of the European Union prevented the suit. But not all hope was lost. In May 2018, after the European Data Protection Regulation (GDPR) came into effect, similar class actions were launched against the five main platforms (Google, Facebook, Apple, Amazon, and Microsoft) by French NGO La Quadrature du Net and Austrian not-for-profit organization None of Your Business. Years later, the delay in addressing these complaints led to allegations that the data authorities played a "complicit role" in protecting large technology companies.[62]

Despite their high media profile, these labor disputes around community management and the extraction of personal data are not entirely representative of the issues we're discussing. In most cases, protest came from multimedia content producers' demanding payment from the platforms for the texts, images and videos they had created. Shortly after the class action suit of community managers at AOL folded, about a thousand bloggers from an AOL subsidiary, the participatory journalism platform *Huff-Post*, joined another class action suit sparked by a union activist from the National Writers Union. The bloggers sought compensation for their "work as unpaid content producers." Among the reasons given was the fact that their contributions had provided a substantial part of the price paid by AOL when it purchased *The Huffington Post*. Although the action failed,[63] it influenced future claims due to AOL's scale and visibility. For example, the claim that YouTubers brought against their platform included the renegotiation of the partnership agreements for monetizing their videos. Following the discovery of huge quantities of racist and violent content on the platform in 2017, the platform became the center of controversy. As advertisers retired en masse, revenue dropped spectacularly, which is known as the "adpocalypse."[64] YouTube tightened the rules of its algorithm for recommending content. In an effort to challenge these new rules and take back visibility on the platform, US creators sued. Because the platform is a springboard for careers in the cultural industries, from music to television, the struggles and conflicts surrounding YouTube have a particularly high profile.

Yet social media content producers on less technically sophisticated platforms have made similar claims. Vine, a short-form video hosting service owned by X/Twitter and shut down in 2017, is a clear case in point. Unlike TikTok, Vine did not have a revenue-sharing component. The act of sharing on this platform had always mattered more than the ideas, the framing, the editing, or the plot—in short, sharing mattered more than all the skills needed to make a video.[65] In 2016, a few dozen influential Viners demanded payment upon threat of leaving.[66] Negotiations between the platform and its users to develop an advertising revenue system similar to YouTube's lasted several months.[67] To everyone's surprise, the platform owners decided that their business model would no longer be viable if users requested payment, and so they announced the closure of the service. Vine contributors' struggle over pay came to an abrupt end too.[68]

The situation has echoes in claims for compensation by Wikipedia contributors. The famous free encyclopedia is no stranger to tensions between volunteer and paid work. Run by the Wikimedia Foundation, a nonprofit organization that employs staff around the world, who handle the technical, administrative, and partnership aspects, Wikipedia content creation and management is delegated to users, who are organized into different categories, including unregistered contributors, registered Wikipedians, editors, and administrators. The site is free for users, has no advertising, and does not pay its contributors. The question of how to distribute the value generated by users' contributions remains unresolved. The problem is first that the encyclopedia functions as though it were a social media,[69] since users create profiles, earn badges, boost their reputation, and discuss edits to each page. Yet although the Wikimedia Foundation is a decentralized nonprofit, and uses none of the techniques for capturing user-produced value directly, it does capture value indirectly, such as when Facebook uses Wikipedia to train its deep learning solutions[70] and to verify information. Moreover, ever since the inclusion of Wikipedia in the Google Knowledge Graph, and Google's access to structured data via the DBpedia data interface,[71] Wikipedia has contributed to improving Google's search engine, functioning de facto as a knowledge and content database. In recent years, Wikipedia data became an important source of text for language models like ChatGPT.[72] An unexpected consequence has resulted from this fusion with the corporate world: marketing, political spin doctors, communication and search engine optimization professionals have entered the community of contributors, and they are imperceptibly changing its nature by tweaking and spinning the entries of their clients.

Wikipedia's contributors thus found themselves producing content and data for major platforms and perfecting those platforms' AI. For Wikipedia devotees, recognition of the value of their contributions understandably became a major concern.[73] Growing discontent met with fierce resistance from the foundation and from some of the administrators and editors. When Wikipedian Dorothy Howard first used the term "digital labor" to describe the work of users, she accompanied it with a proposal to compensate certain super-editors in exchange for their high volume of contributions. This compensation was not to be taken from advertising revenues (which Wikipedia never had anyway), but from the $51 million in donations that the project receives each year.[74] The proposal was debated within the Wikipedia community, with supporters arguing that passion and dedication to the common good can be combined with a desire to make a living from one's online contributions. However, the editors, administrators, and volunteer users closest to the Wikimedia Foundation (as well as those who had paid employment with Wikipedia) were adamantly opposed to the idea. The more peripheral contributors tended to be quite favorable to the payment plan.

Employability seemed the main factor in these different positions in the debate. Unlike the foundation's members, who were opposed to offering compensation, everyday contributors had difficulty otherwise capitalizing on the skills they acquired through their Wikipedia participation.[75] They

therefore preferred to receive payment immediately, instead of considering their contribution to Wikipedia as a professional investment that they
would be able to use in the future. In contrast, other Wikipedians saw contribution as a way to earn money in the future, just like volunteering in an
unpaid internship or other extracurricular activities that graduate students
can take part in to improve their job prospects.[76]

Hope Labor: Boosting Employability by Contributing Online

These legal claims suggest that networked communities do not work together quite as harmoniously as the "pleasure" approach to platforms participation is wont to believe. They also make clear that the most active contributors, who organized and led these labor struggles, did so because they
hoped their online activities would help them gain a foothold in the cultural industries. Many *Huffington Post* bloggers wanted a start in journalism, many YouTubers in the world of television or music, and Yelp commentators in restaurant reviewing. But, as discussed, social media locks
users into an amateur status, working to thwart their hopes for a professional career.[77]

Activity on a platform is thus conceived as a way of developing job skills
and gaining credibility to assist with a successful job hunt and to attract the
attention of recruiters. Kathleen Kuehn and Thomas Corrigan call this activity "hope labor."[78] It can be found in traditional forms (photographers
who showcase their work on Instagram) as well as in more inventive ones
(gamers who hope to build a career in e-sports by showing off on Twitch).[79]

The platforms see the situation quite differently. For them, user participation equates with the willingness of individuals to provide labor regularly while lowering the price of their contribution out of "love for their
subject."[80] This classic form of exploitation in the traditional cultural industries (and in academia), is here transposed to cultural activities on social media. In both cases, the celebration of creativity masks a downward
pressure on pay, which is expressed most radically when contributors work
for free.

Yet every new video, every new image posted by users on social platforms is work. These "hope workers"—just as the Uber drivers or Amazon
microtaskers we met earlier—contribute in the hope of obtaining a stable
job or at least some legal protection. As with Uber Eats couriers or Google
Raterhub reviewers, the onus is on the contributors to train and find opportunities for involvement. But, unlike other forms of digital labor, social media users perform hope labor as a bridge between present "volunteering"
work and future paid employment.

Their approach thus looks less like a disinterested passion and more like a job search, one in which attracting an online audience combines with building a network of potential clients. Online employment resources are now as important as offline personal and professional networks. This is particularly true for finding employment. Almost three-quarters of job seekers have been on the internet to look for work, and the same proportion believe it would be easy to showcase the skills they have acquired on social media.[81] Still, there's a difference between being a passive user only and actively publish online, as well as between using digital resources while job hunting and during the actual hiring process. Online browsing plays a crucial role in job searches, especially for low-income job seekers.[82] But is using search engines and looking through app profiles enough to land a job? The ability to advertise one's skills online through an extensive and varied social media strategy is particularly important for job seekers.[83]

Wishing to capitalize on one's digital activities in order to boost one's employability is certainly not restricted to those wanting to work in the cultural industries. In Japan, "cell phone novels" (*keitai shōsetsu*) became a form of popular literature in the mid-2000s, against a backdrop of economic recession and declining job prospects for young workers. Each creator sends texts that end up constituting chapters in a novel that other users can read, comment on, and suggest changes to. In some cases, these fragmentary participatory novels have found their way into bookstores and risen up the literary charts, exemplifying a literary genre in its own right.[84] At the peak of this craze, the Magic Island portal gave access to over one million texts of this sort. What this pastime, mainly practiced by young women, expresses is the plight of millennials faced with ever more competitive and insecure jobs. Particularly in East Asian countries, exclusion from the job market entails exclusion from social structures, such as the family, political organizations, and local community bodies, structures that had been essential for empowering their parents' generation and ensuring their social integration. The intensive use of cell phone literature platforms by these young people, authors and participatory readers alike, is a paradox: As their professional prospects dwindle, they seek work under any conditions. Even in the absence of financial reward, contributing on platforms becomes a way of being part of a work community and of finding fulfillment in the recognition they receive.[85]

Facebook, the Hard Grind

Content producers on social media therefore "volunteer" their time and skills to break into the world of work. Yet they are unable to challenge the rhetoric of the passionately dedicated amateur that platforms require, on pain of being excluded. The language of pleasure, leisure, and entertainment are all essential parts of this arrangement. Through the "glamorization" of unpaid labor, platforms delegitimize anyone who aspires to professionalization over pleasure.[86] Of course we have discussed the "1 percent" who are active contributors, but the same role restrictions apply to the vast majority of internet users whose participation is limited to comments, sharing, or even just reading. On social media, younger members who suffer from boredom or discomfort are usually pathologized—stigmatized as addicts or being FOMO (fear of missing out) sufferers[87]—while older members are treated as old-fashioned victims of the digital divide—labeled as "digital immigrants" or technophobes. While scholars have recently discredited the notion of digital addiction,[88] that of the digital divide has been made more complex, with a view to including different demographic and social characteristics.[89] If we compare children from wealthy and from working-class families, a new picture of digital appropriation emerges.[90] In her work on the "informational habitus" of young Californian social media users, Laura Robinson shows that, while middle-class children are more interested in experimentation, research, and online games, those whose families are closer to the poverty line prefer online tasks with immediate results, such as sending a message or checking information—often under the supervision of a teacher or adult—and even searching for jobs.[91] According to a 2023 Pew Research Center study, teenagers in the US have easy access to computers and phones, so one in five is on YouTube, TikTok, Snapchat, Instagram, or Facebook "almost constantly." However, use varies according to race and ethnicity: 32 percent of Latinx teens use TikTok on a regular basis, versus 10 percent of white teens. Income plays a role, too. More teens from families making less than $30,000 a year still use Facebook (45 percent, while it is only 27 percent of kids with $75,000 or more in annual household income). The same goes for TikTok. 71 percent of teens from lower-income homes are on it, compared to 61 percent from higher-income ones.[92]

Both parents and children in these lower-income families fear for their future employment prospects and invest more time online. A proportionally higher number of children from lower-income and non-white families perform unpaid digital labor in the platform economy. Their counterparts from more privileged backgrounds can afford to use the internet less intensively, and even to have periods of media detox. Thus, on one side of the digital divide we have families who cap their children's screen time, and, on the other, parents, teachers, and political authorities who encourage children to use the internet to improve their chances of social advancement.[93]

Among the field surveys addressing these questions, one conducted by Sophie Jehel, which focuses on young users, provides interesting insights.

Her research explores how arduous this investment in terms of social media labor can feel. In the words of the adolescents from working-class backgrounds, Facebook is "boring," "tedious," "a hassle";[94] and, like any chore, it generates avoidance strategies. The dominant discourse—in which digital activities are creative and freely chosen—contrasts with a daily life in which information saturation and the difficulty of managing so many parameters of visibility and attention are simply exhausting for users. Educators and parents introduce additional pressure by characterizing these young people as narcissistic and uninhibited. Yet, for this youth, being present on social platforms is indispensable for monitoring their reputation, which involves deleting problematic photos, removing derogatory comments, and taking down false profiles.[95] Being on social media, instead of generating recognition and an extension of the self, risks turning into a nightmare, ultimately sapping autonomy.

In May 2023, the United States Surgeon General Dr. Vivek Murthy released an advisory warning for parents about the growing evidence that social media use is associated with harm to young people's mental health. Young people who display patterns of problematic social media use face double the risk of experiencing depression and anxiety. Social media users are increasingly worried about their emotional well-being. However, let me suggest another perspective on this issue: What if we looked at poor mental health outcomes primarily as a collective effect of work stress, rather than as an individual problem?[96]

While, at first glance, the cause of these problems could appear to be information overload,[97] excessive online presence is an effect of a labor market on which it is imperative for young people to constantly plan for the future. Their attitudes resemble those of employees who are forced to engage with technologies in the workplace. They share the same concerns, including fears of harassment, worries about compromising their reputation, and feelings of needing to "disconnect." The burden on social platform users is expressed in a specific type of exhaustion that is closer to an employee's bore-out than a school pupil's boredom.

For these users, social media is a training ground for a labor market where they may end up with unstable jobs or no work at all. Unlike the content producers who make up 1 percent of social platform users, these teenagers do not project themselves with hope or pugnacity into the world of work through professionalization, but rather through experiencing the labor market's hard grind. They expose themselves to the same risks as workers: time pressures, loss of autonomy and loss of meaning, fragmented tasks and time, and oscillation between over-stimulation and disillusionment. Managing these risks is also that much harder given that these online activities are supposed to be mainly entertaining and social.

Putting Paid to "Free"

Supporters and opponents of the position that social media content producers are fully fledged workers do agree on one point: whether it's called work or not, activity on social platforms is "free." If the opposing camps agree, this is because the concepts of "unpaid" work and "volunteer" work are often conflated, as are "choosing to work for no pay" (when work is performed pro bono) and "being required to provide free services" (having one's salary withheld). Consequently the contributions made to social media for free are overrepresented, and the specific governmentality of labor on these platforms is concealed. A rhetoric of amateurism takes over, and the problematic aspects of this work are passed over—its arduousness, conflicts, and costs. In fact, a huge amount of paid work drives the so-called free uses. Alongside the users who have managed to get themselves paid, artists who have formed partnerships in order to monetize their presence on the platforms, and beginners who hope to become professionals, there is an invisible army of users working behind the scenes of social media as moderators, "click farmers," and reviewers.

A number of global social media services, such as America's Facebook, Russia's VKontakte, or China's Sina Weibo, don't pay their members. These digital platforms can count on a critical mass of users. Due to their positions and size, they are a service that people end up needing—to buy and sell, to meet with friends, to find information. Platforms become indispensable, and so do not need to use monetary incentives to encourage digital labor.

Platforms can still claim to be "doubly free" because they promise two things: that their service is "free and always will be" (to use Facebook's established tagline), and that user-generated content is not remunerated. This dual principle may be suspended or altered in the case of, for example, a business campaign to enter a new market or to retain users. Starting in 2016, Microsoft launched income-generating activities. For the "active" use of the Edge browser (for example, by clicking on content or watching videos),[98] or to use the Bing[99] search engine, internet users earned points redeemable for music or movies. Google pays users for participating in market research. Between 2012 and 2016, the Screenwise Panel Trend program allowed users to earn $5 per quarter to install an extension on their browsers that recorded all their login data.[100] More recently, the Google Opinion Reward application promised 60 cents in Google Play vouchers for each survey answered. There have also been recent developments since popular platforms like Instagram and X have begun to allow creators to monetize their content and sell products. Yet, platforms that allow users to make money from the content they create usually also require them to pay a commission, like a subscription to premium features or a selling fee. These apps adopt a model of "pay to be paid." Overall, these payments eventually balance each other out.

The notion of the "doubly free" internet persists, despite recently launched platforms and social media often offering to pay their users. Many websites and applications offer a range of ways to earn money or gifts: watching advertising videos, solving CAPTCHAs, downloading mobile apps, selling photos, sharing opinions with other members or taking part in competitions. They're usually smaller, short-lived apps that need a gimmick to compete with the big, free ones. Many of them fail, making the internet look like a cemetery for bad business ideas.

Some have tried to offer a mix of microtasks and social interactions, such as Coinbase Earn, which used to pay its users to receive emails containing small requests (answering questions, visiting a website, providing an email address) paid in the form of tokens or bitcoins.[101] One of the latest crashes of the cryptocurrency market killed the service. Other platforms came up with complex payment schemes, some even trying to pay the user for just being online and using their social space. The social platform Tsu, which closed in 2016, ingeniously managed to imitate Facebook in both its appearance and its business model, with two categories of users: *subscribers* who filled in their profiles, published content, interacted with and recruited other users; and *advertisers* who paid to retrieve data and content from the first category. But, unlike Facebook, Tsu sought to return some of the price paid by advertisers to subscribers. For example, data from a profile which advertisers had paid $1 to obtain, generated 45 cents for the subscriber who created the profile (after Tsu had taken a 10-cent commission), and 30 cents (equivalent to one-third of the value of the new profile) for the subscriber who had introduced them. Tsu also paid the subscriber who had introduced the introducer, and so on, each earning one-third of the previous one.

This complex idea, which actually ensured that every subscriber received some payment, was not the cause of Tsu's commercial failure. The company went out of business because its main competitor, Facebook, classified any link, comment, or content relating to Tsu as spam. This effectively eliminated the service from Facebook and its partner sites in the Meta universe.[102]

The example of Tsu turns the spotlight on another essential aspect; the concept of "working for free" on social platforms must be viewed as an ideological construct, while its reality is shaped by the economic and social structures that govern these services. The existence and survival of platforms depends on their ability to attract several distinct groups of users— with each one embodying one facet of a multisided market mechanism.

The pricing structure that a platform sets up will be consistent with its environment, its objectives, and its specific methods. It may, for example, make different groups pay different rates, on different terms, or have one group pay and another one not (the latter case applies to Facebook users, for example).

In addition to the pricing system, the platform can use economic incentives to attract certain groups of users, such as discounts and—particularly important for our discussion—negative pricing, i.e., a payment to platform users. But this is not always necessary: some platforms are able to count on the loyalty of their captive users, and can thus do without economic incentives to draw them in. When the incentive reaches zero, we come to the second aspect of the "doubly free" system: users do not get paid for providing content or for their participation more generally.

The so-called free nature of social media labor is therefore, paradoxically, a manifestation of platforms' systems of pricing and incentives. Indeed, what interests the platforms is not how enthusiastically the subscribers contribute, but the relative positions of the user categories. "Free" work is closely linked to what those who interpret social media activities as work call *exploitation*, in the strict sense, that is, the gap between the value derived by the platform from users' contributions, and the earnings received by the same users. This gap is obviously greater in the category of users to which platforms offer no payment.[103] Nevertheless, the size of this gap can vary if users make a successful claim for payment, join a partnership program, or become creators. Not all users are exploited equally. "Most users," says Robert Gehl, "receive no financial reward for their participation but, of course, some become microcelebrities and build audiences that can be traded in for money."[104]

On the rare occasions that "free" labor is remunerated, it tends to be overvalued financially; individual users' earnings on social platforms are much more polarized than on microwork platforms like Upwork or ZBJ. Microwork income ranges from a few cents to a few thousand dollars, whereas on TikTok and on YouTube, it may range from zero cents to several million dollars.[105]

A further aspect worth highlighting is that, as I have already mentioned, not all tasks performed by users on microtasking platforms are paid. Sometimes they are performed to qualify for another task, sometimes the requester refuses to pay, etc. But no one would call this work "doubly free" or think that microworkers are "volunteering" their content. This reflects the continuity between social media labor and other forms of activity on digital platforms that capture user-produced value. Whenever digital labor is engaged, there is a gap between the value produced and the revenue distributed to users.

The Link Economy

Thus, is social media labor a form of work or pleasure? For the proponents of the former approach, users' contributions are a hybrid of work and leisure that platforms exploit. According to those who advocate the latter, online participation is free and generates pleasure for the user. But I have shown that platform activities can be just as much of "a grind" as a real job—meaning they are neither leisure nor pleasure. I have also demonstrated that platforms' claim to be "free" is merely a business strategy. Therefore, they cannot be viewed either as simple participation or as exploitation in the classic Marxist sense. These findings undermine both approaches.

In either case, those who see social media simply as "pleasure" and "work" seem to miss the point of how people work on digital platforms. Both analyses emphasize *content* to the detriment of *data*, namely personal information. This bias is understandable. When online participation first began to be analyzed in the early 2000s, the internet was mainly conceived as a place for multimedia cultural production. By the end of the decade, these sorts of analyses were obsolete because the "small content producer revolution" failed to deliver the anticipated results. Today, on social platforms and in the current media environment more generally, the recognition of creators as *content* producers is limited. A 2014 study by the Columbia School of Journalism shows that photos, videos, and information posted by users (as "citizen journalists" or witnesses to newsworthy events) are used by the mainstream media, but that their authors are rarely credited, only the platform hosting the content.[106] The 2023 strike of film and television writers,[107] as well as the lawsuits filed by journalists,[108] and fiction[109] and non-fiction authors[110] against OpenAI revolve exactly around this lack of recognition: these works are not treated as products of human creativity, but simply as data for training a machine learning model.

In general, content is not a direct source of revenue for social platforms; when it is, it is almost never through simple resale. For example, gaming platforms such as Steam do not directly appropriate users' content, but take a commission when users sell their virtual creations (by modifying existing games or "mods," accessories, skins, levels, or characters).[111] Others, such as Vimeo, charge a transaction fee on the direct "on-demand" sale of videos to fans willing to pay.[112] Meta does not sell its users' texts, photos, or videos, but it obtains indirect revenue by using them, for example, to illustrate ads that are displayed contextually on its subscribers' walls.[113] Similarly, sites that specialize in graphic arts and photography, such as DeviantArt, must periodically remind users that they do not sell their users' creations.[114]

The early emphasis on the value captured by platforms from users' content production has given way to a focus on the value of *data*. This includes personal information, visits to brand pages (clicks, shares, views), and links between these (forming friendships, the frequency of interactions, time spent chatting). The business models of the larger platforms depend on ad sales brokers (Adsense, Adwords, and DoubleClick for Google; Audience Network for Meta), which collect and exploit personal information such as browsing history, profile scores, lists of friends or followers, geographical location, right down to session replays (the highly detailed tracking of specific user actions and site responses).[115] More recently, these models for value capture have shifted to machine learning. They use data relating to digital usage patterns (server logs, tags on images in an archive, the behavior of users of a site, etc.) as their main sources of examples for training algorithms and measuring their performance. The real revenue streams and opportunities for profitable investment are derived today from the data extracted from behavior on social media. From the platform's perspective, content is actually a loss leader, serving as bait to sustain a lively community for the benefit of advertisers and investors.

Unlike content, data does not lend itself to being equated with conscious work or enthusiastic participation. This probably explains the relative indifference of the "pleasure" approach to data. At most, for them, users leave "traces" of digital data that contribute little to user satisfaction and pass largely unnoticed. Some observers explain that these traces "do not make sense," do not become "a basis for any knowledge," and even that "conferring on them the status of 'information' seems to greatly overestimate their indexical capacity."[116]

Still, metadata such as timestamps, session logs, URLs, and unique user identifiers are valuable assets that platforms exploit both for computational purposes, that is, to calibrate algorithms and train AI, and for economic reasons, when advertisers, data brokers and even governments are given paying access to it.[117] Personal information, and records of online and social media behavior are exploited in addition to the information that platforms explicitly request from users.

This makes for a thriving data economy,[118] as is now well known thanks to the media and research coverage of the vast scale and variety of the personal data and metadata collected by platforms from their users. For example, when Austrian lawyer Max Schrems asked Facebook to retrieve the personal information stored on him by the platform, he received a file of 1,200 pages.[119] A *Guardian* journalist who made the same experiment with the dating app Tinder received the equivalent of 800 pages.[120]

This puts the importance of content in a platform's business model into perspective. Where the content producer sees a photo skillfully depicting a certain sociocultural context, the platform sees nothing but a set of extractable metadata. Computers do not "see" photos and videos in the same way as humans. When you look at a photo, you may see your best friend standing outside her house. From a computer's perspective, that same image is just a data set that the computer can interpret as shapes and color values. When content is first created, its metadata includes information generated by the authors. As it circulates online, new information is added by those who like, share, or flag it. Hence the content is less important than the act of sharing it and the comments, rankings, and filters that it receives. All of these actions produce data.

The involvement of users in enriching content with new data is also of interest to platforms because it allows for sharing communities to emerge within larger digital social networks. Affinity networks enable preferences, interests, and practices to be tracked. Joining a group of music enthusiasts or commenting on a certain television debate with other users, interests a platform only for the social graph resulting from the exchanges—its size, its homogeneity, its composition, or its structure. What social media are practicing is, as the economist Laurent Gilles aptly puts it, not an "economy of goods" but an "economy of links"[121]—especially when these links can be represented as units of a data set.

The skills and creativity of the people involved in online social networks, far from constituting the backbone of these platforms, only become attractive to them when they have been broken down into data points to be recomposed, grouped, segmented, and classified.[122] The flair and particular personality of a creator are irrelevant when data is generated just as copiously through nonspecialist acts such as sharing, viewing, and browsing. As with microtasking, what counts for the platform is clicking on a button or following a hyperlink in a mechanical way—that is, in a way that benefits the machines.

Qualification and Monetization: What's the Value of a Like?

Determining whether contributing to social media can be considered digital labor requires us to assess the value of the personal data extracted. At first, since this data is the object of strictly regulated transactions between the users who produce it, the platforms that capture and monetize it, and the advertisers who use it, its value seems easy to calculate. As with any transaction, the estimated value of the negotiated good varies considerably, depending on the asymmetries of market power, that is, on the ability of the different parties to impose their price. A "like" on Facebook is estimated at between $0.0005 (a value calculated by a downward auction mechanism between a social platform and an ad sales platforms),[123] $0.004 (if you buy 1100 likes for less than $5 on a click farm platform),[124] and $174 (according to a marketing agency's estimate of the purchase cost for a potential customer).[125] The value of a post on X may range from $0.001 (based on its contribution to X's commercial value),[126] to $0.02 (based on the cost of a premium subscription for an intensive user),[127] to $560 (if it is a company preparing for an event).[128]

Whether this wide range is due to the difficulty of assigning a single value to a good in the presence of multiple economic actors and multiple purchasing methods, or, more generally, to a "crisis in the representation of value," as discussed in chapter 2, it's never a good idea to reduce data's value to its contingent price. This is because the market interpretation of value introduces an artificial break between production and circulation, between "estimated value" and "value realized," and between company and market, within platforms that are, in fact, both at once. Any act of producing data on a platform is already an act of putting that same data into circulation. The indicators devised for the preceding century's economy of markets and companies are no longer pertinent: prices, exchange value, willingness to pay, etc. Considering data to be the product of users' labor on social platforms means applying to it the same mechanisms that we have already discussed for valuing human labor on any platform that uses digital labor.

Knowing how much data is worth means looking at its use for qualification, monetization, and automation. The qualification value appropriated by a platform derives from the work done by users to describe content, information, and even other users. The platforms require this to function, since the architecture of social media depends on social norms and processes.[129] That is, algorithms require audiences to evaluate and prioritize information. The real fuel for social media is not so much "user-generated *content*" as "user-generated *content classification*." In order to offer relevant and personalized information, the algorithms that determine the suggested reels on Instagram or the contents of a banner ad on an app are based on the tastes and habits of users. For the latter, the experience of social platforms is full of opportunities for qualification: comments on blogging platforms, film reviews on movie recommendation platforms,[130] likes and subscribes on YouTube,[131] "karma points" on Reddit,[132] hashtags that describe the argument or general feeling of a conversation on TikTok:[133] all of these are examples of activities where users of social platforms evaluate, group, and sort information.

In addition to qualifying content, users qualify reputation. They identify the most influential and popular people in a community and evaluate their credibility in sharing information.[134] Low-visibility users are no less concerned by these practices of reciprocal qualification as high-visibility ones. On the dating application Tinder, for example, the number of individuals accepted, as expressed by the "swipe right" of potential partners, determines the "ELO score" (in reference to the ELO ranking used for chess players) of each profile, which in turn makes an algorithm select and recommend other profiles.[135] Platforms use similar matching mechanisms for friendship or professional contacts.[136] Here, subscribers are not only invited to specify the type of connection with their peers (friendship, family, collaboration, or romance), but also to indicate the intensity of this relationship and its associated feelings.

Likewise, Facebook Reactions and their equivalents on other social platforms are tools for gauging the importance of the social relationships qualified.[137] But the responsibilities for classification that the platforms entrust to users go far beyond the framework of sociability alone. On Facebook, subscribers find themselves assessing page and interface design changes in A/B testing, without giving their consent. They do so not because this activity is consistent with their affects and social affiliations, but because sometimes Facebook gives them a concealed task.[138] In the same vein, the platform have offered for a number of years a feature accessible by anyone, called Facebook Editor, which allows the user to evaluate the quality of the information available about a business, a place, a movie, and a television show.[139] Users dining at a restaurant or simply consulting its website can check and correct its name, address, and the menu. If they agree to make changes, the usual Facebook interface disappears and is replaced by another one, which looks remarkably similar to a microwork platform. A status bar measures the amount of information contributed, a tab gives access to the history of all contributions, a button displays the number of people who have benefited from this information, and a score measures the accuracy of the information entered by the user. It is also possible to earn badges: "novice" for the first change made, "oracle" after submitting 100, "pioneer" for 250, etc. The quantity of work carried out by Facebook editors is not the only thing to attract points. Consistency of effort, and the variety of the information provided are also rewarded. Those suggesting changes when located in ten different cities earn a "jet-setter" badge.

Researchers and artists at the SHARE Lab attempted to map the data captured by Facebook's "algorithmic factory."[140] Some major categories emerge from examining the 300 million petabytes of data that the platform had collected from the activities of its users. The first category corresponded to the activity of content qualification. This consists of static information entered by users for their own profiles (name, age, etc.), or else information drawn from an ever-mobile mass of activities and behavior. The figures are disturbing: 10 billion messages, 350 million uploaded photos, 4,300 million "likes," 4,750 million shares per day, etc. Other personal data is extracted in order to sell it, in what is a billion-dollar market for targeted advertising. Information about the digital devices of subscribers, for example the brand of smart phone or computer used, is also collected (the SHARE Lab inquiry again showed this). Facebook can thus gather information about the income or the level of digital literacy of the owners. IP addresses can be traced back to subscribers' homes and give access to estimated real estate values of users' property. Moreover, the still or movie cameras used to make the videos and photos posted online are used to deduce the identity of their owners by cross-referencing with other metadata. Data collection continues outside Facebook, which tracks users through cookies and "like" buttons placed on pages of other websites. Meta controls the relationship between Facebook, Instagram, Messenger, WhatsApp, and others, and opens up the data across mobile applications. The photos, videos, and phone contacts of their subscribers, and the stored details of the Wi-Fi connections used are all relevant data.

A final area of monetization operates via the Meta Business Partners program,[141] through which the platform makes business deals with companies and large data brokers. These are companies that aggregate and cross-reference information derived from media outlets, ad agencies, public administrations, open data directories and, logically enough, from data collected by internet companies from hundreds of millions of citizens. By comparing its own data sets with those of partners specialized in monetizing personal data (Acxiom, Epsilon, and Datalogix), Meta contributes to tracking the habits of 77 percent of internet users around the world, about 3.59 billion people. This database matching is all the more disturbing because data brokers gather information on the health, political opinions, sexual orientation, and religious beliefs of citizens. They do this despite calls for regulation and accountability from the US Federal Trade Commission.[142]

The Eyes and Ears of Automation

Platforms capture value in more ways than qualification and monetization. While qualification is a necessary condition for platforms to operate, monetization ensures their financial viability, and finally, another form of value, automation value, feeds into investments in technological innovation. This time, the information users produce trains algorithms and builds the huge databases needed for deep learning programs. As computer scientist Yann LeCun, director of FAIR (Facebook Artificial Intelligence Research) and a pioneer in this branch of computer science, explains, advances in this field are not due to recent improvements in scientific methodology, but to the increased availability of billions of examples of images, texts, and sound bites divided into millions of categories.[143] As a technique, deep learning had been around for three decades before the availability of massive amounts of data allowed it to leapfrog to its next stage.[144] It is now possible for machines to recognize patterns in complex quantities of information, such as images or natural language. This profusion of content and data is necessary because, unlike humans who are able to learn from very few examples, machines require vast volumes of data.[145]

But the examples produced by users of large platforms and smaller apps contribute only to the first stages of automation—known as data generation. Machine learning requires other human acts such as data annotation. What this means is that data is arranged in a specific order, cleaned by fixing errors and inaccuracies, enriched with extra information, and formatted for convenience.

This is why some platforms encourage users to add tags to their content. For example, on Instagram, a user assigning the keywords "beach," "ibiza," "umbrellas," "deckchairs," "summer," "2024," "from my hotel room," "sun," "relax" to a photo describes the image very precisely—so much so that recognition software can learn its content, the exact time and place it was taken, as well as the feelings associated with the image. In the case of Meta, this content tagging is used to train the visual pattern recognition devices needed by businesses and the law enforcement agencies for face recognition. Since the introduction of the "real name" policy (banning the use of pseudonyms and encouraging the use of one's legal identity to register on the network), and the purchase of a company specializing in facial recognition,[146] Meta experimented with the identification of users in the photos circulating on the platform. Under the pretext of fighting fake news and revenge porn, the company even launched a "Face CAPTCHA" feature that required users to upload a recognizable photo of their face to confirm the authorship of their account. The photo had to be suitable for the biometric verification algorithms used to detect highlights and model the face based on the faceprint principle (analogous to fingerprints). Faces were then matched with real names by other functions, like the "tag suggestions" feature, which encouraged users to link photos to a friend's profile, or Photo Review, which asked users to annotate photos in which they appeared.[147] Prior to a privacy lawsuit,[148] Meta halted its face recognition feature—allegedly used by one-third of Facebook users.[149] Despite this, other companies have developed face recognition algorithms using biometric data from social media. The controversial provider of law enforcement identification services, Clearview AI, has been extracting forty billion photos from Facebook to train its facial recognition engine.[150] These computer vision processes are sometimes also developed by other Meta companies, such as Instagram or its VR branch Quest, where images abundantly annotated by users are used to recognize goods and target commercial offers.

In other cases, Meta put users to work to improve natural language processing. To date, machine learning has focused on translation, especially when different language communities are practicing it together. Facebook provides automated multilingual versions of its users' posts. External communities (such as the open collaborative translation database Tatoeba) provide sample texts and phrases in multiple languages to calibrate the platform's translation software.[151] Facebook users are then invited to improve the translation by clicking on the "rate this translation" link at the bottom of any multilingual post, whereupon a star rating appears, as well as a "I have a better translation" tab that allows users to contribute another version of the same post.

The cyclical vogue for chatbots is another indication of the importance of users' digital labor and its conversion into automation value. Chatbots are marketed to make appointments,[152] book hotels, or buy airline tickets,[153] or they may send private messages recommending a concert or a purchase,[154] or even simulate the presence of a friend or partner.[155] While presenting themselves as completely automated, chatbots presuppose the delegation of computer tasks such as verifying, and giving training to, human beings. A conversational agent that gives complex information depends on human input. As OpenAI admits, "[users'] instructions will be used to make our models better unless [they've] opted out." Whenever ChatGPT's AI replies to users' queries, they're asked to rate the response. Like Reddit or YouTube votes, they use "thumbs" to rate the chatbot's response. This user feedback is used in reinforcement learning with human feedback.[156] Rather than learning by training, the machine learns by trial-and-error. In this case, the users are the "teachers" who rate the AI's attempts at talking naturally. Aside from users' inputs used for reinforcement learning, OpenAI has been employing trained data from CommonCrawl, an open repository of web data harvested over the years.[157] Founded in 2007, CommonCrawl has listed more than 250 billion pages.

Several AI solutions use this approach. Meta's AI models are trained on the personal data of Facebook, WhatsApp, and Instagram users. The platform even issued a "Generative AI Data Subject Rights" form so its members could access and delete their personal info.[158] Grok, the chatbot developed by Elon Musk's xAI company, was initially trained with internet-crawled data. It generates output based on current data that users post on X. The same platform has performed other iterations of this experiment that have ended in disaster.[159] Back in 2016, when X was called Twitter, the training of Tay, Microsoft's conversational bot, was, according to its designers, initially intended to take the form of an interactive game. Tay was meant to have the persona of a teenage girl, and to develop as she communicated with her peers on social media. She was meant to learn languages, conversational norms, and eventually to form opinions. Some users got caught up in the game and tested the chatbot's limits by suggesting illegal behavior or racist expressions, which (before she was shuttered) Tay promptly reproduced uninhibitedly.[160]

The failure of "M," Messenger's virtual assistant, which we discussed in chapter 1, is another example of how much the automated learning programs of social media depend on the human work that feeds automation—and of how chatbots are not naturally destined to success. The device was initially available to a few tens of thousands of users, with whom it interacted through instant messaging. Facebook, unlike other companies, did not want to hide the human component, and the app's marketing materials declared it to be artificial intelligence "driven mostly by humans."[161] Because Facebook found the unpaid contributions of users difficult to coordinate, it had resorted to hiring microworkers primarily from the United Kingdom, Kenya, and Nepal through the CloudFactory microwork platform.[162] These "external operations teams" were to be the nucleus of a fully automated AI solution. However, despite Facebook's vast financial and scientific resources, after three years up and running, M had still not got beyond 30 percent automation.[163] Thus human microtaskers were actually responsible for satisfying more than two-thirds of the user requests, which ranged from delivering drinks to writing a song. The microtasking costs were deemed too high by Facebook, which permanently discontinued M in 2018. The fate of M not only revealed the limits of the fantasy of complete automation, but it also demonstrated the fact that, if a platform fails to get humans to work for free, as it so often does, then the program is not always financially viable.

Heroes or Cleaners? The Production Chain of Content Moderation

The rise and fall (and rise again) of chatbots highlights the continuity between work for social media and paid or micropaid digital labor. Social platforms do not rely only on their users to calibrate algorithms and perfect deep learning methods. On Amazon Mechanical Turk alone, requests (human intelligence tasks, or HITs) from Google tripled between 2012 and 2017, according to a study published on the MTurk Crowd forum.[164] Around 2019, the bulk of AI training microtasks had moved to SageMaker Ground Truth, the Amazon service specializing in "low-cost machine learning" that allow developers to "prepare their datasets for training their ML models"[165] by quickly sending tasks to MTurk workers for annotation and verification. Requests cover a vast range of tasks—judging the relevance of search engine results, comparing two YouTube videos to decide which is funnier, annotating Android applications, ranking posts by topic, and so forth.

On social platforms, users are not the only providers of digital labor. The content they produce, and the validations and tests they carry out do not exhaust the amount of value-producing work required. Digital labor of a different kind, which is low- or micropaid, is also needed. Platforms thus connect a growing number of microtaskers to users. These two categories of digital labor are linked in the same chain of production and value creation.

An illustration of this shuttling between micropaid and "free" work can be found in Facebook's fight against the proliferation of propaganda, sensationalist news, and hate messages that are usually classified as fake news. The platform emphasized by turns the power of its algorithms, the professionalism of its "external operations teams," and the dedication of its users. Between 2016 (the beginning of the Cambridge Analytica scandal) and 2021 (the year of the Facebook Leaks revelations), it's been a tough road for Meta's moderation. Initially, former Facebook subcontractors revealed that behind the system which suggests "personalized trending topics," previously thought to be fully automated, there lay a team of human operators.[166] The platform quickly laid off the entire team and, to reassure the public and investors, declared that it had replaced them "with an algorithm." However, without human supervision, it rapidly fell prey to manipulation by fake news producers.[167] The platform was then accused of having been largely responsible for introducing electoral bias into the American presidential election and got rid of its personalized topic service.[168] This storm forced Facebook to bring back digital labor. It introduced a "mark this post as fake news" feature, whereby any user could report what they considered to be misinformation.[169] However, the work of fact checking, thereafter carried out by the platform's users, proved insufficient. Snopes.com, the independent platform of volunteer "myth debunkers," and also Wikipedia, were mobilized.[170] External teams, under contract with Facebook, were also added. Companies like Telus, Accenture, Teleperformance, and Sama were called upon to evaluate Facebook's content for Europe, Africa, and South America.[171] According to Meta, both microworkers and users improve the quality of information by compiling a database of contested media stories, which will feed the platform's artificial intelligence with examples, enabling it to take decisions automatically concerning the posted content: Should it appear in a certain group's news feed? Is it eligible for a sponsored post?[172] This emphasis on automatic content decisions should not omit "humans in the loop," who are still crucial to moderation. Besides the "free" social media labor of users, Meta admitted in 2021 there were "over 40,000 people [. . .] working on safety and security, including global content review teams in over 20 sites around the world reviewing content in over 70 languages."[173]

Often, as in the case of fake news, evaluating information for a platform amounts to filtering out whatever does not suit the sensibility of its users or the aims of the platform's inventors. When microworkers on Amazon Mechanical Turk or Remotask have to perform annotation tasks on images with potentially problematic content, their microwork helps to generate automated solutions in the form of moderation. They help to align filtering and recommendation algorithms with regional contexts by complying with local regulations. In Pakistan they can be recruited to class TikTok videos containing blasphemy, in Japan to mark the Google links that point to adult websites, in the US to label images that violate copyright laws on Instagram, and so forth.

Sometimes, and this is particularly relevant to the discussion of social media labor, users are also requested to behave as moderators. For example, Chinese streaming platform Bilibili, with 500 million users, has created a participatory "disciplinary committee." According to reports, 70,000 users serve as "inspectors" by monitoring issues and voting on videos flagged as "vulgar."[174] Users of X, Tinder, and Discord perform the same operations when they block, hide or report messages and profiles they find offensive or inappropriate. Each downvote affects recommendation algorithms and content filters.

Whether the content filtering is paid or performed on a "voluntary" basis, it is typical of how online media function. In the second case, it is similar to what researcher Sarah Roberts has called commercial content moderation (CCM).[175] This "dirty work" is done by digital laborers to make sure the content is compliant with terms of use, legislation, and community standards. This makes for wide international variations. In some contexts, moderated content will be pornographic images; in others, violent, racist, homophobic or misogynist statements; in still others, statements against religion or posts from a government's political opponents.

These moderators face arduous chores and significant occupational risks.[176] The consequences can be disastrous for their mental health and well-being. Some develop forms of post-traumatic stress disorder after viewing extreme images (gore, violence, nonconsensual sex acts)[177] all day long. Like other types of digital labor, moderation has followed the methods of offshoring. Workers are often found in Kenya (like Sama, discussed in the previous chapter, which used to moderate Meta content), in the Philippines, and in India (working for US and Chinese platforms).[178] In Sarah Roberts's words, the offshoring of content moderation is a kind of "illegal offloading of technological waste."[179]

Moderators work almost invisibly in geographically remote places, yet, paradoxically, they are necessary to ensure that the contents posted by users conform to current standards of "visibility."[180] Moderation also fulfills another essential function, which is to further the cause of automation.

In evaluating content, moderators find themselves sorting and enriching it—in short, qualifying it. Qualified content can then be integrated into a corpus that can be used for purposes other than online publication. As such, they evaluate not only content but also data. This is summed up most clearly by the leader of a team of Indian microtaskers in the documentary *The Moderators* (2017): "Basically what you have to do is you have to judge the data."[181]

According to Sarah Roberts, CCM happens at four types of organizations: in-house in big tech companies, offshore in "boutique" local firms, in remote call centers, and on microwork platforms.[182] Yet, unpaid users do a lot of the moderation, too. Microwork becomes intertwined with users' "free" digital labor as they contribute to the data corpuses that advance machine learning. YouTube provides clear evidence of this fact. It enforces strict rules for filtering problematic content (violent or adult), protecting copyrighted material, and classifying by language and country. Its reputation is built on the "automatic" detection of content, and on reporting offenders. Actually, the platform also uses another method, based on a work sequence with three categories of moderators. The first category is that of users who flag possible violations of the platform's rules. After this initial screening, carried out for free, YouTube activates a second category of "super-users," first called "trusted flaggers," but subsequently known as "YouTube Heroes."[183] Their reward consists in points enabling them to access to premium content or special offers. They do not only check the material flagged, but they can also enrich videos by adding subtitles and check descriptions or give feedback to other users.[184] The last category is that of commercial moderators, who are recruited and paid by platforms or by specialized subcontractors.[185] They review dozens of videos per hour and classify them according to set criteria such as commercial videos, music videos, TV show excerpts, but also videos "containing inappropriate content." One of their most important roles is to decide whether a controversial video should be "demonetized" or not, that is, deprived of advertising revenue and demoted by the algorithm. YouTube attempts to reassure its advertisers, who do not want to damage their brand image by having it associated with objectionable videos, that its moderators are human. But to which category will they belong?[186] There are the tens of thousands of paid moderators, a few thousand "Heroes" paid in kind, and millions of unpaid occasional flaggers—clearly, it will be a combination of the three.

Click Farms Recruit Networked Workers like All Others

If we take a closer look at the daily tasks of internet moderators, we notice a surprising fact. When moderators must decide whether content classified as racist or hateful should be removed, they have to take into consideration the fact that social media platforms would lose a lot of their marketability if they gave the impression of restricting and cracking down on the contributions of their users.[187] The right to free speech becomes an ambiguous ally of business interests, which ultimately articulate with society's racist, misogynistic, and homophobic discourses. This presents a paradoxical situation for a moderator.

It is common for platforms to support and even encourage forms of production and circulation of content that, while clearly contravening the general terms of service, contribute significantly to the volume of traffic on their networks. Since social media is designed with an ideal of virality in mind, that is, a continuous buzz of conversations and new surges of content, they turn, whenever possible, a blind eye to high-visibility, but problematic, content. This is precisely the content for which search engines and referencing algorithms have been optimized. This optimization often involves adding a critical mass of "artificial" clicks in order to stimulate the work of "organic" users.

The economics of virality is rooted in structures very similar to those of microwork. Content mills, for example, which produce the blog posts, reviews, comments, descriptions, and tags that support or construct the visibility of brands and public figures are not fed by freelance writers, but by microworkers, who produce no more than a few dozen words each, on various subjects. Their pay ranges from 0.1[188] to 3 cents per word,[189] but they are lured by the promise of higher sums.[190] Often, these aren't genuine translations, but just software-translated sentences and, increasingly, content produced by generative AI.[191] In turn, the texts produced in this way are not intended to be read by humans, but by web crawlers programmed to detect distinctive keyword sets. Workers of these content mills are put to work optimizing the ranking of certain web pages in search engines. For this, identifying given terms in indexed pages is not enough. They must also detect hyperlinks and cross traffic between sites. This is why companies other than content mills are required to manufacture virality, such as link farms and click farms. These semi-clandestine enterprises aim to fabricate committed users and artificial fans.[192]

"Click on content," "increase the number of views of a video," "give five stars" to an iTunes application, or even "create profiles that subscribe to the feed" of a celebrity or company—these are some of the tasks "farmers" perform for a few cents. This work is barely distinguishable from the tasks any normal user would carry out.[193] When one talks of "fake followers," this refers not only to automatic accounts or fake profiles, but also to internet users who subscribe to a page, channel, or account in exchange for a micropayment.[194] In addition to the fact that it is not always easy to distinguish fake followers from real ones, the two categories of users can influence each other. Click farms want to produce wave upon wave of site traffic, and so they "pump-prime" online marketing campaigns, political propaganda, or rumors. This can lead to cascades of information flows that go far beyond the initial core of fake clicks and ends in messages being promoted by "organic" users.[195]

Click farmers and microworkers have the most in common. Some click farms have their own premises in which individuals scan and share content all day long, being paid at rates that are unacceptably low and working under extremely harsh conditions that some have qualified as slavery.[196] Again, the workforce is provided by low-income countries or less advantaged areas of high-income countries.[197] These types of click farm are well known. TV reports, with scenes of police officers bursting into apartments in Thailand or Russia, where dozens of people are clicking away on hundreds of smart phones, are quite common.[198] A large number of click workers operate from home and are also recruited on platforms such as Upwork or Fiverr. In 2011, Freelancer, another platform, paid particularly poorly for the creation of fake profiles, fake reviews, and fake subscriptions: 30 percent of the tasks received around 10 cents.[199] Over the following decade, the prices have increased significantly: in 2021, clients would pay from 25 cents to $100 per review.[200] Reviews on Google, Tripadvisor, and IMDb are actually some of the most expensive forms of fake engagement, since they require customization and linguistic proficiency. Views are at the opposing end of the spectrum. Prices vary depending on what platform is targeted. TikTok views are $0.01 to $0.03, Instagram views can be sold for $0.08, and an Instagram story can reach $0.30.[201]

On some of the lesser-known platforms (MinuteWorkers, ShortTask, MyEasyTask, etc.), between 70 percent and 95 percent of the microtasks also involved creating fake profiles or fake links.[202] The geographical origin of the requests (66 percent from the United States, Canada, and the United Kingdom), and the geographical origin of the "farmers" (38 percent lived in Bangladesh and 30.7 percent were from Pakistan, Nepal, Indonesia, Sri Lanka, and India) also suggest the symbiosis between click farms and more legitimate microwork platforms.[203] The only fundamental difference is that on platforms for fake clicks, microworkers can hope to be paid about five times higher than on Amazon Mechanical Turk.[204]

Although click farms are described as anomalies that deviate from "normal" social media usage, they have direct links to the standard platform economy and are by no means a marginal phenomenon. In 2013, the sale of fake followers on X (then Twitter) represented a turnover of $360 million, while on Facebook, fake clicks reportedly generated some $200 million per year.[205] By mid-2010, mCent (a service that paid users to install apps and make them more popular) was reporting thirty million users and ranking among the top five apps in India.[206] It also attracted $57 million investments from Verizon, among others.[207] At the beginning of the next decade, other indicators revealed how big the fake follower market was: on average 1.2 percent of social media followers of US companies were fake and, if found, the average firm was estimated to lose up to $7.10 million in market capitalization.[208]

Russian Dog Food

The setting is an online discussion forum where digital workers from South America share anecdotes about clients, practical tips, and cautionary tales. Rosa hails from San Cristóbal, Venezuela, near the Colombian border. As a newbie, she's learning the ropes. For now, she likes microtasks. She is experimenting with several platforms, including Microworkers and YSense. Her last task consisted in accessing Facebook and leaving a "like" on the page of a Russian dog food brand. The pay wasn't great, but as a job, it's quick and harmless. Little did she know it would get her blocked on Facebook. Panicked, she asks her online forum friends for help.

"My Facebook profile is dead!" she laments. "I don't know what to do. It's gone."

There is an outpouring of support from forum members. Juan, a veteran microworker with a knack for troubleshooting, chimes in among the comments.

"Rosa, do you speak Russian or did you use a translator to write the comment? Facebook might have flagged it as suspicious behavior. It happened to me once."

"I don't speak Russian, and anyway, it wasn't about writing a comment. The only thing I had to do was click on a link, view a post on a Russian page, and like it."

"Were you using your own profile or a work profile?" Juan inquires.

"My own . . ."

"That's why you were blocked. You are geotagged in Venezuela. And you do not speak Russian, so you do not usually engage with Russian content. They thought you were a bot."

Juan is a geek, but he's also kind of a poet. He explains that liking content for a living "is like practicing Capoeira with Facebook's algorithm."

"The algorithm is watching you," he adds, "and it doesn't like irregular behavior. If you accept a microtask that requires to access Facebook, you'll need to create another account to bypass restrictions. But remember: even fake accounts need to appear real. To avoid detection, they need to be authentic, too."

"Maybe," quips Rosa, "to prove I'm a legit Russian pet lover, I should start a profile where I only talk about dogs in Moscow, and another about cats, and so on. Do I need to have a million parallel lives to please Facebook?"

It's not uncommon for people to be in Rosa's shoes. After analyzing over 600,000 profiles and interviewing hundreds of workers, my team and I found that the overwhelming majority of tasks on the popular platform Microworkers involve some type of fake engagement. Some tasks request that the workers watch YouTube videos and like them. Some workers have to leave reviews on booking platforms for hotels in places like Dubai and Bali, where they have never set foot. Other times, a task may require taking a selfie and sending it to a company that produces face recognition software.

Our research shows that a "Juan" (a thirty-something guy with a degree in STEM), is more likely to carry out these tasks than a "Rosa." However, other surveys conducted by our colleagues show that demographics change based on countries, platforms, and years. Click farms have been mainly observed and studied in Southeast Asia and Latin America. The marketplaces of fake clicks in Indonesia are full of young people who work from cybercafes.[209] Authoritarian presidential candidates in the Philippines use PR agencies to recruit urban, overeducated, marketing pros to create fake news.[210] In Brazil, fake virality is generated by the clicks of thirty-five to forty-five-year-old women who work on local and international platforms, earning less than a penny per task.[211]

Click farms have been described as "parasite platforms" because they rely on social media to survive, but they also threaten them. Facebook in particular has been extremely aggressive in taking down 837 million posts containing spam and 583 million fake accounts.[212] Usually it bans suspicious behavior (for example, users who "like" or even review products not sold in their country) or profiles (for example, accounts that share only third-party content without producing any themselves).

There haven't been many studies that look at the perspectives of workers on click farm platforms. Even less is known about their social relationships, which could help to detect groups and communities that are more likely to work online to generate fake engagement. For researchers, the matter is further complicated by the proliferation of accounts for individuals across social media, commonly done by workers to circumvent platform restrictions.

Who Is an Organic User on Social Platforms?

Brands can buy followers, fans, and even artificial traffic, on social platforms. This type of content is sometimes referred to as "non-organic," meaning "not produced by users with a genuine interest in" a brand, a celebrity, an organization, or a politician. Platforms pretend to take a dim view of these practices, and periodically engage in purges of fake likes,[213] declare that they have installed mechanisms to detect these fake clicks automatically,[214] or even file complaints against users who turn to click farms.[215] Advertisers have repeatedly asked platforms to report on the extent of this artificial traffic.[216] The companies that incur the wrath of Meta usually have very explicit names, such as GetPaidForLikes, but the tech giant's business model is based on a visibility market that has much in common with how click farms function.

On this point, Facebook reminds us that there is only one way to acquire likes for your message: target a certain segment of the population and wait for it to react "organically."[217] Experts in marketing, communication, and IT set out to test this claim by using honeypots and bait-clicks.[218] In computer security, these terms designate traps to lure cybercriminals into launching attacks. In this case, they refer to contents designed not to be interesting to organic users, but to attract fake traffic. The experts in question concluded that automated bot traffic is very limited. The bad news is that most of the activity came from profiles that were probably paid to view, like, and share. These profiles behaved suspiciously in that they were fans of thousands of brand pages, political figures, and celebrities from India and North America. The hundreds of thousands of clicks and shares these false fronts have generated are unlikely to be from passionate consumer groups. Once again, the prime suspects are platforms that recruit users located in Egypt, Iraq, Tunisia, Algeria, Morocco, the Philippines, and Brazil, among others.[219]

These experiments with honeypots yield a further insight: false clicks are not an anomaly or a perversion of the system, but the result of Meta's own business strategy. For a long time, in the name of fighting spam, platforms have been restricting the circulation of shared content. Today it is obvious that, whenever a user posts something on a profile with, say, 5,000 followers, only a tiny fraction of them will get to see it. Every message has a different "reach" based on its time, location, topic, etc. Content is curated by the News Feed algorithm, which chooses what's relevant and hides the rest. This has been accomplished over the years by severely restricting the visibility of posts published by users. The number of views received by "organic brand posts" in Facebook pages has decreased over time. The percentage of users' contacts that saw each new post dropped from 16 percent to an average of 2 percent.[220] After the reform of the News Feed algorithm decided by Mark Zuckerberg in 2018 to "encourage meaningful interaction between people," media and brand name messages became even less visible.[221] Facebook business page owners then worried about the "demise of organic reach."[222] But the drastic restriction on the spontaneous circulation of content on Facebook represents in fact an opportunity for the platform to offer users the possibility of paying a few dollars to get more people to see their content by boosting their posts.[223] The platform promises brands, celebrities, and creators that they'll be able to select (or "target") the users who see their messages. However, if left to its own devices, targeting might not work well. Users have to pay more to the platform if they want to maintain control over how their message spreads. This is called "paid reach."[224] This is how Facebook explains it: "Like TV, search, newspapers, radio and virtually every other marketing platform, Facebook is far more effective when businesses use paid media to help meet their goals. Your business won't always appear on the first page of a search result *unless you're paying to be part of that space*. Similarly, paid media on Facebook allows businesses to reach broader audiences *more predictably*."[225]

Facebook does not directly sell "likes" and does not allow anyone to sell them. Meta's been aggressively suing platforms from Hong Kong to New Zealand for fake ads, fake likes, and fake followers.[226] But the platform does sell visibility, along with the assurance that a specific audience will actually be engaged. If a Facebook page with one million fans refuses to pay, only a tiny minority of them will see the posts. If the owner of the page pays to "target their advertisement," the number of persons able to see the posts will be proportional to the amount paid. This "turn to paid media" blurs the line between users who pay to purchase exposure on click farming platforms and those who purchase exposure directly from Meta.[227] Here is where the logic of click farms springs back into action.

Furthermore, advertisers are growing tired of Meta's promotion tactics. Advertisers have repeatedly expressed concerns about Facebook,[228] while marketing experts have pushed for transparent "viewability" measures to ensure true human engagement.[229] Professionals' doubts were confirmed when Facebook admitted to having overestimated viewership by as much as 80 percent over a two-year period.[230]

*

Because Facebook's business model seems inseparable from the economy of click farming, we must subject its declared values of authenticity and autonomy to sharp scrutiny. The visibility market shatters the illusion of the user's voluntary and carefree participation. New subscribers to social media are now caught up in a system of click production based on hidden work, including the "free" work they are asked to do, and the work done by their micropaid counterparts. The distinction between artificial and organic traffic has become blurred. Consequently, voluntary clicks by users on social media are of the same nature and are part of the same system of economic incentives, as those of individuals who are paid to circulate content, the only difference being that organic users are less costly for the platform. All users are workers.

· PART 3 ·
THE HORIZONS OF DIGITAL LABOR

· 6 ·

WORK OUTSIDE WORK

Platforms are crucial to the production of AI. To function, all platforms require digital labor. And digital laborers themselves sit on a spectrum of taskification and datafication, ranging from on-demand platform workers to social media users. Their labor is sometimes considered freelancing, sometimes temping, sometimes piecework, and sometimes not "work" at all.

Digital laborers struggle to gain recognition as workers who share in the benefits of standard employment. They are not alone. In this sense, they belong to a larger community of people who conduct unrecognized work. This includes domestic work (that relates to household and caregiving duties), "gamified work" (that features elements of play, such as competition or rewards), "consumption work" (that involves purchasing, using, or disposing of goods and services), "audience labor" (the act of producing value through being the public of media and target of advertisement), and "immaterial labor" (that produces goods and services by intellectual or creative activities). In recent decades, these categories have emerged, showing that platform workers' position, conditions, and activities are all affected by their degree of "conspicuousness"—meaning the extent to which their occupations are visible.

The Consumer's Work Outside of Work

However varied, digital platforms have one common feature: they paradoxically set users both to work *and* apart from work. Whenever a platform commands expertise, attention, and loyalty from a contributor, in the same breath it demotes that user to the status of "participant," "consumer," or "guest," but *not* "employee."

Digital labor resembles consumer work, especially when it is framed as part of a "collaborative economy," where platforms thrive on the work of users. The idea of consumer work is not new—in fact, it predates digital labor. Remember the productive tasks that were delegated to customers (and not employees) in some supermarkets, train stations, airports, post offices, and restaurants? Initially, the tasks were physical, such as assembling a piece of furniture, preparing one's tray in a self-service restaurant, helping other consumers by training them, or evaluating the work of employees by ticking the boxes of satisfaction questionnaires (which also transformed the consumer into the employee's informal supervisor).[1] This evolution has reached a new level with the start of the digital economy, as for example with self-service checkouts and automatic tellers.

With the advent of digital platforms, consumer work does not simply complement the standard work of employees but becomes the center of a productive system that relies on an unstructured but loyal workforce. Involving the consumer in the work process fulfills every employer's dream: having a large, passionate, docile workforce—for free.[2]

Consumption is integrated into production as a field of capitalist accumulation (and conflict) in itself.[3] This is evident in the commodification of services in the areas of nutrition, housing, and personal care (as noted in connection with on-demand platforms), with the exchange of intangible goods relating to art and culture (as we have seen around social media content), and with the construction of human social networks (as in the case of online communities). Platforms offer workers standardized tasks—such as leaving reviews and star ratings of services—and pay them poorly. But the same tasks are performed by disciplined consumers outside of these platforms as well. The work of producing cultural content, and keeping the flow of social contact going, has enabled social media to develop. This extension of the commodification of human attitudes and social ties ends up transforming the most trivial activities into productive functions "ranging from remembering the birthdays of relatives to finding a date."[4]

Platforms capture economic value from this work outside of work, which is marginal to consumers' lives. They channel toward production all the "cheap excess capacities," usually found on the fringes of work.[5] The consumer work fulfills the promise of customization that's proper to contemporary economies: consumers personalize their own goods and services using their own work. In addition, consumers help improve product visibility or quality by communicating verbally or symbolically about the good they consume. Yet this consumer work is presented as something that sits outside the sphere of the market, where financial transactions happen. This is what unites consumer work and digital labor. While dealing with users, digital platforms constantly avoid questions of payment. In the age of platforms, "free consumer work" is inextricably linked with the micropaid, underpaid, and unpaid activities that characterize digital labor.

Labor of Love

The activities developed by digital platforms tend to escape the collective agreements, standards, and legislation that govern formal work. They contribute significantly to the rise of nonstandard forms of employment in different sectors and various countries all over the world.[6] This is why they have often—and rightly—been accused of perpetuating and normalizing the informal economy. They are also, however, transforming informal work in several parts of the world, and in unexpected ways. Workers in high-income countries often see digital labor as a route into informal work. In economies where informality is rife, just the opposite may be true: Digital platforms promise to offer new opportunities for informal workers to formalize.[7] The actual impact is less clear, though. Platforms promise greater access to everyone, regardless of gender, race, class, or caste, but at the same time they reproduce surveillance, discrimination, and economic exploitation.[8] Since platforms are driven by technology and competition, they have their own logic. Informal workers and small businesses that provide the same services as they do are affected by their presence. In particular, platforms undercut the more tacit, often more advantageous, ways in which workers negotiate their service fees.[9]

The instability of tasks and of payment, and the absence of guarantees, are reminiscent of the conditions of craftsmen and workers in the manufacturing sector in the early days of industrialization. Construction, transportation, information processing, and cultural production are among the sectors where digital platforms consolidate capitalist power and restrict workers' ability to negotiate their wages and working conditions.

But we should not forget that a significant part of digital labor activities is care work, personal services, and other functions traditionally associated not with production, but with the reproduction of the workforce: accommodation, catering, maintaining social relations, and leisure. From this point of view, digital labor is similar to domestic and parental work, which societal expectations and stereotypes traditionally associate with women.[10] Some authors have therefore referred to the "digital housewife,"[11] or to "digital care work," to characterize digital labor.[12] Regardless of gender, every user becomes a sort of modern incarnation of a "housewife," performing indispensable reproductive work online that, importantly, remains for the most part unpaid.

The use of videoconferencing between geographically distant family members is a particularly good example. The preparation, scripting, and running of a video chat session is a work of active engagement in communication, which is mainly delegated to women in the family context. It's especially true when relatives connect with kids or when platforms are used for social gatherings. For communication to run smoothly on video, children, parents, and grandparents have to perform not only the minimum technical work, but also the "social work" of coordination, presentation, behavior management, and stage setting (that is, creating the material conditions for communication, such as adjusting the settings, moving children closer to the camera so that they are in the frame, and so forth).[13]

Like domestic work, digital labor by its very nature blurs the gendered distinction between the private and the public spheres,[14] between reproduction and production, use value and exchange value.[15] Scholars having examined the affective, cultural, and social dimensions of this labor demonstrate the denial and invisibilization of both domestic and digital work. The "pleasure" approach advocates, with their rhetoric of the "collaborative" economy to describe the involvement of users of digital platforms as a form of participation in a social project, are, so these scholars argue, agents of domestic servitude.

Even with generative AI, there's a clear link between affects and exploitation. As I mentioned previously, OpenAI GPT models use human labor to make sense of data. This use is required during training and is mostly performed by microworkers. However, it also takes place during the reinforcement phase, when the AI interacts with users and, by gathering their feedback, gradually aligns with their preferences and expectations. This second phase can be viewed as affective labor, mainly because generative AI solutions cause their users to feel "affective turbulence." As an example, whenever ChatGPT exposes users to biased and inaccurate statements, this turbulence appears. The users don't think the system is broken; they think it's prejudiced or discriminatory. Because AI biases mostly impact women, queer people, and people of color, this form of labor also highlights intersectional concerns. As a result, marginalized groups of users have the greater burden in fixing and moderating GPT's biases. Some authors even suggest that "those more harmed by linguistic AI are *doubly exploited* as they generate more value through their affective turmoil and their culturally attuned moderation."[16]

Whether it's chatting with AI or joining a subculture of amateurs and creative individuals, digital labor conceals and perpetuates traditional values and hierarchies.[17] The "love" of work had already been unmasked by feminist theorists as "the heaviest of ideological mystifications imposed on a labor relation, namely housework, in order to force women into performing this work without getting paid."[18] It is also at the center of attacks by theorists of digital labor against the representation of users in front of their screens as pure creatures of affect[19] whose dedication justifies exploitation. The artist Laurel Ptak's installation "Wages for Facebook"[20] (2014), demanding that the work of internet users be remunerated, refers implicitly in its title to the influential essay written some forty years earlier by Silvia Federici, "Wages for Housework,"[21] and the connection made is not just of pure form. As these women authors say in concert, what patriarchy and capitalist platforms call "love" and "friendship," respectively, "we call it unpaid labor."

Audience Labor

Despite today's focus on the "attention economy" while studying consumer behavior on the internet,[22] few seem to recognize another important source of value production on platforms: the audience. In fact, the notions of attention economics and audience economics were almost born at the same time. Herbert Simon first theorized the former in 1971,[23] while the latter was developed just a few years later by communication theorist Dallas Walker Smythe. Smythe introduced the concept of "audience commodity" to account for the activity of traditional media audiences.[24] From his perspective, the time devoted to commercial media such as television and radio must be considered as working time when the attention devoted to the programs or associated ads proves to be fundamental to value creation. The capital circulates and accumulates by exploiting audience activity. Audience members aren't simply consumers of cultural products. They are actually producers, not just of attention, but of "ideology and consciousness," which makes of them service providers for advertisers and marketing professionals.[25]

Smythe's reasoning seemed justified, yet it referred to pre-digital media. Let's take the example of free newspapers, distributed at no cost to readers and paid for by ads. In the same way as multisided digital platforms, this form of traditional media runs on a complex pricing model: advertisers pay a positive price, the public does not pay (zero price), and producers are paid (negative price). How to interpret this "zero price," though? It could mean that consumption is free, but for Smythe, it meant audiences produce value that media owners and advertisers don't pay for.

Through the prism of "audience labor," Smythe was the first to understand communication flows as a mode of capitalist accumulation.[26] Other authors expanded on his analysis by saying that audience labor is the activity of watching, which involves "capacities of perception" and aims to creation of meaning.[27] The reference to watching brings to mind marketing-driven platforms like Meta or YouTube, while the reference to creating meaning recalls how microworkers help AI to make sense of data. The logical leap from audience labor to digital labor has been made. According to more recent research, audience labor extends to clicks, shares, comments, views, posts, status updates, and reach, all data that measure and organize visibility.[28]

Digital labor differs from audience labor in two respects, however: the type of activities performed by audience, and the purpose of their work. The audiences of social platforms are not simply assigned to the work of watching, but must additionally perform the "work of being watched,"[29] that is, the construction of an online presence that relies on managing sociability indicators (like attention to participation scores, confidentiality parameters, and contact lists). Platform audiences are thus doubly put to work, insofar as their attention and "consciousness" is captured and as they are bound to create the content and goods that circulate on the social media and collaborative platforms they use.[30] The other difference is that digital labor, in addition to qualifying objects that are subsequently monetized in advertising, also involves performing human computational tasks from which value is extracted by the platforms for training algorithms and perfecting machine learning processes.

Just-in-Time Playbor

Smythe's most important contribution to the current analysis of digital platform work is undoubtedly his problematization of the notion of "free time." His work came in the wake of Theodor Adorno's critique, for whom this modern category is based on the capitalist and consumerist model of work. For, it's impossible to think about "free" time without also thinking about "unfree time," which is occupied by labor and not controlled by the worker. As the philosopher put it, "free time is shackled to the contrary," to the point of being its sinister extension.[31] Smythe, however, departs from Adorno's perspective in envisaging the time spent interacting on communication media not as the time of recuperation and of reconstitution of one's labor power, but as a subterfuge that abolishes the distinction between work and leisure altogether.

This interweaving of work and leisure is a crucial issue in studies on digital labor today. For some, it is the sign of a change in the very nature of work, which is underplaying its character of constraint and subjection. Concepts, such as "playbor," proposed by Julian Kücklich,[32] have emerged to describe a new generation of activities that mix leisure and work. The notion of "playbor" originated in studies of the video game industry, which is now dominated by several large platforms that distribute games and video broadcast game sessions, as well as providing multiplayer universes that connect their audiences. These audiences are part of vast ecosystems of content producers, who generate much more value than what is remunerated by the video game industry and platforms.[33] The work of gamers, e-sport practitioners, modders, and testers is proving to be a vital complement to, and a mirror image of, the formally recognized work of software developers, salespeople, and designers.[34] Unlike Smythe's traditional spectators and readers, digital platform gamers do more than just stare at a screen. They perform important functions of testing, debugging, and character and prop production. They also train the game engine, improve the characters, and help algorithms to create new game patterns.[35] In recent years, behavior data from gamers have been used to make annotations and examples for computer vision solutions and simulations. In an effort to create "population-based reinforcement learning,"[36] Google DeepMind used millions of human players' sessions of StarCraft II to train its AI model AlphaStar. Meanwhile, OpenAI has linked its reinforcement learning techniques to thousands of games in several of its projects (Gym, Universe, Five, Neural MMO, Robosumo, etc.).[37]

All these activities are, according to Kücklich, comparable to productive forms of waged labor, because in both cases the creators of the produced goods do not "own" their own products.[38] Whether or not gamers are the ultimate expression of precarious work in the video game industry is not the issue here. The importance of "digital game labor"[39] in understanding digital labor as a whole lies in the fact that the intertwining of private and public spheres, work and entertainment, leisure and conflict,[40] is gradually spreading to other fields.

Gamification, as discussed in the previous chapter, is the transformation of any human occupation into a game; taking advantage of play's interactive and fun aura, today's "gamification of the world" is flourishing.[41] In digital platform labor, gamification is everywhere. The designers of on-demand work apps have introduced elements such as scores, awards, and competitions; users of microwork services are encouraged to gain badges or to improve their ranking in order to move up the levels; social media adorns their interfaces with gimmicks reminiscent of video games: "favorites" to boost users' participation, competitions around the number of views and subscribers, rankings of trending topics, and best friends lists.[42] This is how platforms foster engagement and prosocial attitudes, ultimately to encourage data production and to make users perform higher quality qualification and automation tasks.[43]

Gamification is not confined to digital labor, however, and one could argue that playbor in the digital sector simply reflects a more general movement, which has taken root in traditional firms over the last few decades.[44] It involves a management philosophy revolving around personal development, creative emulation, user-friendly workspaces, horizontal relations, teamwork, and the conversion of objectives into "challenges" and games.[45] In their seminal book, *The New Spirit of Capitalism*, sociologists Luc Boltanski and Ève Chiapello mentioned two types of critique of society, both of which relate to work: the artistic critique (which demands freedom, autonomy, and authenticity) and the social critique (based on solidarity, security, and equality).[46] Playbor can be interpreted as an adaptation of the artistic critique of work, with a view to renewing workers' commitment and encouraging them to use their initiative.

While traditional companies and digital platforms may share the same managerial paradigm, they do not apply it in the same way. In the workplace, play tends to occupy the place of discipline and effort—at least on the surface. On digital platforms, the boundaries are blurred to the point of altering users' own awareness of their activity. In an indeterminate zone between leisure and production, playbor is something like mandatory fun.[47] The philosopher Ian Bogost consequently describes gamification as "exploitationware," a program in which a worker's effort is not elicited through direct and proportional rewards—a promotion or a salary increase, for example—but through opaque incentives such as stars, points, access to a new part of the platform, etc.[48]

Platforms not only confuse users as to the nature and purpose of the activities they perform, but they also disturb their relationship to the time spent on them.[49] Since gamified digital labor passes as a hobby, it can seep into every nook and cranny of daily life, without limit. This is particularly evident on social platforms, where users compete to share content. But it also applies to microtask platforms, especially those that use gamification as an incentive to keep users engaged or motivate them to work while they're away from their main jobs. Just as platform labor takes place outside of circumscribed places of production, its working hours go beyond the legal limits set in standard jobs. Even when digital labor is considered part-time, or minimal-time, it's tendentially a continuous-time occupation.

According to the critic Jonathan Crary, capitalism in the age of the internet produces just-in-time lives and thus sleeplessness, which is to say "the state in which producing, consuming, and discarding occur without pause, hastening the exhaustion of life and the depletion of resources."[50] In many ways, his words echo the remarks made by Netflix CEO Reed Hastings: The main competitors for the video-streaming platforms are not other platforms or cable television: "When you watch a show from Netflix and you get addicted to it, you stay up late at night. We're competing with sleep."[51] This dramatic image of a sleep-deprived humanity is an allegory of internet culture, in which users, distracted by attention-grabbing ploys and encouragements to accumulate and produce information, cannot get even a wink of sleep. Crary lists gambling, porn, and online video games as examples of activities that change our relationship to time through gamification. With the illusion of mastery, victory, and appropriation they provide, games stimulate impulses and appetites that intensify the production of information twenty-four hours a day.

Digital labor consumes life, such that living itself becomes a moment within work. According to Crary, "This time is far too valuable not to be leveraged with plural sources of solicitation and choices that maximize possibilities of monetization and that allow the continuous accumulation of information about the user."[52]

Is Digital Labor Immaterial?

Platforms' distortion of users' relationship to time leads us to a final theoretical antecedent of digital labor, namely immaterial labor. This concept, popularized by the philosopher Maurizio Lazzarato in the 1990s, can be applied to all the work performed in the cultural industries that is designed to add value, share, and recommend content, in the context of cognitive capitalism. As with consumer work, domestic work, audience labor, and playbor, there is no distinction between productive and reproductive work from this perspective. Cultural, relational, cognitive activities—and time itself—aren't outside the market, they've been absorbed by it.[53] To speak of immaterial labor implies shifting the focus onto value creation, and not through a tangible transformation of reality, but through increasing a commodity's informational content. In this context, production requires activities that are not normally recognized as work, to inject culture, norms, and modes of consumption into the products that circulate on today's markets. One example is social media users' value-enhancing activity of qualification. Another example, more relevant to automation, is microworkers organizing data to fit them into human cultures and values.

Immaterial labor differs from digital labor, however. Despite being a social activity, sometimes requiring nonspecialized skills and abilities, immaterial labor remains an eminently "intellectual" labor. In this respect, it has a lot in common with the Marxian concept of the "general intellect," discussed in chapter 5. In Marx's "Fragment on Machines" and subsequent post-workerist developments, the general intellect is described as "general social knowledge." These are occupations with a higher level of creativity than digital labor. Immaterial labor occurs within the framework of the general intellect, performed by a class of workers that philosopher Paolo Virno calls "mass intellectuality." In Virno's view, this social group is nothing like, and is actually the polar opposite of "a new labor aristocracy." Yet he also argues that mass intellectuality is "the depository of cognitive competences that cannot be objectified in machinery."[54] In other words, it's nothing to do with the repetitive labor of an assembly line, objectified in fixed capital and embedded in automation. By contrast, digital labor is not dissimilar to the repetitive labor embedded in automation, as we have already seen.

The fact is that when post-workerists were formulating their analyses, they were faced with an utterly different socio-technological reality, one that was still dominated by the traditional media and cultural industries. The best-known figures of immaterial workers at the time were not the app-based couriers or the data annotators—who did not yet exist—but those involved in marketing, audiovisual production, advertising, fashion, and photography. For the emerging digital sector, one could cite software, website, and multimedia content producers. All these examples come from universes in the advanced tertiary sector that draw on specialized professions with high visibility, where standard employment contracts were still the norm, although the subcontracting and self-employment were also pervasive. The notion of immaterial labor is thus based on a relatively "noble" vision of the activities involved. As a consequence, studies are more likely to focus on the creative work of the data scientist or the storyteller who sells a fantasy of automation than on the more prosaic reality of the users behind their screens.

The concept of "immaterial labor" comes closest to that of "digital labor" in the way that technologies are conceived in their relation to their human users and designers. Influenced by Deleuze and Guattari's *A Thousand Plateaus*,[55] theorists of immaterial labor posit that humans are vulnerable to "machinic enslavement" if they don't control technologies.[56] This subjection operates not only in the places and at the times of day that correspond to the specific activity we call "work." It is effective also where no standard employment relation exists, in work not recognized as work, and it is capable of "furnishing surplus value without doing any work (children, the retired, the unemployed, television viewers, etc.)."[57]

Digital labor can be analyzed the same way but turns work into an algorithmic process for the purposes of information, communication, and management. This does not mean the end of the strictly "material" part of work. Neither does it amount to the replacement of a "manual workforce" [*main d'oeuvre*] by a "brain workforce" [*cerveau d'oeuvre*], to borrow the economist Michel Volle's expression.[58] The Marxist dualisms of tangible and intangible, intellectual and manual, skilled and unskilled labor already came under scrutiny in the 1970s,[59] and their conclusive inadequacy was shown in the 2000s by internet studies, which also criticized their most recent incarnations (real/virtual, material/disembodied, etc.).[60] Digital labor is no more intangible than is the work of a lawyer or a warehouse worker—especially in the context of the computerization of all professions. Click farmers in Venezuela, Facebook moderators in Kenya, data annotators in India, and Uber users in Germany are all confronted with highly concrete issues and concrete tasks, for which they use their bodies, their locations, their senses, and of course their fingers—their *digiti*.

Discovering Inconspicuous Digital Labor

By examining the earlier concepts of consumption, domestic, audience, and immaterial labor, I've tried to pinpoint the characteristics of digital labor, and to leave behind some of the dichotomies (between work and nonwork, tangible and intangible) that make it hard to understand work on digital platforms. There is, however, a more fundamental opposition revealed by the analyses of these concepts. It runs between *work that is immediately recognizable as such*, and other activities that are mediated by the digital platforms and transformed into "silent," *inconspicuous*, or merely "implicit" forms of work.

The invisibility of domestic labor, the casual nature of consumer and audience labor, the ambiguity of playbor, and the immaterial components of cognitive capitalism all illustrate how hard it is to make work both visible and valuable. The problem of recognition applies to all kinds of work, according to sociologists Susan Leigh Star and Anselm Strauss seminal 1999 contribution "Layers of Silence, Arenas of Voice." When people voice their labor, they're making it visible and explicit to each other; when it is silenced, it is not recognized as "real work."[61] The oscillation between these two poles helps to define what exactly counts as work. Ambiguous cases, where it is not self-evident that the activity is work, include domestic labor and care. What a society recognizes as work can be related to long historical processes of professionalization, which, made visible, can become careers. The definition of what constitutes work can shift in such a way that certain activities classed as work are gradually deprived of this status, and silenced.

In the last quarter of the twentieth century, when Star and Strauss were writing, there may have been just such a shift. With rising unemployment, waves of business restructuring, and concerted outsourcing policies, the hope of obtaining a steady office job was receding, undermined by the trend in subcontracting and the increase in nonstandard forms of employment. Since the dawn of the twenty-first century, these trends have become even more pronounced. In particular, there has been a push toward work from home—especially noticeable during the global COVID-19 health crisis. Flexibility, casualization, and generalized temporary and contingent work have grown, fueled by a financialized economy and pressure on traditional employment models. From a career perspective, this has blurred the lines between phases of professional life where one performs activities that "count as work" and phases where work becomes invisible and resembles idle communication, information consumption, or vanity projects.

Labor on platforms accelerates this development. Faced with these tensions, negotiations become necessary to decide which activities can be classified as work within the ever-expanding gray zone of digital labor.

Ironically, the recent changes in the job market present an opportunity to discuss what work is.

According to Star and Strauss, visibility is crucial in this regard.[62] Visible work is harder to ignore. Making no mention of work, condemning it to silence and invisibility, is therefore tantamount to taking a step toward its disappearance. One of the most common tactics for "silencing" work, according to Star and Strauss, is to make the worker "unseeable," which may happen when "large-scale networked systems," which bring about a change in task allocation, are used to give the impression that work is performed by machines, whereas it is in fact delegated to other, hidden, human beings.[63] Analyzing the technologies available in the late 1990s, Star and Strauss attributed this concealment of work to the emergence of information management systems in companies. If we transpose this analysis to today's digital platforms, these theories expand to an unprecedented extent. A contemporary example of a large-scale system, the algorithm of a ride-hailing app that distributes tasks to the drivers, in the meantime makes coordination work and price negotiation invisible. On e-commerce platforms, recommendation systems suggest purchases, but they also hide the fact that human data annotators have checked prices, removed duplicates, and updated products descriptions. Last but not least, large language models put the work of data scientists and platform tycoons front and center, conveniently overlooking microtaskers' pre-training work and users' reinforcement learning.

On digital platforms, the distribution of what counts as work is therefore inseparable from the distribution of its visibility. Both aspects—the category of work and its visibility—influence the definition and value of the data and content that circulate within informational infrastructures. A lack of recognition of work, for instance, can lead to a lack of visibility of data—and vice versa. A study by sociologist Jérôme Denis supports this conclusion. In a startup that distributes open data, for instance, non-machine-readable data isn't considered data at all, and labor to produce it doesn't qualify as work.[64]

When we review the lack of recognition of digital labor in terms of the division between visible and invisible labor, we can better understand the macro-social context in which those who perform hidden and alienating tasks have difficulty entering the employment market. Automating, networking, and aligning business processes with information and communication technologies all contribute to muzzling and erasing work.

Despite this, Star and Strauss's analysis fails to incorporate the fact that platform work is divided into multiple tasks, some of which struggle more than others to surface. Uber Eats couriers juggle visible deliveries and less visible activities like bike maintenance and messaging customers through their app. On Amazon Mechanical Turk, crowdworkers perform visible microtasks, while in the background they read, look up information, and navigate the complexities of payments. Volunteer YouTube moderators oversee visible content, and they also handle emotional burdens behind the scenes.

This shortcoming is partially mitigated by the link the authors make between "invisible work" and "shadow work," a notion introduced by social critic Ivan Illich.[65] This "shady side of industrial economy" is most apparent in "the preparation for work to which one is compelled"[66] This activity is not only the shadow cast by "real" work, nor is it work that has been relegated to the shadows, due to conflicts over how work itself is defined. Rather, it is the portion of every activity that is pushed out of sight.

Digital labor oscillates between visibility and invisibility, leading us to the notion of conspicuousness. The concept of "conspicuous consumption" was formulated at the end of the nineteenth century by the American sociologist Thorstein Veblen to convey the social importance of the economic activity of consumption not to fulfill needs, but as a spectacle for others and to signal one's social status. In the 1970s, legal scholar Jethro Lieberman drew directly on this notion to propose its opposite, "inconspicuous production," which refers to production performed outside the field of vision.[67] In many jobs, "unskilled chores" constitute the bulk of the daily workload, whether performed by an affluent professional or a menial worker. This held true way before the rise of platforms, as Lieberman's explored a range of occupations such as lawyers, barbers, morticians, dentists, and keepers of mental wards. He highlighted their search for a balance between the conspicuous and ritualized part of their work (meeting with clients, carrying out work in public, etc.), and another, unobtrusive and ordinary part, which was performed out of sight. In this pre-internet context, inconspicuous labor consisted mainly of routine, and sometimes arduous, tasks that could be complex and difficult to communicate.

In principle, then, each occupation can be characterized by a ratio of conspicuous to inconspicuous work. Lieberman does not seem to believe that the ratio is higher for creative and specialized professions or lower for salaried jobs, clerical work, and administrative tasks. But, in keeping with the culture of his time, he is convinced that routine tasks that represent the inconspicuous portion of any job "can be accomplished by anyone with a modicum of training if routine enough, in fact, it can even be adapted to the machine."[68] Still, those who design automated processes, algorithms, and robots will find it difficult to make machines do their tasks, particularly the elementary and commonsensical ones that AI can't simulate. The more inconspicuous the production, the harder it is to automate.

Maybe the way that these authors characterize automation within the socio-technological contexts that preceded digital platforms is no longer relevant today. Lieberman thought that machines would replace inconspicuous work. Star and Strauss blame large-scale technological systems with "silencing work and continuing invisibility." The conspicuous/inconspicuous divide provides another perspective, showing that platforms require invisible labor to produce what passes for automation today. This view places inconspicuous digital labor in conversation with anthropologist Mary L. Gray and computer scientist Siddharth Suri's concept of "ghost work."[69] In contrast to the notion outlined here, these authors restrict the scope of application of their concept. According to Gray and Suri, ghost work refers to a workforce of uncredited data trainers, employed in a particular industry to achieve a specific goal—help develop AI solutions.

Following Illich and Lieberman, inconspicuous digital labor may be defined as the invisible taskification and datafication that occurs within every job. Next, we should ask: To what extent does digital labor now affect other forms of labor?

Hyperemployment

The emergence of digital labor is not only an internal process linked to the rise of the information and communication technologies, but also the consequence of how standard employment has been affected by late twentieth-century trends in corporate restructuring, outsourcing, and subdivision of work tasks, and by early twenty-first century advances in financialization. The redefinition of work in terms of its visible and invisible components does not necessarily imply that the quantity of activities "outside of work" increases, nor is the main issue that of the increase in working time, as was the case with immaterial work. Rather, what is at stake is a shift to a system which the philosopher Ian Bogost has called "hyperemployment."[70] This refers to the obligation of workers to integrate into their productive activity periods of digital labor able to be mobilized at any time of day and from any place, in public spaces, at home, and not only in traditional workplaces. As we have already mentioned, exogenous shocks such as the COVID-19 health crisis have accelerated the hybridization of presence- and remote-based work. It's true that many companies, especially in the tech sector, have very publicly recalled their staff back to the office at the end of lockdown.[71] But the availability of technological innovations has made it easier to work from home and on public transportation.[72] Multitasking, like when working and riding a train, functions both ways. Sure, digital technology makes work permeate family time. At the same time, family and domestic chores are often done while remote working, like cooking or cleaning during a Zoom meeting.[73]

Due to the abundance of devices kept on our person and in our own homes, users today may receive a request, a demand, or an order at any time of day or night. This constant availability, often interpreted as addiction or compulsion, or as a social requirement, reveals the massive increase in the number of productive tasks distributed by colleagues, superiors, clients, work partners, and so forth.

"Hyperemployment" relates to the extension of the forms and methods of digital labor to standard employment. Increasingly, formal employment is affected by technologically enhanced insecure work situations, as well as challenges relative to "work outside of work." A common thread is shared by employees, platform workers, and even the unemployed. In fact, they are all "hyperemployed." The Greek word "hyper" ("over") means that digital labor has been superimposed on any other activity workers do. Any person employed by a traditional company with a formal contract can be seen as an *inconspicuous part-time digital worker on one or more platforms*. The traditional dependence of employee on employer has not so much ended as taken on a multiplicity of different forms.

Bogost focuses on the example of email, conceived as a task allocation mechanism that disregards schedules and local situations. Initially introduced to coordinate people using the same computer resources in a collaborative context, email was designed to be a work-related correspondence tool. In fact, email functions as an unending list of tasks—a request for information, or the delegation of a piece of work. Each message conveys an obligation from a supervisor, a customer, or a colleague to act, and it threatens recipients if they do not reply and react. It embodies and epitomizes two crucial aspects of digital labor in all its forms: It changes any occupation into a just-in-time activity, to be carried out "on demand"; and its eminently social nature mobilizes relational skills to produce multimedia content (the texts and images of the emails and their audio and video attachments). Bogost underscores the work discipline imposed by email, as its primary function "is to reproduce itself in enough volume to create anxiety and confusion. The constant flow of new email produces an endless supply of potential work."[74]

Email has been around for long enough to be considered a universal technology. It is "the plumbing of hyperemployment."[75] But, of course, the philosopher's analysis applies to other tools. Imagine receiving a group chat on an instant messaging app after you wake up, or having to accept a late-night video call from a colleague in another time zone. There is no shortage of potential work here either.

I don't want to imply that the notion of "hyperemployment" should be used as a criticism of the introduction of information and communication technology into the traditional work environment. But, rather than increasing productivity at work, these technologies encroach on the time meant for free and personal enjoyment. A number of businesses have gone digital, particularly hastened during COVID-19. From mom-and-pop shops to big corporations, they started using a variety of tools to coordinate their employees' work, from email to Zoom to Microsoft Teams. These are off-the-shelf solutions made by big platforms. In fact, they *are* the big platforms—and employees are now using them.

Employees do not have to contend with the "free" work of social media users or with the underemployment of the on-demand economy or microwork, but they do experience a loss of autonomy as the result of "feeling overwhelmed" by email, chats, video calls, LinkedIn comments, and Instagram stories. This easily turns into a "feeling of resignation," since it seems impossible to work in any other way. Salaried employees and workers in insecure online gigs share the occupational risks of communication burnout and an escalating workload. There is only one difference between despair of hyperemployment and that of unemployment, Bogost notes, namely that despair seems acceptable for the latter. Indeed, society celebrates hyperemployment for the supposed freedom it offers—freedom to carry out unpaid work constantly from the comfort of employees' own cars or toilets.[76]

· 7 ·
HOW DO WE CLASSIFY DIGITAL LABOR?

As we discussed in the previous chapter, scholars have explored unrecognized, invisible, and unmeasurable work for decades. Little matter if it happens at home as unpaid domestic labor or in social contexts as immaterial cultural contributions, there's a common thread—work is positioning itself "outside of work." The reason is that every occupation has both visible and invisible aspects. This dialectic between visibility and invisibility is evident in digital labor as well.

Yet there's one thing that makes digital labor different from other types of unrecognized labor. In the shift from firms to platforms, contemporary organizational practices are embracing it to promote flexibility. Companies that hire flexible workers, potentially, don't have to pay benefits, train employees, or pay employment taxes. These workers do not create a constant cost to the company. Ideally, on platforms and in their ecosystems, every individual is a one-person startup and work tends to occur outside any traditional firm.[1] As part of larger efforts to restructure and be more agile, the same traditional firms encourage salaried workers and redundant employees to start their own startups. This phenomenon of companies spinning off bits of themselves into separate companies is known as "hiving off" or "starbursting."[2] It shows how widespread platformization has become. Indeed platforms do this all the time. Take, for instance, Uber's decision to spin off its autonomous vehicle business into Aurora (the startup mentioned in chapter 3) or how layoffs at big tech companies usually lead to the creation of smaller software companies.[3]

Is this the dawn of "work for oneself," freed from subordination, as some claim?[4] As the revival of the term "gig work" over the past decade suggests, are we headed to a return to historic labor hire arrangements, home-based work, and piecework, characterized by the tension between insecurity and control? This is the fundamental question that digital labor raises.

Escaping Formal Employment to "Open Up" to Work

Despite prophecies since the 1990s announcing the advent of a "jobless" or "post-work" world due to (or thanks to) information and communication technologies,[5] regular employment continues to occupy a central place in our societies. According to data from the International Labour Organization (ILO), salaried employment is buoyant at a global level, although, after rising from 1990 to 2010, it has stagnated in advanced countries. However, it is increasing in emerging and developing countries.[6] Before the COVID-19 crisis hit, the industrialized world was experiencing rapid employment growth. According to *The Economist*, "most of the rich world is enjoying a jobs boom of unprecedented scope."[7] The health crisis has upset these predictions, causing a worldwide spike in unemployment and bringing an end to the longest economic boom in US history. According to the OECD, the pandemic has triggered one of the worst job crises in a century.[8] The recession was short-lived, though. By the end of 2021, little more than eighteen months after the start of the pandemic, employment rates in the EU and the US were back at pre-crisis levels.[9]

If this occurs on an aggregate level, we should not expect similar consistency on a smaller scale. Individual employment histories are more unstable, oscillating more frequently between periods of employment and unemployment, or, in lower-income countries, with phases in the informal sector. Moreover, global competition is eroding wage bargaining and encouraging companies to subcontract or recruit under nonstandard and less protected contracts.[10] These trends, as sociologists Zygmunt Bauman and Ulrich Beck already argued, contribute to a "political economy of insecurity."[11] Richard Sennett has emphasized both the "hidden wounds" and the emotional cost of professional instability, and individuals' coping strategies.[12] The value placed on flexibility, independence, entrepreneurship, and controlling one's own future by taking ownership over projects creates a paradoxical situation. Professional stability is at once coveted as a way to deal with economic risks and belittled as submission to the hierarchical order of the old industrial world. The individual is made to feel that "being a 'good' or successful worker today means avoiding commitment, being mobile and being flexible."[13]

Might the platform economy allow workers all over the world to shake off the suffocating shackles of wage labor? This idea has advocates. According to the current reimagining of labor, worker rights would improve, earners would be given more flexibility, and entrepreneurship would be encouraged for the masses. Work on platforms allegedly delivers on the promise of turning all workers into "venture laborers" who profit and take risks at the same time.[14] In addition, this reimagining has it that job instability is no longer a risk, but a chance; that it no longer undermines identity, but offers possibilities of personal fulfillment by enabling one to experience a variety of situations and settings. This reversal of perspective is based, however, on a very restricted definition of work; for workers, it is supposed to only consist of "meaningful" activities, designed to give an existential dimension to their professional lives.[15] There is no room in this definition for the routine tasks of the microworkers, the drudgery of food delivery couriers, or the traumas experienced by social media moderators. Who are the losers of digital transformation? Those for whom extreme flexibility is not a life choice. For them, digital labor is a modern-day assembly line; it can hardly be considered as a path to self-realization.[16] Once again, the platforms celebrate self-made men, the techbro entrepreneurs, and ignore digital laborers, who suffer from pointlessness, repetition, and discrimination. Certainly, the desire for freedom among workers is no less intense than that of tech startup founders, who see themselves as escaping the nine-to-five. But clearly, they are no more likely to gain this freedom than they are to achieve stable employment.

Theorists of both the "work" and the "pleasure" approaches discussed in chapter 5 argue, for different reasons, that work and leisure have become indistinguishable. We see echoes of this convergence in the aforementioned notion of playbor. There would seem to be a correlation between the desire for autonomy and openness that platform-based gig workers and freelancers seem to be engaged in, and the notion of work as side hustle. This notion refers to an income-generating work performed alongside full-time jobs, allowing workers to reaffirm their dignity in the face of the mounting disparagement of the working classes. Working on a platform is regarded by some analysts as a way to pursue "self-enhancement, self-transcendence, openness-to-change, and conservation work motives."[17] Digital platforms lend new momentum to the side-hustle culture, transformed and transfigured into a new paradigm of "open" work—open, in reality, to having multiple flexible jobs. Can this theoretical framework account for the instability, conflicts, and distress of Uber drivers and Amazon Mechanical Turk workers? The siren song of the platforms is seductive, as is the image of empowered workers with agency to fulfill their passions. But this vision ignores the exhaustion that comes from multiple side jobs as well as the unavailability of choices from which these workers suffer. The insecurity and small gigs that come with digital labor mean that workers lose the traditional social protections that come with standard work arrangements.

It might seem like an individual concern, one that workers can address by negotiating between their own autonomy requirements and needs for stability. Yet we should worry collectively about how people will make a living in a platform economy. The tax structure, retirement funds, social safety net, and access to health care are based on laws and institutions that assume stable employment and full-time jobs in formal organizations that pay a regular paycheck and benefits—at least in high-income countries. Risks associated with these rapid changes in how we work aren't just person-specific; they're systemic.

Exalted Workers and Microtaskers: A Dual Labor Market

The fiction of platform work as path to worker emancipation ignores reality—workers lack choice. When confronted with the real working conditions of digital laborers, some theorists dismiss them—imagining their precarity and exploitation as anomalies that better training and minimal regulation can correct. In their perspective, "flexicurity" would strike the right balance between stable and flexible work arrangements.[18] One thing that has been overlooked is that, if unrecognized digital laborers are subject to precarious working arrangements, this is because, over recent decades, recognized workers in the tech sector have set the example. Engineers, designers, and developers are all perceived to belong to the class of 'innovative' workers that embrace entrepreneurship. On-demand workers, microworkers, and moderators are considered 'mundane' laborers,[19] and they are made to follow the example of their colleagues who are deemed to possess more skills and visibility.

Platform economies intensify pressures on "downstairs" workers to emulate the "upstairs" ones, but such pressures were actually developed already in the 1990s and early 2000s. Flexibility and job mobility have been a staple of tech work for decades. Job-hopping was very prevalent among the venture laborers of the dot-com boom. To develop professionally and create new profitable businesses, these workers would move from one company to another, with periods of unpaid work, unemployment, and public subsidies. The phenomenon of venture labor, as defined by sociologist Gina Neff, "created a vicious cycle—taking risks seemed to be the only way to get ahead—encouraging entrepreneurial behavior from people in the industry, which in turn signaled to others that taking risks was a good idea."[20] Paradoxically, for these high-profile workers, career mobility and unpaid work were ways to deal with uncertainty. Workers could stay on top of the market by frequently changing jobs, building their portfolios, and working at big companies either for free or at a discount.

Around the same time, economist Bernard Gazier characterized the "heroes of the electronic frontier" as the reincarnation of the "exalted ones," the elite workers of the first Industrial Revolution, who were more qualified and less submissive than the majority of their colleagues.[21] Known as the fastest and best workers—who only work three days a week, choose their employer, and don't risk unemployment—only the most qualified workers made up this class. They express a "desire for autonomy and their profound allergy to forms of brigading of collective labor and the moral order one attempts to impose of them."[22]

The entrepreneurial spirit displayed by the recognized and visible tech workers is supposedly a link with their historical counterparts, the "exalted workers" of the Industrial Revolution. But is it true that underpaid and unpaid platform workers, like the elite ones of the nineteenth and twentieth centuries, can choose their working conditions, their pay, and even the type of work they perform? Certainly, some workers in computer engineering, electronics, robotics, graphics, data science, and software development do have these choices. But, as we've seen, Amazon Turk's microtaskers, Facebook's moderators, and app-based delivery couriers all lack similar agency.

The exalted and "innovative" laborers are (or at least aspire to be) independent and self-employed, to be living out "hacker spirit" in a renovation of the first capitalist entrepreneurial spirit depicted by Max Weber.[23] Tech workers are advocating freedom from work, passion, and self-determination in hackerspaces popping up all around the world, from HasGeek House in Bengaluru to Noisebridge in San Francisco.[24] There, individuals looking for new horizons and opportunities try to establish innovative forms of collective organization. Unlike the trade unions and the worker struggles of the past, these spaces of innovation aren't rooted in the same antiauthoritarian and social justice tradition; rather, they are rooted in libertarianism and digital utopianism.[25] The emphasis is on entrepreneurship and a laissez-faire economic model, as well as on sharing skills and knowledge. Although this alleged "antiestablishment innovation" is clearly aligned with market capitalism, it also lets hackers take on the role of the labor force of the future, capable of expanding workers' autonomy.

Despite their political posturing, these attitudes and possibilities remain the prerogative of highly specialized and recognized workers, driven by the desire to compete and to display virtuosity in one's chosen field. Analyzing the career paths and personal attitudes of the users of TaskRabbit or Instagram in the same terms as the highly paid data scientists and engineers at Google is problematic at the very least. Arguably, just as traditional labor markets are polarized into stable and unstable jobs, insiders and outsiders, protected and unprotected professions,[26] so, too, do platforms sustain dual labor markets. There is an increasing pay and recognition gap between "exalted" developers and platform-based microtaskers, between those who can choose when and where to work and those who have no control over their remuneration or working conditions. Of course, this expanding gap depends on many factors. Worker specialization is also determined by those who run the gigs, and whether they recognize workers as experts or as mere hired hands. Performing conspicuous tasks, rather than inconspicuous ones, is also an effective way to get recognition. The level of automation associated with a task, which employers use as an indicator of worker productivity, affects recognition. The primary market of the highly skilled workers offers flexibility and access to well-paid, conspicuous, and highly automated tasks, whereas the secondary market involves individuals with less regarded qualifications, whose access is restricted to low-paid or unpaid, relatively inconspicuous tasks that mostly support smart technology and autonomous systems, but that are not themselves automated.

The creation of these tensions and divides looks a lot like what big firms used to do in segmenting and breaking up the labor market, so they can create artificial fractures in it. Eventually, working conditions end up fragmenting and diversifying work so much that workers can't establish a basis for collective struggle against capital.[27] Career development opportunities for digital laborers will always be limited, especially for those from marginalized groups.[28] In the tech sector, there seems to be a huge gap between the global masses of microtaskers and the elite of "exalted heroes," and no system in place to bridge it. Self-fulfillment and recognition in labor proved to be hollow promises.

When Exalted Workers Become Microtaskers

In this dual labor market reconfigured by the platform economy, those who innovate with automated solutions should, in principle, enjoy the same security as "insiders" in traditional labor markets. In reality, they find themselves on shaky ground. In the twentieth-century firm, employees' salaries were protected from fluctuations in price and demand, and shielded by employment contracts that did not obey market forces (indeed so much so that some philosophers detected positively anti-capitalist traits in these wage labor structures and the potential to subvert the institutions of capital).[29] In contrast, digital platforms join the market and take part in its characteristics as a market–corporate hybrid. Far from attempting to neutralize price and stock fluctuations, they abide by them, adjusting their prices according to the slightest variation in supply and demand. Uber's algorithm, discussed in chapter 3, is typical of this mechanism. Platforms also push market segmentation to the extreme to maximize their profits, by differentiating between classes of users (as discussed previously in the case of Facebook). In the upper echelons of the labor market, tech workers can obviously romanticize the spirit of free enterprise. Instead of subverting capitalism's institutions, they reinvent them by embracing a business mindset. The fact that they rationalize out of clairvoyance or blindness doesn't matter as much as realizing that not everyone gets the freedom and fulfillment they find in their own work.[30]

As early as the 1990s, developers and computer experts embraced the imperative of flexibility, and became accustomed to seeing themselves as freelance and adaptable multitaskers. Strong competition in international markets from developers from emerging economies able to compete with their European and American counterparts by manufacturing IT products with standard functions ended up restricting their professional opportunities. This happened at a time when the major employers in the North (IBM, Deloitte, Accenture, EY, and Capgemini) were confronting the new global giants in Asia (for example, India's Tata and Infosys), which employed innovative tech workers at lower cost, and so drove salaries down.

Suddenly, autonomy ceased to be a prerogative that these highly skilled workers enjoyed by default and instead became a problematic goal for them to strive toward. Working conditions deteriorated, and obvious forms of dependence on clients set in, which compromised their quest for emancipation.[31] The image of a new generation of workers who are hyper-flexible of their own free will looked less and less credible. Flexibility, especially when it takes the form of moving between different projects or clients, does not imply that the workers concerned are managing their career paths as they wish. It is the companies, that is, the employers, who lead the dance. The flexibility and adaptability of workers, far from being signs of a new attitude toward work, are explicitly written into their contracts.[32]

Now fast forward to today, and it is hardly surprising that the "exalted ones" themselves look more and more like microtaskers. This close connection between experts in digital services or freelance hackers, and click-workers or on-demand platform workers becomes clearer when we discover that both categories are subject to the production chain constraints, controlled by digital platforms and large IT consulting and service outsourcing companies.[33] Neither structure differs significantly from the other, as outsourcing companies adhere to the platform paradigm, and include digital platforms in their global production chains.

Platforms now sell on-demand code development. Two online platforms I already mentioned, Fiverr and Upwork, offer programming services starting at $3 an hour. Others, like Gigster, connect clients with freelancers in Asia or Eastern Europe. There are even websites like Codealphabet or Codementor that supply one-to-one live tech support and debugging; it can take only fifteen minutes and cost as little as $8. Also thriving are platforms like LDTalentWork, which offer inexperienced developers a way to improve their skills by working for a small fee (which can be as low as $4 an hour).

As a final piece of the puzzle, there are social platforms for coders, such as StackOverflow and Microsoft's GitHub, which allow "volunteers" to contribute their coding skills for free. For years, these unpaid software developers have unknowingly fine-tuned OpenAI's machine learning tools, through publicly accessible code on GitHub.[34] Additionally, OpenAI has hired approximately 400 computer programmers through other platforms in Latin America and Eastern Europe. In 2022 and 2023, such programmers provided data and documentation for OpenAI's Generative models of code, pretrained on large corpora of programs. According to reports, some of these coders are working in five-hour shifts and not even managing to get paid by OpenAI.[35] It is ironic that software developers are limited in their view of risk when using tools like ChatGPT. They will not lose their jobs because this solution is capable of writing code, but they will instead join the ranks of the underpaid or unpaid toilers that train the model.

This scenario portrays a world where the exploits of "exalted" code developers are imperceptibly morphing into cheap digital labor, and where worker autonomy is compromised through the fragmentation and standardization of tasks.

The Return to Archaic Employment Relations

Could these signs of taskification signal the end of the free and entrepreneurial tech worker era? In fact, common sense seems to suggest the opposite. In the tech industry, the flexible, business-minded, independent worker isn't threatened. On the contrary, many see the era of formal, dependent employment—with its stability and social guarantees—as ending.

Has the emergence of digital platforms irreversibly blurred the line between worker and entrepreneur, with their less protected and more flexible work? This question can be answered from another, more critical perspective. There's some cause to argue that digital labor brings back archaic methods of recruiting and managing workers. Such work arrangements have been banned in many countries. Several terms in different languages refer to regimes of "work for hire." In French, "*marchandage*" (literally "haggling") is the act of an intermediary employing a worker who's actually working permanently for a different company. In Italian, "*caporalato*" (literally "gang master regime") means the illegal hiring of worker contractors in agriculture. Often used in a pejorative manner, these terms have become synonymous with irregular work, unofficial contracts, and a lack of social security benefits for economically dependent workers.

Originally, they described traditional ways of managing workers before wage labor became widespread. Between the end of the nineteenth century and the first half of the twentieth century, standard employment contracts as dependent employees, or unitary contracts, became increasingly common. There is a close relationship between them and the rise of the organizational model of the firm during that period. But, at the beginning of the twenty-first century, as firms progressively morph into platforms, and the paradigm of the firm becomes less relevant, dependent employment schemes are called into question. Previous work arrangements resurface, such as those that were used to recruit hired hands in the agricultural sector or freelance artisans in urban centers.

In the eyes of advocates of independent and autonomous work, especially in the tech sector, the advantages of a gig economy seem obvious; this reversion to old work practices isn't necessarily bad either. After all, even when they weren't qualified to join the exalted workers' elite, nineteenth-century laborers, artisans, and journeymen were free to negotiate the price of their services, to leave their employer for better wages, and their only obligation was to achieve results. Workers were often described as "masters in their own home" because, once hired, they were free to decide how, where, and with what resources—including human resources— they would carry out the work. The distinction between worker and entrepreneur was therefore not fixed. The historian Alain Cottereau even speaks of "a principle of bilateral free will," to indicate that, when agreeing to perform the work, both workers and employers would act entirely willfully and on an equal footing. There was thus little difference between the persons who demand a service and the persons who provide it.[36] Workers in this regime were considered to be "entrepreneurs of their own work."[37] The expression means that they took on a more businesslike attitude toward choosing, performing, and getting paid for their job. They viewed the companies that hired them more as ad hoc business partners, rather than as their employers.

Today, digital platforms champion precisely this employment model, in which the roles of entrepreneur and worker are fluid. Uber, for example, presents its drivers as partners "who are their own bosses,"[38] despite the dissenting opinion of many public bodies and the courts.[39] Portraying drivers as small business owners, Uber even encourages them to drive rental cars during their off-peak hours. Similarly, some microwork platforms offer advice to their taskers on how to develop entrepreneurial projects, as in the case of the impact sourcing platforms presented in chapter 4.

Another aspect of the nineteenth-century "work for hire" logic that is rehashed by today's gig work and freelancing is more rarely mentioned: historically, workers employed a system of interlocking subcontractors whereby each worker could hire others to carry out an assignment. This brought with it a cycle of oppression within work units. As a result, workers could make bids for jobs, and then quickly assign them to unaffiliated, or less experienced workers, forming a chain of subcontractors. The hierarchies between an employer and a worker, in relation to others hired by the latter at lower rates, created an environment of mutual exploitation.[40] It was precisely due to the abuses committed by these workers-outsourcers with regard to their fellow workers that the system of *marchandage* was abolished in France in 1848, while the *caporalato* wasn't banned in Italy until the early twenty-first century. Both were described as an illegal intermediation of labor that exploits workers by using subcontractors.

The system of mutual dependence that one finds on digital platforms today reflects the eminently social nature of labor—the bonds of dependence between members of the same community are very common. But it also mirrors the subcontracting chains of yore. Instagram influencers, for example, rely on their ability to engage their followers and influence their behavior. The work they perform for the platform consists of creating videos, images, and texts, which circulate thanks to the social media labor of members engaged online, who share and qualify this content. Likewise, within the small communities created around the online labor platform Upwork, microworkers employ other members, acquaintances, or even family members. After a top-level agreement that sets out all the tasks to be performed, the work enters an unofficial labor exchange, where the chain of subcontracts extends ever further at the local level.[41] This phenomenon of reintermediation, which I discussed in chapter 4, also happens among Uber drivers. In both high- and low-income countries, many people sell or rent out their app accounts to drivers who are effectively both their business partners and labor suppliers.[42]

These practices reflect a continuity between the forms of contingent work that are currently flourishing on platforms and forms of work typical of the early phases of industrialization. Other examples of these latter forms, ones that were prevalent in the United Kingdom and in Asian countries like India, Indonesia, and Japan, included the "putting-out system."[43] In this system, central agents would negotiate the work with the client, then dispatch it to subcontractors who worked from home, hence the names "domestic system" and "cottage industry." Without any regulation of wages, hours or working conditions, this system organized the production of goods in small workshops ("sweatshops"), particularly in the textile industry. This system is reappearing with digital platforms, in the production of home-cooked meals offered via on-demand apps, in transcription or translation services carried out as piecework remotely, or in makeup YouTube tutorials shot in bedrooms. Some of these workers even describe their work as part of a "global digital sweatshop."[44]

In the nineteenth century, it was common for work to be performed and paid for by the piece, per unit, by the task. Today, microwork and on-demand labor most often follow a similar arrangement. Several authors maintain that services like Uber or Amazon Mechanical Turk are modern examples of piecework.[45] Platforms fragment computational tasks by reducing them to actions for which, allegedly, no qualifications are required, and which are remunerated not based on the time spent performing them, but on the number of requests satisfied. When content producers monetize their contributions on social platforms, one can also call it "digital piecework," paid by the unit.[46]

So, to a certain extent, the spread of subcontracting chains and the fragmentation of tasks on digital platforms points to a revival of the sweatshop system. Despite the "exalted" fantasy of self-determination at work and of a balanced power relationship between employers and digital laborers, the old relationships of dependence and subjection remain.

The Persistence of Subordination

One point that formal employment and platform piecework have in common is that both involve a *bond of subordination* to a main employer. The concept of a subordinate who works for a central power is much older than work-for-hire arrangements. It originated during the early medieval period.[47] Paradoxically, the modern employment contract that supplanted the work-for-hire model, mentioned while discussing the *marchandage*, reintroduced some feudal features. In particular, it replaced the mutual consent and "bilateral free will" of employers and workers with a set of legal articles that forced workers to perform their tasks for the employer, who had several prerogatives over them. However, unlike the feudal system, the wage system was not based on coercion, but on "protected subordination," a relative subjection of the worker in exchange for protection by the employer. This principle was progressively generalized, beyond wage labor, to other legal frameworks for work. At the end of the twentieth century, a certain number of protections specific to dependent work were granted to licensed professionals (doctors, lawyers, engineers etc.), and more broadly to self-employed workers. This mechanism was based on the principle that legal scholars describe as "allegiance in independence." A reverse dynamic set in at that point, marked by a gradual decline in the discretionary power of managers and business leaders, and an increasing autonomy of employees.[48] As a result, self-employed workers are better protected, whereas salaried employees enjoy a little less subordination; clearly, the two kinds of workers converge.

These trends were amplified by digital platforms. Increasing digitization and quicker production flows gave rise to paradoxical forms of subordination bonds, both constrained and free. When food app Deliveroo informs the aspiring courier, in a friendly and perfectly nonhierarchical manner, that "[they] can stay connected for as long or as little time as [they] wish,"[49] or when Mechanical Turk presents microwork as tasks "to complete whenever it's convenient,"[50] this rhetoric of independence is clearly designed to attract workers, but it hides a technical system that controls and determines their activity. If couriers and Turkers really worked whenever they wanted, their profiles and fulfillment rates would suffer.

Digital labor involves a specific form of subordination, yet it brings with it none of the protections enjoyed as a counterpart by formal employees, such as the stability of a job or financial cover for certain occupational risks, nor, more broadly, does it engage a company's liability. Subordination in digital platform labor does not equate with the economic and social dependence of employees who are "ranked below" (subordo) their employers but is closer to the sense of the Latin word subordinatio: "placing someone under the orders of a superior." It mainly describes a relationship of subjection aimed at carrying out a task. No symbolic power—religious or political—establishes this subjection; rather, it originates from an authoritative source of "orders" (ordines) which pertain to commands (imperata).

Analyses of on-demand work,[51] microwork,[52] and even social media work[53] have brought to light the fact that all forms of digital labor are based on series of injunctions and instructions that users receive from services, apps, and interfaces. The dependence of users on platforms that act as technological intermediaries can be described as technical subordination.[54] Workers, who are supposed to be independent, experience the platform's requirements as constraints on their behavior, a situation described in legal terms as "subjection."

Direct forms of subjection consist of the rhythm and order of execution of tasks prescribed by the platform. The majority of digital applications and services include "triggers" in their interface, in the form of prompts, reminders, and calls to action. Specialists in persuasive technology design and ergonomics induce the behavior desired by the platform immediately through these messages, messages that tell people to "act on impulse." As behavioral scientist BJ Fogg explains, "when Facebook sends me an email notification that someone has tagged me in a photo, I can immediately click on a link in that email to view the image. This kind of trigger–behavior coupling has never before been so strong."[55] The same is true for taking a photo, entering a query, accepting a contact, or attending to a smartphone alert. There are many examples of the orders transmitted by triggers: LinkedIn sends emails to its subscribers, reminding them to connect to their profile, to "compare [their] salary with that of other professionals," or to "find out how [their] profile stands out." Clickworker sends notifications to its microtaskers to let them know that they need to take urgent action on a pending task. These alerts, always couched in the imperative ("Connect!" "Click here!" "Accept this task!"), embody what the philosopher Maurizio Ferraris defines as "calls to order."[56]

In addition to these notifications, which are typical of the transformations of contemporary capital and its strategies of attention capture,[57] there are also more marginal constraints, similar to those that are often evoked to prove a relation of subordination in labor law cases (obligations of time or place, dress codes, and accountability). For digital laborers, this means carrying out the tasks that might be prescribed. In addition to the use of proprietary software or systems that cannot be modified or audited by users, platforms may require specific access arrangements. For example, Microsoft's UHRS microworkers are required to log in using their Outlook account, produced by the same company. Additional obligations include complying with rules relative to personal appearance. This is equivalent to the dress code of some companies or establishments. Express delivery apps require couriers to carry bags and, in most cases, to wear the platform's logo. Some of these rules are not about branding, but about how users present themselves in general. Ever since social media such as Facebook and Google applied their "real name" policy in the mid-2010s, the ways their users present themselves to others have been drastically restricted, for example, by not allowing pseudonyms or certain types of profile pictures.[58]

With the generalization of platforms and AI tools in the workplace, all these obligations, which are designed to streamline the productive tasks that digital laborers are required to perform, have been integrated by the traditional firms. Even the work environments of contemporary formal employees are loaded with alerts, notifications, reminders, invitations, and pop-ups that appear on their screens. Stimuli and demands requiring an immediate response have become ubiquitous, wresting control from employees, and forcing them to proceed at a rhythm and according to priorities defined by others.[59] This generates what has been described as a "regime of dispersion at work."[60] This regime represents a new way of coordinating work. That's especially true in clerical jobs and among knowledge workers that were traditionally organized by schedules, action plans, and roadmaps.[61] Today, this is also true for Amazon warehouse workers, who follow instructions second by second rather than following established routines or knowing their inventory by heart[62]. The generalized situation of scattered attention is becoming increasingly prevalent in the workplace.[63] In this respect, digital labor hardly differs from contemporary transformations of work found in any company, in any industry sector. The much-vaunted autonomy that was to be ushered in by platforms as firms, especially in the wake of the pandemic crisis, embraced digitization, is a far cry from the reality of technical subjection.

Digital Labor as Unfree Labor

My argument so far is that modern forms of labor, whether they're mediated or not by platforms, carry the echoes of past servitudes. Freelancing and gig work are exploited by intermediaries like work for hire did in the past; formal dependent employment inherits the semantics of subordination of feudalism and serfdom. Work conceived as enslavement is an unbroken thread running through the history of our societies. Wage labor is no exception. To have to rely on wage labor in the Middle Ages was invariably a sign of deteriorating status. Despite widespread misery, wage labor was connected to the plight of vagabonds, convicts, and prisoners, and so it had a deep social stigma.[64] At the dawn of the Industrial Revolution, only domestic servants such as lackeys and valets were bound by "employment contracts," a sign of their inferior status.[65] In contrast, the work-for-hire and putting-out system that emerged in the proto-industrial era seemed to give workers freedom and dignity. So when, starting in the nineteenth century, *marchandage* was displaced by dependent employment, it imposed a work discipline in which the employer could "sanction" disobedient workers for their "insubordination."

In the debate about social media labor, the "pleasure" approach sees platforms as escaping previous relations of domination. Those who support the "work" approach see them as a return to servitude, far from the freedoms of exalted workers.[66] Some online labor intermediation services brazenly appeal to the idea of slavery as a way of emphasizing the reliability of their services. For example, the Philippine service MicroSourcing, which provides moderation and data preparation services for Western companies, boasts that the work is carried out by "virtual captives."[67] Its website explains, unabashed, that the rates of pay of virtual captives are generally much lower than offshore labor, but that foreign customers retain full control over the entire operation, including its personnel, infrastructure, and processes.[68]

The notion of captive is not always a simple metaphor. In the United States, prison labor now comprises data entry, proofreading, and document preparation services[69] intermediated by microtasking platforms like the ones we have discussed. Finnish startups have turned prisoners into data workers, scanning text and labeling words as part of an experimental program.[70] In Russia, these activities are sometimes presented as return-to-work programs. In Number 7 state penitentiary in the Omsk district, for example, the inmates are required to produce YouTube videos (sketches, musical performances, and tutorials).[71] More often than not, both the prison administration and the platform monetize the digital labor the prisoners perform. In China and other Asian countries, some prisoners are constrained to get into gold farming: in online video games, they collect objects, earn virtual money (which can be converted on auction sites), and create customized characters that the prison administration can then sell.[72] Forced digital labor of prisoners (or, in some cases, of illegal rural migrants) has become an extremely lucrative business in China.[73]

Some digital platforms obtain contracts to provide digital communication services to prisons, and then reuse (and monetize) the content and data produced by prisoners and their relatives. One such platform, JPay, aims to become "the Apple of the US prison system." It enables inmates to access sites for downloading music, books, and email software, or sites for money transfers and video communication with families. These services are paid for by the prisoners, especially the video visits, which involve more than 600 structures in forty-six states. On all of these services, the prison takes a 20 percent commission.[74]

The Electronic Frontier Foundation has expressed its concern that the platform captures personal information for monetization, and it warns inmates and family members on the outside who use JPay, that they are transferring their rights on all records to the platform, "including any text, data, information, images, or other material that [they] transmit through the service."[75] In some cases, prisoners have used illegal methods to reappropriate some of the value they created and exchanged on JPay. As an example, 364 inmates in an Idaho prison hacked the tablet provided by the platform and transferred nearly $225,000 to their own accounts. These can be viewed as extreme cases of users trying to defend their economic and human rights and reclaim control of their data and content in a situation of unequal bargaining power with a platform.

The Panopticon of Production

Platformization and prisons are intertwined, so it's important to understand how this pattern of punishment and control permeates other sectors. Since platforms clearly put their users under surveillance, the supposed freedom they grant them may seem suspect too. New software solutions and platforms have been launched to coordinate mobile workers and digital nomads. "Bossware" apps and platforms are everywhere as a result of COVID-19's incentives for remote work.[76] It's hard to trust their brand names. The CleverControl app promises "total control over employee monitors." StaffCop is another, but with law enforcement vibes. Even Microsoft lets businesses track worker activity on their platform Teams. On a dashboard, managers and supervisors can inspect employee activities: text chats (even deleted ones), recorded calls, video meetings, application usage, file creation, IP addresses, and so on. They additionally record workers' movements and habits so closely, and at every moment of the day, that some of these solutions have been compared to electronic bracelets.[77]

Platform workers are subordinated both in terms of the constant calls to action they receive and the way their behavior is continuously monitored and evaluated. Digital labor is part of the long history of workplace surveillance, but it's different because it doesn't require an actual "place" where attendance can be tracked.[78] Employee monitoring in the workplace used to be strictly confined to the division of labor within the factory, with a view to checking product quality and the number of hours worked. Preventing "time theft" was a priority for employers, and employees could be sanctioned and prosecuted for failing to meet their contractual obligations.[79]

On digital platforms, however, surveillance arises primarily from the need to coordinate workers and teams that are not in one place, and not active at the same time. The platform monitors users everywhere, more frequently, and in much greater detail than traditional employers. In fact, surveillance is part of a bigger trend called algorithmic management. Today, it is used to handle all the traditional employer duties, from hiring to managing day-to-day operations to terminating workers. Algorithmic management is rooted in the platform economy. Through increasingly sophisticated algorithms, platforms match consumers and workers. In order to compute routes for couriers, prices for drivers, and nudges for social media audiences, these systems have to monitor their users constantly. Platforms have inspired modern companies: now they take screenshots of the employees like Upwork does with its online workers, track their GPS data like Uber does with its drivers, etc.

While these examples refer to standard salaried jobs, the underlying logic of algorithmic management and automated surveillance applies to platform work as well. The productive effort of digital laborers is quantified through performance metrics (likes, scores, ratings, stars, and also the volume of followers, shares, and contacts), which are often associated with gamification (winning badges, goodies, triggering animated sequences in exchange for personal data and contributions), competition (comparing scores between different users of a platform to rank them according to their performance), or with self-assessment (activity reports, analysis of the most productive time slots, etc.). In fact, these are all control methods that are associated with a high level of technical subordination. Both for standard employees and for platform workers they can lead to sanctions. On a platform, when users are too slow to earn followers or accept tasks, a notification calls them to action; if they still fail to improve, the service becomes less accessible. On the other hand, if the results improve too quickly, or deviate abnormally from the average of other platform users, the suspicion of fraud or cheating can also lead to exclusion. The algorithmic devices for managing users' workflows thus have a powerful disciplinary effect, especially because productivity is monitored in real time.

In a regime of digital labor, attempts to define the limits of work break down on a social as well as a spatial level. As users produce more data, services, and content, the boundary between public and private needs to be negotiated and redefined with the platform, and even with other users. The legacy of nineteenth-century American case law, which viewed privacy as a "negative liberty" (the proverbial "right to be let alone,")[80] has given way to a more active approach, based on exchanging and interpreting digital signals,[81] or preserving the integrity of a social context.[82] The definition of what is private or public is never a given. It results from what can be described as an iterative and dialectical "collective negotiation,"[83] whose outcome depends on the power relations between workers and platforms. Yet despite the demands of workers that their privacy be better respected, and in the near absence of collective instruments for defending their rights in this area, mass surveillance and permanent tracking programs are actually increasing.

Platforms are not developing these systems only for themselves. They also sell them to other companies. For example, Microsoft has launched 365, Facebook has marketed its collaboration software Workplace, and Google sells its business solution Workspace. Their successes have attracted other companies. Digital corporate networks or applications for keeping in touch and collaborating with colleagues, such as Cornerstone OnDemand and Salesforce Chatter, are now competing with the giant platforms in the work monitoring market. The number of applications has grown so massively[84] that Julie Cohen speaks of a "surveillance-innovation complex"[85] and, famously, Shoshana Zuboff has popularized the notion of "surveillance capitalism."[86]

The role of surveillance on digital platforms is quite different from the situation in traditional companies. Previously, the time clock, the supervisor, and the manager checked workers' productivity and attendance. Today, tracking users' activity is a platform's core business. Surveillance is not only a means to ensure that the amount of work remains steady or increases, but also an end: it generates valuable data for the platform. As I have argued in the previous chapters, platforms extract *qualification value*, which increases knowledge of workers' preferences (the most sought-after content, the tasks best suited to each person's skills, their closest delivery points); the resulting behavioral analyses can be sold on to brands or institutions, to generate *monetization value*; the data extracted from users can also be converted into *automation value*, used to train algorithms and AI solutions. Increasingly intrusive algorithmic management infrastructures make surveillance and production inseparable for users.[87] On platforms, management costs are shifted to the workers, who are forced to monitor themselves and others.[88] However, in line with the spirit of digital labor, this surveillance is passed off as participation based on reciprocal disclosure. It can therefore be described as "participatory surveillance,"[89] no longer under the watchful eye of Big Brother, but under that of "millions of Little Brothers and Little Sisters" who carry with them "the tools of [their] own transparency."[90]

There are countless examples of this "participatory surveillance": X users tag other subscribers to track them more easily, Uber Eats users rate their delivery couriers, etc. This participatory surveillance is at work in traditional firms, too. Human resources management systems and corporate social media harvest data for employee surveillance. This is how the application Betterworks is used. Managers can use it for "transparent tracking" to measure the progress of projects and the involvement of workers. Supervisors can "evaluate employees fairly by providing data-driven insight into [their] performance," and the platform encourages emulation and mutual monitoring between employees through one-to-one meetings to boost "collaboration and alignment."[91] Each person's participation is recorded on a table accessible to all staff, and productivity is mapped by a chart that visualizes goals and progress by team or department.[92] This "soft surveillance" is self-imposed and carried out collaboratively, both on platforms and in firms on their way to platformization. Rather than quashing the user's will, it feeds on it to monitor and track. Panoptic surveillance gets completely reinvented with participatory surveillance. Digital labor ultimately establishes itself as "voluntary servitude"[93] or "mandatory volunteerism"[94] rather than as a way of emancipating workers.

General Terms of Use: A Labor Lock-In?

A decisive moment in the user's consent to being monitored is the ritual acceptance of the terms of service (TOS). TOS vary greatly from one platform to the next, and they are not strictly speaking contracts of employment or service provision. They do, however, stipulate who has the right to use a digital service, how one can unsubscribe, and who benefits from the information that circulates in the form of data, content, profiles, and so forth. From a legal point of view, these are known as adhesion contracts, "take it or leave it" agreements whose terms are not based on equal rights since one party imposes conditions on the other.[95] Platform owners, keen to limit their own liability while maximizing users' productivity, are careful to differentiate TOS from employment contracts. Yet several clauses closely resemble those in contracts: TOS rigidly define users' roles—driver and passenger, requester and microworker, advertiser and subscriber—

and prescribe certain product quality criteria (standards for the acceptability of content on social platforms, accuracy rates in microwork, ratings with on-demand services). More importantly, they determine how value will be distributed, and name the end owners of the data, content or services produced, and the conditions of the producer's remuneration.

For the platform owners, the balancing act of the TOS's clauses aims to ensure that these terms of use cannot be employed by workers, microworkers, creators, moderators, and simple users to file a lawsuit for job reclassification. Platforms go to great lengths to make it clear that they do not solicit, or profit from, the work done by their users. For example, the delivery service Uber Eats stresses that the delivery couriers are "delivery partners" and that the company is just a "technology services provider" that connects restaurants, couriers, and consumers.[96] Mechanical Turk defines itself as a place where requesters and Turkers can enter into an agreement, while each labor provider operates "in their personal capacity as an independent contractor and not as an employee of a requester or Amazon Mechanical Turk or our affiliates."[97] The Howtank platform, which connects brands with "helpers" who agree to provide after-sales services, states that participation "cannot be interpreted as an employer/employee relationship with Howtank nor a partnership, joint enterprise or any other association."[98] These clauses are not inherently problematic if users are able to retain autonomy when performing their tasks and when producing services and data. But given that their activity is subjected to the forms of technical subordination and participatory surveillance that I pointed out, this independence becomes wholly illusory.

In some cases, however, TOS may fail to protect platforms from legal action, such as against claims by users to be reclassified as employees, or to obtain monetary compensation for past pay, or for social insurance charges and future social benefits relating to the work performed. The TOS may well reflect the platform owners' wishes, but not what the law lays down. The platforms' legal teams are aware of this, and tend to introduce "enhanced independent contractor clauses."[99] For example, the on-demand housework service TaskRabbit compels its users to protect and defend the platform from any claims related to the misclassification of a user as an independent contractor, including potential claims associated with employment-related laws such as termination, discrimination, and benefits.[100]

Even though platforms may refuse to accept their status and obligations as employers, they have no qualms about imposing subordination, surveillance, and other restrictions, as detailed in the very same TOS. They shackle their users to their application, a behavior that can be described as "labor lock-in." The expression comes from the concept of "vendor lock-in," where companies try to keep their service providers loyal, so that they don't switch to competitors ("competitor lock-out").[101] Thus some services insist that all tasks be performed on proprietary applications or software. Amazon Mechanical Turk requesters cannot request work from Turkers outside the service's website, and Uber and Airbnb users are not allowed to contact each other, except through their respective applications.

As well as requiring users to behave like employees, the TOS promote a semi-subordinate relationship. This is evident in how the platforms handle users' personal information. Many platforms make it difficult for subscribers to transfer the data of their accounts to other services, despite the best efforts of the law, unions, and user associations to have the so-called right to data portability recognized.[102] The General Data Protection Regulation (GDPR) in place since 2018 recognizes this right for European citizens.[103] The California Consumer Privacy Act of 2020 contains the same provision, while other important pieces of legislation, like India's Digital Personal Data Protection Act of 2023, do not include it. It doesn't matter if the right is recognized or not: if platforms do not work together with users to enforce legislation, these initiatives won't succeed. In the past, solutions like Myspace Data Portability (2008) or Google Data Liberation Front (2010), allegedly designed to enable users to access their data, proved to be misleading. Format incompatibilities and flaws in the saved content and databases relating to platform profiles make it difficult for users and workers to utilize the contents—and more importantly to export them to another platform. Clearly, Uber is a case in point. There's no data portability for its drivers, and the platform doesn't allow them to access their data. A series of lawsuits, referrals, and court decisions have stipulated that the company must disclose more of the drivers' data.[104]

Reputation data, ratings, and "certifications" are particularly crucial for on-demand workers, microworkers, influencers, and social media moderators. They might want to transfer this information between platforms. Restricting the portability of data, platforms create obstacles to workers' freedom of movement. Platforms lock workers into providing services, while at the same time denying them the protections and guarantees that accompany the employment relationship.

Digital Labor: Real Work for Surreal Pay

Digital labor thus occupies a gray area between work-for-hire and salaried employment. Like older types of work contract, it is an activity without a fixed location, based more on horizontal relations and cooperation that eventually turn out to induce cascading subcontracts. Like salaried employment, however, it involves subordination, surveillance, and unequal rights between the workers and the owners of digital services. The boom in digital labor, which is meant to respond to a general desire to work autonomously, should be reconsidered in the light of these characteristics.

As discussed previously, the tasks performed in digital labor may well be atomized and concealed, but they nonetheless involve real work. Users of all platforms *produce value*. They calibrate applications, train algorithms, create content, and provide services. Above all, they produce the data that sustains innovation in areas as varied as smart cities, insurance, or health.

Moreover, digital labor *is not part of the informal economy*. Digital labor is clearly associated with informality in low-income countries,[105] but it retains some distinct characteristics. In contrast to informal labor, digital labor involves some form of contract. It is defined in contracts that, although primarily intended to distinguish it from employment, impose rules concerning the nature of the tasks to be performed, the way in which they are to be carried out, and the attribution of the resulting products. TOS establish an asymmetrical relation in which obligations are imposed on the users, whereas the platforms are released from any responsibility toward them.

A further point in support of the argument that digital labor is real work that relates to formal employment concerns *surveillance*. According to the ILO, by the mere fact of logging on, digital platform laborers are "subject to extensive and intrusive monitoring of their performance, similar to that applied to traditional employees."[106] In a similar judgment handed down by the Paris Court of Cassation in 2020, it was found that "at the moment of connecting to the Uber platform, there is a bond of subordination between the driver and the company."[107]

Furthermore, when users are assigned productive tasks, they enter a *relation of subordination*, involving direct or indirect forms of subjection. Some of these are comparable to salaried employment, yet the digital laborers do not enjoy the benefits of a stable job. As the numbers of formally independent but economically dependent workers increases, so intermediate statuses of "para-subordination" have emerged, particularly in Europe: we find "co. co. co." ("coordinated and continuous collaboration" contracts) in Italy, TRADE ("economically dependent autonomous workers") in Spain, *Arbeitnehmerähnliche Personen* ("quasi employees") in Germany, and so forth.[108]

Last, there is the issue of *remuneration*. Although some have argued that because digital labor is often free or very poorly paid we can't consider it work, labor law provides no such exclusion. Employers have been held liable for accidents during unpaid activities such as volunteering. Forced labor has recently been defined as an activity "without remuneration or in exchange for remuneration that is clearly unrelated to the amount of work performed."[109]

On the other hand, remuneration does not automatically imply an employment relationship. Rather, in the world of platforms, payment differentiates between platform activities according to the degree of task visibility. The more inconspicuous the task, the less likely the performer will earn income. On-demand drivers receive payment for the time spent behind the wheel, not for the time spent on the app managing their reputation. Microworkers are paid for tasks performed on behalf of a requester, not for finding assignments and projects, preparing for them, and acquiring "certifications." Similarly, on social media, users such as influencers, streaming game players, and adult performers may be able to monetize their production and performance more easily, but they are rarely paid for the endless clicks and data management required to align with algorithms or execute mind-numbing tasks. The number of lawsuits for concealed labor filed against tech companies that derive their success from the data and contributions of their subscribers has skyrocketed—in light of new data protection laws and current trends favoring on-demand workers' rights.

Digital labor exists on a continuum from conspicuous to inconspicuous activities, and from well-paid to unpaid work. This is unlike the experience of nonstandard, nomadic, and freelance workers. The working conditions of platform workers are so divergent that, by design, no collective consciousness of their shared situation can spontaneously emerge. Moreover, the language of free contributions on platforms clouds the reality of technical subordination in digital labor, abetted in this by a perverse syllogism that states that if the activity is unpaid, it is not work—and if it is not work, it must be something like leisure, consumption, participation, or sharing that cannot be paid. However, these platforms request from their users *tasks that generate value*, within a *contractual framework* that entails some degree of *technical subordination*, and involves *continuous monitoring* of users' activity. At the same time, platforms deny these conditions to avoid paying workers and any obligations toward them, including the social protections inherent in our redefinition of work.

· 8 ·

SUBJECTIVITY AT WORK, GLOBALIZATION, AND AUTOMATION

Digital labor is fragmented and volatile. Platforms and users have an ambivalent relationship, resulting in a peculiar form of subordination. Who are users? Are they workers? Amateurs? Partners? Consumers? Platformization raises questions about what type of subjectivity it implies. The concept of subjectivity in political philosophy refers to how a member of society comes to be regarded as a subject. With platforms and AI transforming capitalism, how do workers gain agency, both individually and collectively?

Is This Exploitation?

Classic Marxian theory distinguishes exploitation from theft. Uncompensated or undercompensated platform users, for example, are victims of "wage theft." However, they are not automatically exploited. Throughout history, exploitation has occurred within specific relationships between classes and production. Is there a class today that can extract labor and value from others? To ascertain whether or not it does, the concept of exploitation must be updated to reflect today's realities. This updating can take two forms.

One is to expose the material basis of the supposedly immaterial economy wherein platforms and AI operate. From mining the critical minerals and rare earth metals needed for technologies to function, to the disposal of electronic waste and discarded devices, the digital economy exposes workers, especially in the Global South, to squalid working conditions and very low wages. Far from applying only to digital laborers, this Marxist-inspired approach to exploitation encompasses the entire production chain of the digital economy but focuses in particular on its industrial stage.[1] Kate Crawford's *Atlas of AI* summarizes this approach best.[2] Her essay argues that three factors contribute to the production of goods and services in the global AI economy: data, labor, and earth. The real exploitation associated with today's technologies comes from the way these factors of production are obtained. AI needs data, of course. Labor is also necessary in order to inject human judgment into AI. Then there's earth, the natural resources we need to build the devices—their cases, batteries, screens, and circuits. The journey on the way to reaching the material sources of the tablets that play songs suggested by Spotify algorithms, or to accessing Dall-E to generate images automatically, most likely starts in a lithium mine in Chile, or a semiconductor foundry in Taiwan. The further we get from the final user, the less immaterial this economy looks.

The second approach aligns more closely with this book's focus—reviving the humanist critique of the fragmentation of work, which figured prominently in sociologist Georges Friedmann's 1956 classic *The Anatomy of Work*.[3] The book criticizes assembly-line production for breaking work down into small bits (the French term is *miettes*: "crumbs," "scraps"), making it meaningless and empty. Accordingly, digital labor can be viewed as a series of repetitive, standardized, and atomized microtasks that only platform owners can grasp as a whole. Digital laborers are thus like the workers of the first Fordist era, for whom work becomes a pointless and meaningless activity. But we are witnessing a sinister evolution. Machines, as means of production, oppressed Henry Ford's factory workers; today, digital platform workers are themselves the cogs in the machine that threatens to replace them. In other words, the fragmentation of their work is not a consequence, but a precondition of automation.

A collective subjectivity is needed to express awareness of one's own exploitation. This poses a problem, since platforms rely on a fragmented labor force to accomplish fragmented tasks. Digital laborers struggle to perceive themselves and others as such—or, at the very least, they see only the visible parts of their own and others' work. Within each productive ecosystem, user–workers are often divided into antagonistic categories (like passengers vs. drivers, requesters vs. microworkers, or creators vs. moderators) due to the specificity of different types of digital labor. As Friedmann argued, individuals cannot build solidarity if they do not perceive themselves as mutually interdependent, linked by common interests and objectives.[4] Now, platforms prevent workers from building solidarity functionally (individuals have difficulty perceiving the similarity of their conditions), organically (they are unable to grasp the complementarity of their roles), and even emotionally and morally.

The lack of solidarity between social actors is cause for disappointment. The internet was originally described as a community-building tool. Over the past few years, only platform owners have advocated online communities, mainly for profit. Mark Zuckerberg, for instance, entrusted big platforms to connect atomized social actors through AI in his infamous 2017 "Building Global Community" manifesto.[5] That a platform tycoon would take on the responsibility of keeping society together is all the more ironic, given that the main reason for today's breakdown in solidarities is the unequal distribution of gains from platforms. In general, only a few people can earn a living as a platform worker, so those who have no other source of income feel job insecurity acutely. For example, on some microwork platforms, as mentioned earlier, a small elite of "heavyweights" with exceptional reputational scores manage to earn comfortable sums by taking on many tasks and subcontracting them to other microworkers with lower ratings. It is difficult, then, for the contractor and subcontractors to see themselves as part of the same group. On a related point, digital workers may overlook the collective dimension in how platforms appropriate the fruits of their labor (tasks, content, and data), and their exclusion from the resources their labor generates. They do not own the algorithms, databases, or AI solutions that they train, voluntarily or involuntarily.

Exploited Ever After

Another classic concept, alienation, remains relevant. In her book *Work and Alienation in the Platform Economy*, author Sarrah Kassem examines both Amazon marketplace and Mechanical Turk[6]. As she explores the platform economy, she talks about four areas of alienation: the detachment of workers from their labor activities due to algorithmic management, the estrangement from the products of their labor as work becomes quantified in metrics, the blurring of work-life boundaries eroding human essence or species-being, and the superficial creation of community among workers that masks deeper social issues. Particularly criticized by Kassem are social media platforms like Meta, which promote a "fetishized work culture" that creates a false feeling of belonging, equality, and inclusivity.

By promising self-fulfillment through "voluntary," "participatory," or "collaborative" digital labor, platform owners obfuscate the actual relations of production and remove them from consciousness. To address digital labor only from this perspective may seem reductive, especially since many have, as we've seen, associated digital labor with users' empowerment. Indeed, the meteoric rise of digital labor is inseparable from the cognitive and affective rewards with which social media and other platforms attract workers[7]—the assurance they are practicing the "work of the future."[8] Some authors even suggest that worker mobilization by social media to qualify products, services, and information could harbor a transformative potential that would break with a market theory.[9] Bombarded by the "participatory" commercial rhetoric of platforms (which arguably also shapes these authors' positions), users may indeed be quite willing to live "exploited, happily ever after."[10]

As a matter of fact, digital labor, like other work regimes that preceded it, can be understood as an activity that seeks to balance exploitation with empowerment.[11] The sociologist Eran Fisher argues that users' active role in value creation on platforms increases their exploitation, but that this work remains empowering because of the way their labor fosters self-expression, social network creation, and a sense of community. In this respect, platforms "empower individuals by contributing to their objectification."[12] In this bizarro Marxist dialectic, users have to relinquish their power of not being exploited by a platform if they want to avoid alienation and feel empowered using that same platform.

Fisher's argument focuses on how platform users are empowered individually. He doesn't talk about collective empowerment. Research into collective empowerment has largely focused on how networks can stimulate self-production and exchange based on reciprocity and self-management.[13] But the collective control of workers' activity has not guaranteed protection against platforms' exploitation. It has also not slowed down the fragmentation of the workforce on platforms. A better understanding of how digital laborers construct their subjectivities is essential to evaluating whether a concrete political project can be derived from this empowerment.

The Class of the New: Capital's Allies, or a Digital Proletariat?

When we attempt to imagine an entire class of digital laborers, we encounter the contradictory realities of a "digital proletariat"—reminiscent of the interchangeable, impoverished, and disempowered laborers of the first industrial era—and at the same time an inner circle of "exalted workers," who have more in common with other social groups such as platform owners and designers.

Historically, new social classes have often emerged as the result of technical, economic, and cultural developments. Richard Barbrook's anthology *The Class of the New* lists the main occurrences of the notion of "new class," from 1776 to the present. During the first Industrial Revolution, the new political subject—the proletariat—encompassed all manual laborers.[14] There were internal divisions, however, namely around the technically skilled specialized workers, who had features in common with the capitalist class. Friedrich Engels described them as a "labor aristocracy,"[15] and Lenin attributed their existence to imperialism, which "has the tendency to create privileged sections also among the workers, and to detach them from the broad masses of the proletariat."[16] The mistrust of this second group within the new class signaled a century-old problem: how to account for the intellectual skills of a political subject that defines itself through manual work.

To answer this question, theorists like the English socialist William Morris (1885) considered that intermediate collective subjects existed between the proletariat and the bourgeoisie. It is the case of the intellectual proletariat "whose labor is 'rewarded' on about the same scale as the lower portion of manual labor, as long as they are employed, but whose position is more precarious, and far less satisfactory."[17] Antonio Gramsci proposed the notion of "intellectuals of the urban type" who "have grown up along with industry and are linked to its fortunes" (1934), and who form a link between the proletariat and the entrepreneurs.[18] Due to their ambiguous position, these intermediary figures sometimes played unusual roles in this theory. Thorstein Veblen considered the engineers to be the class capable of overthrowing capitalism and instituting a "soviet of technicians."[19] After the war, this class was mainly associated with the closed organization of civil service or private company employment. William Whyte's "organization man"[20] (1956)—the white-collar worker who renounces his individuality and becomes a conformist model employee totally dedicated to his company—or Serge Mallet's "new working class," which operates and supervises machinery rather than working in production (1964),[21] provide examples of this category. The idea of a new class in the pay of the capitalists, and bridging manual and intellectual skills, thus generated a vast literature.

However, this rich history has been virtually ignored by scholarship on work and politics in the digital revolution. Arthur Kroker and Michael Weinstein coined the expression "virtual class" for the social group of visionary capitalists, specialized developers, engineers, and computer scientists: an intelligentsia whose prosperity depended on their own efforts to expand the uses of digital technologies.[22] The "creative class" popularized by Richard Florida in the early 2000s included highly educated architects, designers, and financiers.[23] Ian Angell's "new barbarians" are venture capitalists, new media experts, and startup founders.[24] The values of the "intellectual" classes of the nineteenth and early twentieth centuries were described as closer to capitalists' values than to those of a labor aristocracy. A "class of the new" has emerged in the digital age that separates itself even further from the working class. It rather resembles a digital bourgeoisie[25] and is even considered to be "unproductive," appropriating the value produced by other social classes.[26]

Current discussions have abandoned this new intermediate class that straddles the categories of worker and intellectual. We see in them a return of the concept of proletariat applied to platform workers. A tutelary figure for this new generation of technologists was André Gorz, whose most significant contributions came in the early 1980s. Even so, the philosopher thought that automation gave way to "non-classes of post-industrial proletarians."[27] Researchers and policymakers have appropriated and updated the concept. Neologisms multiplied, to convey certain nuances within a shared understanding of the concept. Futurist Alvin Toffler coined the expression "cognitariat" in 1983,[28] which philosopher Franco (Bifo) Berardi later applied to the first internet workers. He pointed out that "work in the net economy" necessarily retains a bodily dimension, and has a significant impact on physical health.[29] Nick Dyer-Witheford used the term "cyber-proletariat" to describe the situation of underpaid, insecure, poorly trained, and deskilled workers in the high-value-added tech sector.[30] Ursula Huws refers to jobs involving remote assistance in practical production as pertaining to the "cybertariat." The tasks, performed through a screen, mainly involve standardized information processing. Individuals that occupy these roles tend to develop generic skills, resulting in high occupational mobility.[31] For Guy Standing, workers in the era of the internet are a "precariat" of unprotected and hyper-flexible labor, characterized by nonstandard working arrangements. He compares this with the millions of undeclared workers in informal economies in the Global South, and with the migrant workers who cross borders and come to work in global markets.[32]

Vectoralism, a Class Enemy

There may seem to be little in common between these different notions, but they are in fact facets of one and the same phenomenon and can be brought together within a larger theoretical framework. To get this broader perspective, let us consider the following: Regardless of what you call the digital proletariat, who do you think it stands against? Unlike the social groups that previously dominated corporate and market capitalism, digital labor's adversaries have emerged from financial interests and technology. This new group rose to power when big business began investing in automation and information. Creators of platform and venture capitalists investing in data-driven startups began to form what McKenzie Wark has called the "vectoralist class." This class appropriated and controlled, Wark argues, not the means of production, but information flows (that is, "vectors"). Sociologists had already insisted on Manuel Castells's idea that flows are now more important than physical location for information capitalism.[33] Wark and others backed this idea by arguing that a distinct group of social actors was emerging. The new group provided a variety of logistical services, controlling the movement of information rather than the movement of goods. The vectoralists thus corresponded to the social milieu of platform creators and owners. By appropriating knowledge and know-how through patents, intellectual property rights, and software for capturing data, the vectoralist class plays a central role in contemporary capitalist accumulation.

Just as industrial capitalism's ruling class did not establish their power by buying land, differentiating themselves from the earlier landowning class, the vectoralist have not established their power by buying material assets such as those that made the manufacturing and processing industries powerful.[34] Amazon was able to become the world's largest seller of books without owning a network of bookstores;[35] Uber has disrupted the transportation sector without purchasing a fleet of cars; Airbnb has revolutionized the hotel industry without any lodgings. Surprisingly, the vectoralist class has willingly entrusted ownership of the means of production to users, who could possess the equipment, raw materials, and skills to provide services, without this giving them a political or economic weight comparable to that of those who control the vectors. "Bring your own device," goes the old rallying cry in business organizations. Essentially, it means "bring your own means of production, your smartphone, your car, your Wi-Fi connection, and work for our platform." This lowers the costs for platform owners. And yet the workers, despite having all the material assets at their disposal, are still not granted any access to the means of profit generation—information flows.

However, the new vectoralist class has one feature in common with the industrial bourgeoisie that preceded it. It, too, has an interest in "disposing of the workers" by imposing on them flexible contractual arrangements and real-time adaptability, to align with the rhythms of constantly shifting economic transactions. As McKenzie Wark explains: "A line of economic activity becomes a *vector*, in the sense that it can in principle be deployed anywhere. Connect a supplier of materials to a site of processing with a vector. If the supply becomes erratic, move the vector to connect a different supplier. If the labor at the processing site becomes difficult, move the vector again, connecting the new supplier to a new site of processing. If the capitalist firm doing the processing demands too much in profits, switch to another."[36]

This passage can be seen as an explanation of Uber's surge pricing algorithm, or of any AI tool that finds the best route for a package, or of a just-in-time platform that matches products, friends, and manufacturers. In this light, we can understand infrastructures as enabling the vectoralist class to carry out its core activity of dematerialized logistics. Because management is dematerialized, it appears on the market, transformed, as automation. It's like with "contactless" delivery apps: a food order shows up automatically at a customer's door, but the process appears automatic because a platform hides the worker who delivered it, the chef that made it, and so on.

What the platforms mainly do is relocate productive tasks and manage virtual supply chains, locally and globally. These chains, in turn, produce data to train algorithms, artificial intelligence solutions, and machine learning models. Thus, automation consists essentially of platformization, and platformization consists essentially of outsourcing.

From the point of view of the vectoralist class, outsourcing has two aspects, one *intensive* and the other *extensive*. "The intensive vector," Wark argues, "is the power [...] to monitor and calculate. And it is also the power to play with information."[37] Intensive outsourcing is especially prevalent on social platforms, where users play with information while being monitored. A platform outsources intensively when it encourages, records, and manages user participation—and uses the participation to generate value. This is what Meta does when it wants users to contribute by testing, translating, and debugging new tools. By accumulating data, the company can develop smart systems (or at least systems that pass for smart). By snooping into users' data and increasing their technical subordination, platforms perform "on-site" outsourcing of data tasks to their own users.

When outsourcing is about work performed far away, it is *extensive*. "The extensive vector is the power to move information from one place to another. It is the power to move and combine anything and everything as a resource," concludes Wark.[38] A prime example here is assigning microtasks to click workers, moderators, or "virtual captives" in low-income countries, to annotate, sort, and transcribe information in text or image form. Moving data strategically around the world ensures an unequal exchange of information, which accounts for the prosperity of a class that is located for the most part in the Global North, or, as Wark loves to emphasize, in the "overdeveloped world."[39]

Digital Colonialism and iSlavery

Like Christopher Columbus hoping to reach the Indies by sailing West, platform owners hope to achieve artificial intelligence through digital labor. There is more to it than just a comparison. It's time to address how automation relates to global trade and wealth expropriation. Digital studies have been exploring the implications of "postcolonial"[40] or "decolonial" computing[41] for a while. My own work contributes to this analysis, since I wrote about it in an article for the *International Journal of Communications*.[42] My point was that digital labor and coloniality are connected because they both involve the consolidation of global dependencies.

Indeed, other authors have explored how users, workers, and platform owners are in asymmetrical positions when it comes to wealth and power. These authors mostly use colonization as a metaphor. Just as bacteria colonize a host, platforms and AI are said to "colonize" areas of human experience that weren't previously part of economic activity.

This trend in scholarship kicked off in 2013 when philosopher Roberto Casati claimed that "digital colonialism" refers to the "automated production of norms" introduced by tech companies. A major factor in this dynamic is the belief that technology is necessary for everything in every aspect of life.[43] Although Casati stresses the need to escape systematic data extraction, he does not address the business models, and value chains that shape contemporary labor-intensive digital platforms. Here, "colonialism" describes a set of aggressive policies and discretionary decisions hardcoded into information and communication technologies.

In 2016, technologist Dmytri Kleiner used the term "digital colonization" to describe the transition from a decentralized internet to a closed, centralized network ruled by oligopolies.[44] As far as the idea of colonization is concerned, here it's mostly about enclosing land: just like in the colonies of the nineteenth century, public property was privatized, as the commercial internet took over telecommunication networks and online communities.

The same year marks the publication of the article "Data colonialism through accumulation by dispossession."[45] According to this argument, big data has become a commodity. It reveals a shift in power in which individuals lose control over the information they generate. As capitalism "colonizes and commodifies everyday life" to an unprecedented level, data takes over human existence. Colonization here is mostly a figure of speech, since commodification bears the majority of the burden of proof.

Regardless of the specific definition, it seems contentious to consider American Instagram influencers or European Airbnb guests using the historically charged vocabulary of "colonialism." Importing concepts is never simple, but to what extent is this term useful when describing click workers' conditions? I'll start by abandoning the metaphorical trappings of "colonialism" to treat the term as a conceptual bridge to a global analysis.

From this perspective, we can argue that the Nigerian digital worker who spends all day driving around the streets of Nairobi to collect data to improve Google Maps—just like the individuals in Hyderabad who, for little pay and no social security cover, provide the audio transcriptions necessary for voice assistant apps—consolidate existing colonial relationships. For the information vectors of platforms are, concretely, globalized flows of labor supply. However, the pride that a proletariat of insecure laborers from the South may feel in working for a big tech company seems to demonstrate that the dependence on the North is more symbolic than economic.[46] Paternalistic platform owners take advantage of this sentiment to force workers into accepting smaller payments.

Besides, the vectoralist class has no qualms about employing the term "colonialism." Marc Andreessen, one of Silicon Valley's leading investors and a member of Facebook's board of directors, drew a parallel between the Palo Alto company's penetration of the Indian market and historical colonialism: when Indian citizens, in 2016, rejected Facebook's Free Basics, they had, in his view, adopted an "economically catastrophic" anticolonialist stance.[47] Overall, when discussing the digital economy, conceptual borrowings from the semantic field of colonialism are frequent. There are frequent echoes of the colonization of the American West, as in the expressions "colonization of cyberspace,"[48] "the electronic frontier,"[49] or "digital native."[50]

Media sociologists Nick Couldry and Ulises Mejias change the perspective completely in their book *The Costs of Connection*, published in 2019, by no longer using colonialism as an analogy or metaphor.[51] In their view, colonialism is a historical process that eventually resulted in industrial capitalism, which in turn gave rise to platform capitalism. After all, the primitive accumulation of wealth that resulted in capital investment during the Industrial Revolution stems from a worldwide aggregation of resources, using the Global South to extract wealth and value. So colonialism isn't just about figurative language. Rather, it is an important historical lens to study how platform capitalism came into being.

This concept can't be used metaphorically, because it's absurd to expect historical colonialism to repeat itself in a mirror image. It's important not to minimize the differences between data colonialism and historical colonialism. Data colonialism builds on more than two centuries of capitalism, so exploitation is equally responsible for it. As with historical colonialism, data colonialism is bound to have long-term consequences.[52]

Couldry, Mejias, and I all argue for a "decolonial turn" in digital studies, as a helpful way to imagine strategies of resistance to capitalist data and technology appropriation. Indeed, an increasing number of authors have argued that unlimited automation and the digital labor of platform users, the mining industry in developing countries, and equipment assembly in emerging economies are all part of the same international division of labor.[53] The way in which multinational corporations exploit the cheap labor of low-income countries underlines the continued relevance of the Leninist definition of imperialism as the "highest stage of capitalism."[54] According to media studies scholar Christian Fuchs, the exploitation made possible by digital relations of production ends up equating insecure work, unpaid work, and slavery. The latter exists as much in agriculture as it does in mining, manufacturing, and in the click farms that contribute to the platform economy.

Following a similar line of argument, communications scholar Jack Linchuan Qiu describes an international system he dubs "iSlavery," which extracts freedom from the labor it exploits.[55] From his study of Foxconn, the well-known Taiwanese manufacturer of Apple's iPhone, he concludes that the lack of recognition of platform workers simply mirrors the undignified working conditions in the digital sector more generally. Data sweatshops and click farms are located at the end of a chain whose intermediate links are Asian factories and warehouses, and the rural communities that supply millions of workers to industry and services. In contrast to the image of liberation that smart technologies conjure, the situation of the working class in emerging and developing countries is rife with human rights violations, rigid work discipline, serious health risks, and high suicide rates among workers.[56]

There is nothing "disruptive" or emancipatory in the digital platform economy. Its culture of exploitation and contempt follows in the footsteps of previous unfree labor regimes.[57] Through the provocative notion of "iSlavery," Qiu denounces a society dependent on a global mode of production in which high-value-added activities such as information flows and technological innovation require an army of users and workers who, consciously or not, provide their unacknowledged labor. This workforce therefore exerts downward pressure on the costs of the final products in the digital economy, producing content, data, and services for free or for a microsalary. This has repercussions all along the global value chain, and right to the very bottom, in factories like Foxconn. One can observe an exponential rise in unpaid and underpaid labor at all levels of the global economy.[58]

The landscape of the platform economy is therefore uneven and polarized. The geographical relationships reproduce known political and historical patterns, as shown by the analysis of North–South digital flows. The countries where data and tasks are bought more than they are sold, that is, where the balance of demand for digital labor is positive, are located in North America and Europe. The balance is negative in the lower-middle-income and low-income countries. They provide labor, data, and value to platforms in the North.[59]

Do these tangible realities prove the colonial nature of the digital economy? While the latter is clearly the result of unequal trade flows and forms of exploitation, not all asymmetrical power relations are (neo)colonial. If we transpose this concept carelessly, neglecting the specificities of the past colonial trajectories that have shaped the global division of labor markets, we run the risk of diluting and dehistoricizing the experience of colonialism itself, as well as missing what is new in the current dynamics. Last, the geography of colonial relations of domination only very imperfectly matches that of the dependent relations established by the platform economy. Admittedly, it is mostly microtaskers from former French colonies, such as Tunisia and Madagascar, who are driving French artificial intelligence,[60] but buyers of Indian digital labor are not limited to the countries of the former British Commonwealth, just as Filipino moderators of social media do not work exclusively with the country most influential on their last "insular government," namely the United States.

What I've said so far shouldn't lead us to believe that there is a rigid polarization between the North that specializes in "immaterial work" and a supposedly "pre-digital" South that focuses on the transformation of raw materials—or raw data. This assumption makes us underestimate the importance of the contribution of emerging countries to the platform economy and its culture and their involvement in network governance too. Two good examples are Brazil's Constitution for the Net, the *Marco civil da Internet*, ratified in 2014, and the ambitious (and controversial) innovations developed by Indian governments, such as Aadhaar digital identification system based on biometric data. Other important centers of aggregation of digital labor flows emerge, like Russia and China. The former has aggressively launched its own platform oligopolies like the search engine Yandex, which owns Yandex.Eda for food delivery, Yandex Taxi for ride hailing, and Toloka AI data annotation. China has achieved such intense datafication over the last few decades—and attracts so much digital labor both locally and internationally—that it's become one of the two distinctive vectors of digital platform colonialism worldwide.[61]

Similarly, the way in which users "at the bottom of the data pyramid" experience digital services does not differ substantially from the experience of their counterparts in the North. Everywhere, users are exposed to the same tensions between work and leisure, to similar developments in social media and in work apps, to constant prompts to produce services, ratings, and personal information.[62] It is, moreover, highly problematic to consider that the populations of formerly colonized countries would passively undergo the digital transformation without taking part in it, especially when its driving force comes from those who have historically subjugated their populations. To judge by the stiff competition between African and Indian microworkers to secure the most lucrative AI training projects, or the international success of platforms like China's "super app" WeChat or Nigeria's e-commerce giant Jumia, middle- and low-income countries are far from lacking in agency. Entrepreneurs, user communities, and the public authorities in the South are all vying for dominance in the global ecosystem of digital platforms. Although the United States maintains its lead in terms of stock market valuation in the sector, northern countries are not the only drivers of the digital economy, and middle- and low-income countries do not simply provide physical input and semi-finished products.

Online Outsourcing or Remote Work Migrations?

Platformization is partly a result of colonial dynamics resurfacing and persisting across the globe. Platform development is also influenced by downturns in working conditions and pay worldwide. The changes have certainly been relayed by automation and digitalization. This situation can't be entirely attributed to the strategies of the "vectoralist class." Analyses should factor in the transformations of capitalism itself, in light of evolving economic conditions.

Digital platforms took off at the same time as the debt and financial crises of the late 2000s, which resulted in concerns about mounting unemployment, as well as in wage stagnation, cuts in social protection, and inequalities.[63] The global increase in the number of active participants in the labor market, due mainly to demographic growth and the transition of several countries to a market economy, was accompanied by the integration of emerging countries into global value chains.[64] In the digital sector as elsewhere, multinational companies reacted to the expansion of the labor market by increasing pressure on wages through competitive practices worldwide. However, attempts to lower the price of labor through business relocation were hindered by dissuasive tax policies and the high cost of overseas investment in physical premises. At the same time, the option of importing foreign labor was hampered by increasingly draconian—and, in some countries, downright murderous—anti-migrant policies.[65]

Platformization has been touted as a solution to these difficulties over the last decade. It contributed to the free movement of global labor both by providing jobs with low entry barriers for migrants in countries of destination and enabling "cross-border remote migration." Platforms for delivery and transportation provide clear evidence of this. Even though they're location based, they sit at the intersection of immigration and the on-demand economy, and migrants make up much of the labor force for gig apps.[66] The fact is, however, that platforms that aren't location based also support the claim that digital labor is mostly migrant labor. In the previous chapters, I talked about Venezuelans microworking for US companies, Madagascar workers impersonating AI in Europe in real time, and French reviewers of voice assistants for US companies through a Chinese platform. These are all examples of "remote" work migration.

Only a few decades ago, the labor supply was rooted in certain physical locations, while capital was mobile. In the platform economy, by contrast, labor is geographically dispersed, and distributed along constantly shifting digital supply chains.[67] So whereas, in centuries past, labor was imported, today *a workforce is transported virtually* by means of digital intermediation services that operate as "techno-immigration systems."[68]

Let there be no mistake, however. Digital platforms offer both on-site and remote migrants earning opportunities. They may well allow labor to circulate. But they do not "liberate" workers by opening borders. American and European countries' migration policies have by no means facilitated immigrant labor, and digital labor thus plays the perverse role of *facilitating remote exploitation without providing local protection*. In the age of automation, platforms have no need to relocate workers to make them produce value at lower cost since microtasks can be subcontracted and assigned to the lowest bidders anywhere. Companies in the Global North can tap into a segmented migrant workforce shaped by these technological solutions, while not offering workers the same benefits they would have received if they had been admitted to the target country. The same competitive pressure applies to wage workers in high-income countries (always under threat from automation, which is actually remote outsourced digital labor) as it does to digital labor suppliers in low- and middle-income countries (in a global race to the bottom for remunerations and working conditions). This is why real wages are declining in the North as well as in the South. Platformization and global shocks like the COVID-19 pandemic, have resulted in the first ever negative global wage growth since the 2008 subprime crisis.[69]

Digital labor is thus de-territorialized work, and, as nothing but data flows, it is both concealed and beyond the reach of national regulations. Digital laborers are *foreigners at work*, or more precisely, foreigners *to work* itself, at least in terms of formal employment that they could aspire to in the target country. Because they cannot access the rights conferred by the citizenship of a host country, they are considered outsiders to the labor market.

The Working Class Goes to Palo Alto

The existence of this remotely outsourced labor is, moreover, denied by the platforms that use it. There is therefore a twofold obfuscation, which is the main reason no class consciousness has developed among digital laborers in different countries and in different situations. Developing class solidarity is a challenge because work processes are segmented into tasks. Different kinds of digital labor (local or remote, visible or invisible, paid or unpaid) make it hard to form a common subjectivity. That's the ultimate "disruption" brought about by tech giants. Not only work, but also class conflict is on the chopping block.

When platforms employ the term "class," they do so to depoliticize it, so as to redefine it without any reference to conflict, and in accordance with their worldview. Facebook filed a patent in February 2018 for "user classification." What is the best way to identify users, and how to classify them within a specific audience category to target them with information, advertisements and other triggers.[70] This invention tries to do something quite straightforward: classifying users, in the sense of "allocating them to a class," without knowing their income, which is usually the main metric. Unlike other services like banks, Facebook doesn't know the income of its users, so they have to use other data they've collected. Which explains why in this Facebook patent, demographics like age and gender, as well as educational attainment, matter more. The discriminating element is ultimately how the internet is used (times and duration of connection), and the nature of the equipment owned (types of smartphones, computers, tablets, and network speed). In short, this patent defines the social class of an individual in terms of their capacity to contribute on—and for—a platform.

The categories selected by Facebook redeploy the basic divisions between three social classes, retaining their usual names: "working class," "middle class," and "upper class." But the algorithm is not designed to analyze the social stratification of the users based on their data. The connection that Facebook makes between socioeconomic background and online contribution is meant to assist a particular machine learning system, which will predict to what class the user belongs. Three types of input are used: the profiles entered by the user on Facebook, their behavior (that is, their clicks) and a "training data store." The latter is a database made up of information that has been enriched, completed, annotated, described, and documented to provide a statistical model from the answers to the question "to which of these three social strata does a certain individual belong?" The patent also provides some details about who trained the data. In some cases, it was retrieved "from a global database of training data accessible [on the internet],"[71] most probably prepared by and fed to the system by microworkers. In other cases, Facebook users themselves not only provide their photos, profiles, and information, but also qualify them and make them compatible with the social classification algorithm. Their reactions, likes, and shares help to sort out the content and feed the algorithm. Thanks to their contributions, the users themselves develop this machine learning module, which "can periodically retrain the models using features based on updated training data."[72]

What is interesting about this patent is not so much the technological feat it represents or the vision of society it supports. There is something devious in the way it articulates *work, class,* and *machine learning.* Membership of a social class comes to be closely correlated with one's contribution to improving smart technologies. Workers train an algorithm that classifies them. The system thus identifies classes of users *in themselves* but prevents them from becoming aware of their existence as classes *for themselves.* The distinction between class "in" itself and "for" itself is another classical

Marxist concept. A class is defined either structurally (by the place of its members in a system of ownership) or politically and culturally (by the fact that its members are conscious of belonging to a political subject).[73] The opaque statistical treatment of the work performed by users on the platforms and by microworkers while feeding the database precisely prevents members of this class from recognizing themselves in it. Indeed, the platform would argue, should user-workers know their own classification or understand how their data contributes to it, they might consciously influence the algorithmic outcome.

This Facebook patent epitomizes the vectoralist view of society and technology, according to which automation will eliminate all distinctions between social classes. However, this elimination will not happen in keeping with Marxist principles. In the patent, there is no place for a progressive "class of the new," which emerges due to technological change, and which bears a historical mission of emancipating itself and the whole of humanity. Could the digital proletariat be described as *a class without a mission*, freed from a purpose?

Whereas other class subjects embodied a promise of freedom from servitude, digital laborers seem to be able to expect deliverance only in the form of a threat—that of being replaced by machines. Their destiny, the philosopher Roberto Ciccarelli argues, is governed by a renewed regime of *auctoramentum*. First formulated in Roman law, this bond is defined by the promise to sacrifice one's life to accomplish a task assigned by another, as in the relation between the gladiator and his master.[74] Arguably, the same self-sacrifice is reflected in the relationship between the platform and its users. "Those who are about to automate salute you" the valiant click workers proclaim, as they gaze toward their future of (self-)destruction at the hands of artificial intelligence. Since their existence as a class is only transitory—or so the argument goes—the development of a class consciousness, of a class for itself, is meaningless. From this perspective, the perspective of their class enemies, the digital proletariat has no need for self-reflection, for organization, or for collective action because it represents only the remnants of a world of human labor doomed to disappear.

This negative outlook is based on the expectation of complete automation in the future. Yet complete automation seems to be unattainable. And we may legitimately question the ideological role played by this scenario, which bolsters the digital proletariat's false consciousness.

Artificial Intelligence: A Not So Manifest Destiny?

As the political scientist Bruce Berman noted more than thirty years ago, the debate on artificial intelligence is also a political debate, where scientific advances are intertwined with myths that emanate from belief systems deeply rooted in the backgrounds of experts and associated interest groups.[75] A particularly striking example of the ideological dimension of scientific narrative can be found in a resonantly prophetic text on the future of artificial intelligence penned by Edward Feigenbaum, a pioneer in the field of expert systems, in the early 2000s: "I call computational intelligence the manifest destiny of computer science." He added that he held "no professional belief more strongly than this."[76] Contemplating some memories from his school years, he dwelled on the key themes of "how the West was won"—the US's expansionist ambitions—the "great visionaries like Thomas Jefferson," and the brave settlers, who left the majestic Appalachian Mountains to reach the "the far ocean at the continent's western edge." By analogy, he stressed, "computational Intelligence *is* the manifest destiny of computer science, the goal, the destination, the final frontier." AI will therefore stop at nothing to reach the level of human intelligence: "More than any other field of science, our computer science concepts and methods are central to the quest to unravel and understand one of the grandest mysteries of our existence, the nature of intelligence. Generations of computer scientists to come must be inspired by the challenges and grand challenges of this great quest."[77]

Feigenbaum's prophecy echoes the thoughts of a fellow pioneer in the field of artificial intelligence, Marvin Minsky, who declared, in 1961, that an era of intelligent machines was just around the corner. Thirty-five years later, the MIT professor was forced to admit "we really haven't progressed too far toward a truly intelligent machine. We have collections of dumb specialists in small domains; the true majesty of general intelligence still awaits our attack."[78]Proponents of this discipline can't believe that AI's destiny is less manifest than it seems. There's nothing superstitious or misleading about these statements: they reflect certain scientists' aspirations, not scientific facts.

After all, the ideology they express justifies the endeavors of computer engineers, scientists, and industrialists. Declaring that they are conducting research into simulating human intelligence is one way for technology producers to be at peace with the work they do. They can represent themselves not as a vectoralist class whose function is to manage click traffic around the globe or set up subcontracting chains that end in digital sweatshops, but as a noble elite that contributes to the progress of humanity by devoting itself to cutting-edge innovation. Such a heroic and prestigious vision of their task does not extend to the work of the individuals who produce, train, and qualify data. Yet the fantasy of complete automation, sustained by the designers of digital technologies, still requires workers to close the gap between a reality made up of computer solutions that are inevitably less efficient than expected, and the promise, repeatedly unfulfilled, of the advent of machines capable of simulating human cognition.[79] When technological limitations prevent the full automation of certain activities, human beings are recruited as users, customers, participants, or cheap labor to compensate for the shortcomings of the system.[80]

For human digital labor really does contribute to advances in artificial intelligence, in several ways. Human contributions greatly improve the *accuracy* of automated solutions. Drivers who correct the routes suggested by their GPS system, or social media users who influence the automatic content filtering algorithms perform a work of correction comparable to that of hundreds of thousands of microworkers who remedy the inaccuracies of automatic transcription software. Artificial intelligence also benefits from the *speed* of human work. When workers are involved in perfecting artificial intelligence, the latter develops analytical, decision-making and selection capacities more quickly than in other cases. Last, in order for the functions of an AI device to be applicable to a large number of situations, or even to be universally generalizable, huge numbers of human beings must provide huge volumes of examples for the automation processes to review a large range of cases—this is known as *scalability*.[81] The most important AI successes of recent years have relied primarily on the masses of training data that human contributors have tagged and prepared. Between the 2012 release of the convolutional neural network AlexNet (which proved its worth on ImageNet, a database of images prepared by human annotators)[82] and the 2016 consecration of Neural Machine Translation (a machine that is trained on corpora of corresponding examples in different languages), the computational input required to run these machines has relied on the vast amounts of data produced by microworking human users.[83] Recent efforts like ChatGPT, of course, have transcended all previous initiatives. Following the examples presented in the previous chapters, it should also be clear just how much digital labor has gone into this technology. The previously mentioned CommonCrawl, which is used to train large machine learning models, harvests data from Wikipedia, Reddit, blogs, and so on, to get content produced by the social media labor of millions of anonymous internet users. Add to that thousands of microworkers in Africa, Asia, and America busy annotating the content on online labor platforms and you'll get a sense of the digital labor involved.

Digital labor can thus be construed as a way of channeling the attention, social inclinations, and energy of individuals into a certain type of work, on a gigantic scale. It is work labeled as "useful" by AI developers—indeed, it is the *only* kind of useful work—insofar as it puts human beings into the service of machines[84] to further the ends of smart technologies. The perfect embodiment of this "useful work" is the *semblance* of AI—an "artificial

artificial intelligence"—that Amazon sells on Mechanical Turk, at the same time as it incorporates click working into the narrative of AI as a prophecy soon to be fulfilled.

No matter how many promotional releases tech companies issue, we're no closer to fulfilling this prophecy. Artificial intelligence is just a point on the horizon that keeps receding as we move toward it. Keeping hope alive and converting obstacles into trials of faith requires denying the invisible labor of the platform users, microtaskers, reviewers, and drivers, while whipping up the zeal of marketing professionals.

Over the centuries, the dream of governing human behavior through technology has taken many forms, and artificial intelligence is just one of them. It has been kept alive by a large number of research fields, among them cybernetics, information theory, game theory, systems analysis, operations research, and linear optimization.[85] If we step back a little, we can see that the rhetoric of smart systems has much in common with other discourses on the scientific organization of work, and it is intimately linked with policies that aim at transforming the production process. As in classical Taylorism, automation serves to entrench workforce hierarchies: the design, decision-making, and planning tasks carried out by computer engineers and scientists are overvalued, and even overhumanized,[86] while the remaining activities, starting with those performed by click workers, are reduced to the realization of dull and irrelevant tasks. Taskification and datafication occupy the same place as work steps sequencing and timing did for Taylorism. They are *not* major technical advances, but rather another turn of the screw in the capitalist division of labor, in which the workforce is envisaged as idle, careless, potentially recalcitrant, and needing to be kept under control.[87]

So platform capitalism is one step further along the path of the fragmentation of work and labor. Moreover, under the utopian banner of artificial intelligence and complete automation, it adds a demobilizing and intimidating message of workforce obsolescence: "Be taskified, and so, in the long run, work toward your own extinction." This is the real innovation introduced by contemporary smart technologies.

A Lesson Never Learned

The ambitious research programs of the pioneers of AI in the 1960s and 1970s have little in common with today's digital activities, as surveyed in this book. Far from the dreams of machines endowed with the traits of human beings—the ability to reason, to solve problems, to learn, to create, to play (hence the emphasis on chess)—AI has resorted to "narrow" or "weak" smart technologies. Recent emphasis on generative AI and its creative traits is a massive overestimation of the capabilities of machine learning models. Over time, users realize these tools aren't all they're cracked up to be. When one tries to make generative AI think outside the box and come up with new ideas or solutions, it becomes apparent that it is not designed to be creative and adopt multiple perspectives.[88] In fact, it tends to follow preexisting data and rules, even reproducing biases that are engrained in them.[89] Other times, AI's unauthorized imitation of protected or famous images raises doubts about its originality.[90] Also, the fact that two users with different competence levels in "prompt engineering," figurative arts, or literature can get drastically different results[91] shows just how much human agency is involved.

Other examples emerge in fields such as medicine and finance. Watson, IBM's star AI system, may well assist doctors or financial advisers by "understanding" questions formulated in natural language, but it proved to be limited to a simple keyword search in a corpus of data, which it matches with the terms contained in the question.[92] Its lack of sophistication disappointed the professions that subscribed to it, and they rapidly discovered that there is little substance behind IBM's marketing rhetoric.[93] Watson's methods are derived from the techniques of machine learning that, within the broader field of artificial intelligence, rely on a theory of statistical learning: in order to learn, software must be given enough data to be able to detect patterns in the information, typical distributions of certain observations, like trends and average values of certain quantities. This information is used as "lessons" that the machines should be able to reapply in due course.

Machine learning is likewise rooted in beliefs and axioms that are not only epistemic but also political. There is, first, the idea that a machine learns from its users, and second, that users must produce an increasing number of examples to allow the machine to acquire new notions. To convince themselves that machine learning is feasible, computer scientists have had to abandon the idea that a machine is a device programmed once and for all with certain knowledge; and users have had to get used to taking on the role of "teachers" helping the machine to develop new competences, rather than behaving as owners of it. These teachers do not need to be experts in the subjects taught: No need to be an artist to teach a Stable Diffusion model to produce an image in the style of Rembrandt, and no need to be a specialist in urban geography to improve a GPS. It is not even necessary to master the mathematical methods that organize the teaching—understanding a regression, a matrix, a vector, or an optimization is wholly superfluous. Given a sufficient number and variety of examples, anyone can be a machine learning teacher, as long as the software's double premise is respected—that learning is incremental, and that it takes place through a series of approximations. Once the system has started to learn from the user how to order information efficiently, its performance will keep on improving as long as it is fed examples, and the greater the input of examples, the better the learning machine's performance. Every search engine result, every "like" on Instagram, every order on Uber Eats produces new signals that are fed back into the learning cycle. As more and more data is injected into the system, the results get increasingly precise.[94]

The role of digital labor is most obvious in so-called supervised learning. The teaching method in this case is based on questions for which the answer is already known, such as recognizing a celebrity in a photo or transcribing a handwritten document with a word processor. This form of learning draws on two types of databases, the first to train the machine, the second to test it. The data needs to be "prepared"—sorted, cleaned, integrated, and identified by humans. The digital labor of platform users can be seen at every stage, since they are the ones who produce the content processed by the machines, who prepare the data, and who also check the solutions proposed.

However, supervised learning is only one type of machine learning. Reinforcement learning and unsupervised learning are two others. The former, used for example in ChatGPT, provides for the machine to learn on its own by receiving evaluations of its performance by users. Every thumb up and thumb down on OpenAI chatbots is part of a reward system.[95] Effective reinforcement learning models, however, are partly combined with supervised learning, and as such they need digital labor, because they rely on annotated examples with positive or negative rewards. MidJourney, for example, practices reinforcement learning on the Discord social media platform by asking users to do unpaid digital labor by ranking images in exchange for "free" GPU hours.[96] Computer scientist Jean-Gabriel Ganascia points out that, for both supervised and reinforcement learning, "the same problem arises: who annotates, rewards, or punishes? A teacher is necessary, which means that the machine is not totally autonomous in the sense that it does not spontaneously give itself its own rules, since it follows the lesson that humans teach it."[97]

There is also another type of machine learning, called "unsupervised," in which artificial intelligence finds the solution to its problem by processing unknown data that has not been annotated by humans. For instance, an AI tool could analyze millions of uncaptioned images of fruit salads and categorize all the fruits according to criteria that were not established beforehand. In any case, the lack of supervision is always conditional: People have to take millions of pictures of fruit salads, and then determine whether the results are useful or irrelevant for human purposes. Unsupervised learning has only patchy results at present and is still largely exploratory. The remarkable results of machine learning in recent years are mainly in the field of supervised or reinforcement learning, and thus depend on a massive mobilization of human digital labor. Artificial intelligence is currently able to synthesize what others know already, but it is incapable of developing completely new notions, languages, or formalisms. Unsupervised learning, which is supposed to be able to develop new ways of describing reality, has not yet reached the level of technological sophistication, autonomy, and consistency that would allow it to create new knowledge spontaneously.[98]

Similarly, the "algorithmic black box"[99] effect, often invoked to urge caution in the deployment of technological solutions based on deep learning or neural networks, requires an even more intensive contribution from microworkers and other providers of digital labor. Once a neural network, to name just one type of AI, has managed to differentiate the image of a tree from that of a car, scientists are unable to immediately determine what mathematical model made this possible. They must spend a large amount of time breaking down the steps in their machine's calculations. In some cases, the simplest way is to reverse the learning process: Starting with the result that the machine proposes (a certain series of images, or a certain ranking of results from a search engine), it is possible to test different paths that could have led to it, using pretrained networks.[100] So here again the learning is supervised, and therefore the need to resort to digital labor.

The problem is one of complexity. Whereas a traditional mathematical model may have a few dozen parameters, a neural network will have billions. GPT-4 reportedly has 1.7 trillion parameters, and this has led to an actual AI race. A new scientific consortium to create advanced AI is called, quite simply, the Trillion Parameter Consortium.[101]

Unsupervised learning provides results without necessarily explaining how the machine obtained them, nor does it give precise indications of their level of relevance and usability. An algorithm that divides all the goods in a certain catalog into n groups, without using categories established by producers or consumers (brand, price, etc.), may require several days of modeling to sort through thousands of guesses. Since the reason for a given distribution is not self-evident, it is also difficult to assess the quality of the results. Whereas, in the case of supervised learning, the relevance of the solutions (known in advance) can be assessed, in the case of unsupervised learning, the absence of a benchmark results in multiple and often incompatible validation methods. "Success" then becomes a question of evaluation and arbitration, played out between different stakeholders in the machine learning process.[102]

The Pipe Dream of Full Automation

As automation increasingly relies on machine learning, the quantity of data and digital labor it requires may also be its downfall. Those who believe that progress in AI depends on more powerful algorithms, more complex networks, and a greater number of parameters overlook a crucial ingredient, namely high-quality data. Machine learning engineers have several sources of data at their disposal, the most reliable being the huge volumes of information structured and qualified by users of applications and social platforms, or annotated by the hordes of microworkers all over the world. No other data source can reach this level. Producing synthetic data through generative AI is gaining traction in the scientific community. The process here involves using algorithms to generate information that mimics real-world data. They come with the promise of clean, deanonymized, unbiased datasets tailored to a company's or institution's needs. Most importantly, the datasets will purportedly be cheap, insofar as they do not require human annotation. Unfortunately, synthetic data is usually not reliable, resulting in potential errors in downstream tasks.[103] The quality of artificial data—for example, computer-generated faces to train a facial recognition system—is often disappointing; and, among other shortcomings, open-source databases risk being overadjusted, memorizing overused and overprecise training data, such that the machine proves incapable of generalizing models to new situations.

Since machine learning only progresses if human data production increases, no truly autonomous machine can emerge that would dispense with human-treated input altogether. So if computing really does have a manifest destiny in the form of perfected artificial intelligence, *it will also be irrevocably tied to digital labor*. The Sisyphean task of machine learning has the following form: the more machine learning is applied to the automation of human work, the more it requires human work to feed it high-quality data.

In a text published by the American Association for Artificial Intelligence (renamed the Association for the Advancement of Artificial Intelligence), another of the founding fathers of the field, Nils Nilsson, suggested replacing the Turing test with the "employment test." The Turing test, as is well known, is a way of evaluating the efficiency of a computer, based on its ability to simulate human interaction. When a user is unable to tell whether the entity they are interacting with is a human or a machine, the machine is declared to have reached "human-level intelligence." For decades experts have argued that this method was never a good measure of machine intelligence.[104] Nilsson's employment test, therefore, stipulates that "AI programs must be able to perform the jobs ordinarily performed by humans." As such, Nilsson continues, "progress toward human-level AI could then be measured by the fraction of these jobs that can be acceptably performed by machines." The criterion for success in AI development is therefore *its capacity to replace paid and visible jobs, not invisible digital labor*.

Nilsson makes it clear that reducing the cost of labor is the ultimate goal of the endeavor: "Systems with true human-level intelligence should be able to perform the tasks for which humans get paid."[105] Inconspicuous work, for which humans tend not to get paid, is not even considered: it lies outside the scope of the test.

Yet it is precisely this relegation of workers to poorly paid, unpaid, and invisibilized tasks of data production and supervision of machine learning that is problematic in the age of automation and platformization; and it is precisely the disciplining threat of the "Great Technological Replacement" that keeps the digital laborers on their toes. As we have seen, digital labor outcomes are sold to companies as though the results were automated, but the individuals who perform the tasks day in day out are not recognized as workers. Rather, they are treated as amateurs or hired hands who perform an activity that does not really contribute to the technological innovations they help develop, unlike visible, lauded specialists and scientists.

Furthermore, the need for digital labor is considered to be temporary only—machines need it today in order to learn to do without it tomorrow. In machine learning, relying on past data introduces a major challenge to the prospect of achieving full automation. Because algorithms learn from historical data, they struggle with entirely new situations. It is therefore always necessary to update the data, to adapt the output, and to align both with current social norms—and human workers participate in these processes. In contexts that change, as all contexts do, machines may not be able to make complete and autonomous decisions.

Yet this goal of automation, forever deferred, still has immediate subjective effect on platform workers. They are forced to await the deus ex machina that will put an end to their condition as a "digital proletariat," but they are discouraged from creating the tools by which they could transform their situation from within, and, if possible, to their own advantage. The messianic expectation that automation will abolish human work is never fulfilled, and instead the great providential design of AI proves to consist of disconnected sets of routines governed by no particular grand plan.

The success of digital labor has thus been to transform work into a multiplicity of disparate tasks, but it has failed to create an autonomous intelligent system. This is due to a *logical impossibility*. Processes simulating human intelligence are quicker and more precise, and they can be applied to more areas, but they will inevitably lag behind human cognitive faculties, as in the paradox of Achilles and the tortoise. While, at any given moment, AI is rushing to attain the level of human intelligence, human intelligence never stops moving, however slowly, so the distance between them can never be closed. For the human intelligence that machines would ideally reproduce does not stand still. It adapts to new practices, and so any artificial replica must be constantly updated—a process in which, now as in the future, humans will be always involved.

So we are all—the hundreds of millions of on-demand and microworkers, the billions of users providing paid and unpaid social media labor— embarked on a long, a very long, career as AI trainers. If we recall the economist Ernest Mandel's devastating rebuttal of automation penned some ten years before the web existed, and almost twenty before Facebook, we might even say it is interminable:

> Under capitalism, full automation, the development of robotism on a wide scale, is impossible, because it would imply the disappearance of commodity production, of the market economy, of money, of capital, and of profits. [. . .] So what is the most likely variant under capitalism is precisely the long duration of the present depression, with only the development of partial automation and marginal robotization; both accompanied . . . by large-scale pressure to extract more and more surplus value from a number of productive work days and workers tending to stagnate and decline slowly, i.e., growing pressure to overexploitation of the working class (lowering of real wages and social security payments), to weaken or destroy the free organized labor movement and to undermine democratic freedoms and human rights.[106]

CONCLUSION

WHAT IS TO BE DONE?

Despite promises of emancipation and social connection, today's technologies leave digital laborers exploited and alienated. Cogs in the machines of their own threatened extinction, these workers perform atomized tasks that sometimes aren't seen as work at all. To counter this reality, we need a concerted political effort to recognize digital labor, and to mobilize and organize digital workers. As a society, we must acknowledge that a human workforce powers automation—it provides real work that produces real value.

According to philosopher Axel Honneth, recognition occupies a central place in social and subjective life[1]. Developing AI solutions depends on users and workers on digital platforms, who seek recognition socially and politically. The struggle for recognition, equality, and protection is real and is crucial for fostering an equitable wealth distribution within broader social structures.

Efforts to achieve this recognition take two tacks. The first seeks to extend to digital workers the labor protections of recognized employment; the second aims to rethink the relationships between workers and the infrastructures of data collection and processing. Still in its infancy, this second approach focuses on a governance of the commons, with a view to imagining new ways of sharing resources and reviving the original political impetus behind platforms' creation.

Bringing Digital Labor Back into the Fold

To combat the unlimited exploitation of digital workers, the first strategy is to extend to digital workers the rights and protections that employees enjoy in traditionally recognized employment. The viewpoint expressed here is far from unanimous. While the traditional left is open to it, more radical social movements see it as particularly dangerous because it could encourage wage slavery and undermine workers' self-management.

There are important differences between formal employment and the activities of qualification, monetization, and automation that workers perform on platforms. There are also, as we have seen, several common features, in particular subordination (either contractual or technical), and surveillance (whether top-down or participatory). Generally speaking, if a relationship of dependence exists between platforms and laborers, legal recognition of digital laborers' work can be legitimately demanded. Despite the platforms' denials, workers must prove their position of subordination in order to get fair working conditions and compensation.

The political movements that reaffirm the appeal of the historical paradigm of *protected subordinate employment*, want digital labor to be recognized as a form of work with the same rights as traditional salaried work. Individuals facing life's vicissitudes need protection against the social risk that comes from workers' accepting a form of subordination. Since digital labor is already (technically) subordinate, there are millions of platform workers worldwide who are seeking this recognition by becoming salaried employees. However, by definition, subordination to a formal employer's economic interests is limited to the realization of social objectives, and does not carry the stigma of indignity of other forms of labor, past and present.

From this perspective, extending the social protections of salaried work to self-employed and freelance platform workers would be a first step in the right direction. Other claims go further, calling for digital laborers to be fully reclassified as employees of the platforms, implying also that a real employment contract would replace the TOS. On-demand drivers, AOL moderators, and Google reCAPTCHA users have all asked, at different times and with varying outcomes, to become officially recognized employees of the services and applications they use and to which they have contributed.[2]

This is how transportation services and delivery workers in Europe, the United States, and Asia succeeded in obtaining compensation from the owners of on-demand applications that claimed to be mere intermediaries. The package included minimum hours, compensation for on-call time, and cover for illness and occupational accidents.[3] The importance of reclassifying service contracts as open-ended employment contracts has been highlighted by several recent court rulings and legislative initiatives, some of them already mentioned in previous chapters. Across Europe, 371 initiatives and court cases related to platform economy activities have been reported. They concern ride-hailing platforms like Uber and food delivery apps like Glovo, Just Eat, and Foodinho. Many platforms are fined, most are required to hire as formal employees, or under the status of "workers" or "coordinated and continuous workers," which gives them limited rights.[4]

All of these initiatives revolve around the legal presumption of employment that was the guiding principle of the directive on platform work that European countries adopted in March 2024, in spite of strong opposition from platform lobbyists and promarket governments.[5] In compliance with the directive, platforms now have the responsibility of proving that workers are actually independent rather than subordinate. Additionally, worker representatives have the right to monitor the algorithmic processes governing their activity, thus being informed of the nature of the technical subordination to which they are subject.[6]

Previous attempts to move self-employed workers into waged and protected labor have been contested before and after they are enacted as laws. The Spanish "riders' law," for instance, turned freelance riders into staff back in 2021.[7] But companies have reacted by shutting down operations or refusing to comply. There have been mixed results with other collective bargaining agreements between companies and unions. For example, Danish platform Hilfr has created a dual workforce consisting of self-employed workers (FreelanceHilfrs) and formally employed workers (SuperHilfrs).[8] But it's been undermined by the Danish Competition and Consumer Authority.[9]

A crucial issue needs to be underscored here: all these cases of successful reclassification have to do with location-based labor platforms. But what about digital laborers on microwork platforms? For them, the way forward has been litigation and class actions, which have resulted in better pay or protection.[10] In the rare cases that microworkers have sought reclassification as workers, results have been mixed. As mentioned in the preface, the Brazilian data annotation company Ixia, creator of the InteliChat business solution, was ordered by a court to reclassify its users as employees. The company had recruited thousands of microworkers to enhance its AI system, not hiring them as employees. The São Paulo labor court successfully regulated the situation of these workers, securing 130 million reais ($26.6 million) in compensation for collective damages. In France, a similar case didn't end well for the workers.[11] A 2020 decision by the Douai Court of Appeal, which directed the microworking app Clic and Walk to hire its alleged 700,000 users as employees, was overturned by the Criminal Chamber of the French Supreme Court in 2022.[12]

Bringing digital labor back into the fold of formal employment is not an easy task. In this regard, unions have an important role to play. This attempt has led to original experiments in creating "platforms to protect platform workers," such as the German Fair Crowd Work, founded under the auspices of the German union IG Metall, where users can rate the platform's terms of service and denounce possible abuses by e-commerce platforms, mobile apps, requesters.[13] Since the end of the 2000s, social media platforms have also witnessed "user strikes"[14] and even the occasional creation of consumer unions[15] to protest against unfair TOS, and opaque methods of data monetization by third-party companies.

From the mid-2010s, trade union involvement has significantly helped organize initiatives to defend the rights of precarious and nonstandard workers. In the United States, unions in the transport sector, such as the International Brotherhood of Teamsters, are a good example.[16] In the United Kingdom, on-demand workers have made progress by joining the Independent Workers Union of Great Britain (IWGB).[17] And Spain's "rider's law" was the result of a tripartite collective bargaining agreement where trade union CCOO (Workers' Commission) and UGT (General Workers' Confederation) negotiated with the Spanish government and two employer confederations.[18]

All these mobilizations have a common objective, but it is ultimately quite limited. They do not seek to challenge the power of the platforms, but instead the platforms' denial of their role as employers with digital laborers as their subordinates. If the relationship is recognized, they hope, a regular employment relationship can be established, with improved working conditions and remuneration. This strategy is most effective for conspicuous digital labor, that is, activities closest to the traditional conception of work and employment. The upshot: this strategy often fails those who perform digital labor that involves an increasing number of hidden tasks. Unfortunately, these inconspicuous tasks constitute the bulk of current automation-oriented activities. To what extent will trade unions make visible this work, concealed in the interstices of our daily digital lives and behind the interfaces of our mobile apps?

This first approach doesn't deal with the global dimension of digital labor either. Outsourcing is increasingly moving work abroad in the context of platformization. Client companies and requesters can access massive numbers of users, who can perform remote tasks on platforms, and benefit from cheap, vulnerable workers. There's little coordination between trade unions in different countries, especially between the Global North and the Global South. One of these examples is the collaboration between UK activist group Foxglove and Kenya-based ACMU (African Content Moderator Union), formed after a lawsuit for unfair dismissal filed by 184 content moderators from Sama, a company that provides data annotation services to Meta, YouTube, TikTok, and OpenAI.[19]

It is not enough to anchor social protection to a single country, for operations that are easily offshored, such as moderation, data annotation, or microwork; this only displaces the problem geographically. The digital labor of on-demand applications, while ostensibly confined to a city or region, may also be relocated to other countries with lower labor costs and fewer legal protections. While on-demand drivers in high-income countries might receive better social protection as long as they are formally employed, digital laborers who coordinate transport logistics or offer data processing will not, particularly if they live in informal economies in lower-income countries.

This situation has prompted NGOs, workers' organizations, international bodies, and researchers to work together on certification principles for good practice on pay and working conditions for digital labor on platforms.[20] One of the first attempts so far at the international level is the "Frankfurt Declaration," published in 2016 by a network of European and North American trade unions. The objectives were, among others, to harmonize minimum wages, to decide on jurisdiction in the case of disputes, and to review social protection provisions.[21] Lawyers can help in other cases by leveraging existing regulations on social responsibility and employer due diligence. Corporate leaders need to consider the potential negative effects of AI solutions developed while partnering with digital labor platforms. It is not only a matter of environmental costs, although they are certainly important, but also of social costs. Human rights may suffer when outsourcing is done by exploitative platforms. AI solutions built on human labor require active policies to identify and reduce the risks associated with supply chains. So far, two initiatives have emerged to chart these supply chains and evaluate the due diligence practices of European companies. The first is a study on the impact of the German and EU Supply Chain Act and Due Diligence Provisions, implemented in 2023 by the German Federal Ministry for Economic Cooperation and Development. Another, conducted by the NGOs Intérêt à Agir, examines whether French companies comply with human and workers' rights rules throughout their data supply chains and in low-income countries where they operate through platforms.[22]

There are other risks: social movements are also seeing their ideas and methods co-opted by promarket forces. For instance, the liberal project to generalize a system of microroyalties, which is already operational for certain services, stipulates that in return platforms may exploit any contribution made by their users.[23] Others recommend organizing markets to sell personal data piecemeal.[24] The advocates of these measures, who are often from libertarian backgrounds, tend to ape the language of social conflict, and characterize themselves as oppositional forces to the digital oligopolies. They wield a romantic vocabulary of "powerful strikes" and "data labor unions,"[25] but their measures ultimately aim at generalizing the piecework system already current in microwork, in click farms, and on certain applications that monetize information from social media users' profiles. Their noble proclamations look like pure cynicism. Far from uniting digital workers around the defense of their rights, they individualize the notion of recognition and once again deny the idea that digital labor is real work.

Another Platformization Is Possible

The corruption of platforms' emancipatory message is a perfect illustration of how capitalism manages to convert critiques of labor relations into profit. The digital economy exploits individuals' aspirations to autonomy and their freely produced content and contributions to extract value. Is it still possible, then, to turn platformization against itself? This is a second strategy of resistance. It brings us back to the history of the concept of the platform, outlined in chapter 2, whose philosophical sources go back to the English revolutionary Commonwealth. A political program, in the eyes of the seventeenth-century revolutionaries, "made into a platform," aimed at abolishing dependent labor and private property, while advocating a shared governance of the commons.

Born in the mid-2010s, the "platform cooperativism movement" sought to reclaim the original political philosophy of platforms.[26] This international movement argues for collective ownership of the digital means of production, and for a cooperative rather than collaborative economy, free from monopolies, exploitation, and surveillance.[27] It wants to create a people-centered internet that would promote principles of social justice, a social economy based on solidarity, and ecological sustainability. Its idea of work and, more specifically, of digital labor draws on the substantial tradition of the mutualist movement.[28]

The book by the same name published by media studies scholar Trebor Scholz in 2016 lays out the principles of platform cooperativism. These principles are mostly consistent with the strategies of labor unions that work to extend the status of employment to digital labor. Cooperative platforms must provide their members with decent wages and job security, legal protection, portable health and welfare benefits, and the right to disconnect. In addition, Scholz has added the collective ownership of platforms by those who generate most of their value, and the need for workers to be involved in programming and managing their production flows, in order to establish a regime of "co-determined labor."[29]

Several cooperative platforms proved the viability of this model: On-demand worker cooperatives were formed, such as US app Coopify, a mutualist alternative to TaskRabbit;[30] municipal accommodation services such as AllBnB, a local substitute for Airbnb that pays a dividend to residents from rental profits;[31] the open-source application CoopCycle, which delivery couriers use instead of Foodora;[32] the co-owned content producer collectives Stocksy United and Resonate,[33] royalty-free versions of Flickr and Spotify, and more.

There is one critical point to be made right off the bat: many of these platform cooperatives no longer exist, having never really gotten off the ground. In his research, Rafael Grohmann coined the term "dead platform co-ops" to describe a phenomenon in which platforms struggle to scale up, leading to their failure. In a study focused on the cooperatives in Latin America and Europe, he analyzed platforms from the early stages of the pandemic, before they went out of business. There is something to be learned from understanding why they fail. Most of the time, it's because people can't build a collective organizational culture, because resources and business plans are lacking, worker turnover is high, no consumer base exists, technology is expensive, and organizing a large workforce is hard.[34]

Despite this movement's heterogeneity, it retains a twofold objective throughout: to reform platform capitalism; and to contribute to an ethically responsible platformization of traditional cooperatives. This twin purpose is a strength, but also a potential weakness; that is, if the movement does not manage to turn the tide in the hegemony of capitalist platforms, and provides merely an alternative running alongside them.[35] Faced with the expansionist approach of its opponents, the movement could well remain a niche phenomenon, or even be co-opted by big tech,[36] as has already occurred in the case of the Platform Cooperativism Consortium, which received a direct grant from one of the tech giants.[37]

Platform cooperativism is for protecting and conserving "global commons." A common global resource like the atmosphere, oceans, and outer space can be accessed and used by anyone. The germane notion of "digital commons" includes open-source software, creative commons-licensed content, and collaborative online platforms that foster the creation and distribution of digital assets and information.[38] However, the political potential of this notion, and the associated collective actions, are not followed up consistently by the movement. There has been no concrete manifestation of an integration of a commons economy with the anti-capitalist spirit of alternative platforms. To facilitate this, individual activists and political entities are starting to think about the possible consequences of sharing the most fundamental resource of the platform economy: data. Despite being generally seen as individual contributions (starting with the widely discussed "personal" data), or as information that benefits proprietary platforms, data can be seen as resources meant for sharing (referred to as nonrivalrous goods in economics), which ideally makes them the substance of an "informational commons."[39]

Digital Labor "In Commons"

In order for digital workers to gain recognition through data sharing, they take on the role as bearers of "bundles of rights" that link together those wanting to access the commons.[40] Unlike today's rights, which are built on excluding others, enjoyment of the commons should be adapted to the situation of each group of users, who will have differentiated prerogatives, obligations, and allocated resources. This requires that platform users develop a shared governance of data considered as an economic asset. Data may then be redistributed in the form of rights or resources. Scholarship on "social protection relating to personal data," in the context of the generalization of platform work, is a good illustration of this approach.[41]

Data is created and transformed by users online. Platforms allow them to express themselves and produce economic value at the same time. Preserving the integrity of users' information on platforms, although often framed as protecting privacy or consumer rights, ultimately ensures safeguarding their rights as data workers. Users must be able to benefit from the wealth they help accumulate. Their rights do not stem from private ownership, as in the example of libertarians who seek to sell data on the aforementioned piecemeal basis. They do not stem from co-ownership of means of production either, as in platform cooperativism. Instead, they derive from a protective system based on resource allocation.

Rights, as I am arguing for them, would not be based on private property (as in the case of the privatizing appropriation of data with a view to its commodification), nor on co-ownership of the means of production (as in the approach of platform cooperativism), but on a system of protection based on an allocation of resources. The idea is similar to Robert Castel's concept of "social property": data can be viewed as a set of collective goods made available to non-owners.[42] In societies where formal employment is prevalent, this set of goods already exists. It includes all forms of social protection, social housing, health services, and public infrastructures and services. Despite the fact that they are not owned by individuals, these goods and services may be accessed by those who are entitled to them by other rights (citizenship, tax payments, eligibility to certain benefits, etc.). Today, these can be enriched with new rights that derive from the production and sharing of data. Instead of promoting political–economic regimes where data is a capital asset owned individually, this perspective is compatible with more radical forms of "digital socialism."[43]

But just thinking digital assets can be socially owned doesn't mean they will be. To make this possible, workplace democracy is key, as is insisting on popular oversight of technological decisions. The convergence between the interests of personal data producers and collective interests needs to be defended through organized action. If data falls into the public domain, this could preclude the lawful owners of said data from licensing it or profiting from its property. By using this legal framework, individuals and groups of workers could prevent platforms from commodifying the eminently social wealth that digital labor creates. Such measures would ensure the recognition of work by removing the barriers and enclosures around data and information on the internet, that were created by platform capitalism.

Among the tools available to achieve this are class action lawsuits by users against platforms to regain control of the contents, services, and information that platforms capture.[44] However, this legal instrument will be toothless if the negotiating power of workers is not strengthened. A first step toward this would be to treat platforms' terms of service not as covert employment contracts, but as collective agreements between platform owners, user groups, and institutional actors, whose terms result from progressive dialectical adjustments.[45]

Each platform must negotiate its own collective agreement, in order to incorporate fair labor standards into the algorithms themselves.[46] The FairTube campaign, which began in 2019, was an attempt to achieve this goal; with help from the union IG Metall, it led to the creation of a YouTube creators' union in Germany.[47] Agreements can be extended to an entire sector, for example to microworking or music streaming platforms, conforming with the so-called principle of favor. Throughout Europe and South America, this principle is found in many legal systems. The principle means that it is not possible for corporations and workers to make an exception to labor law if the exception hurts employees. The extension of this principle to the context of digital platforms would allow for the standardization of TOS, a standardization that would have to adopt terms more respectful of users' rights. The extortion of users' consent to the free use of their data would no longer be possible.[48]

Other opportunities for collective action exist, particularly when large platforms are used by residents of large cities around the world. Platforms such as Uber and Airbnb have had some negative impacts on urban settings where they operate in terms of traffic, pollution, and gentrification. But they are the center of conflicts over the use of Uber and Airbnb data, too. So these platforms have attempted to share their data with city authorities. After all, this information itself is produced by citizens. A platform that captures this data can distribute it to the local authorities, which can then use it to improve town infrastructure, develop new policies, and improve services.[49] At the moment, these actions are nothing more than facades. Private profit drives platforms' commitments to transparency, data sharing, and redistribution of informational assets.

Imagine, however, if these initiatives were implemented in a spirit of genuine data sharing. There is nothing stopping us from considering a more radical departure, such as collectivizing data. Data could become the *direct, indivisible,* and *inalienable collective property* of its users. Instead of data generated by individuals and captured and commodified by corporations, collectivized data empowers users by giving them ownership rights over the valuable information they produce through their online interactions and activities. It also subjects personally identifiable data to an important restriction: As a collective good, it cannot be enjoyed exclusively. Think about the big proprietary databases we use today to train big machine learning models. They are now owned by a few big platforms (Microsoft, Meta, Alphabet) and a few top-level universities (Stanford, Princeton, Tsinghua).[50] Collectively, they could be made available for everyone to use, but cannot be turned into closed systems.

By allowing individuals to control how their data is used, shared, and sometimes sold to third parties, this model gives users more control and agency. Additionally, it prevents data from being privatized and enclosed. Like a community lake that can only be fished if it's protected from mindless predation, this data is under the watchful eye of a shared governance system that fosters open access and knowledge exchange.

As a collective asset, data should be shared equitably among users, not concentrated in the hands of a few tech giants. Collectivizing it isn't just about economic justice, it's about reimagining our relationship with digital platforms.

This outlook has been inspired by more people than Nobel Prize–winning economist Elinor Ostrom and other theorists of the commons. It actually comes from a long and often marginalized philosophical tradition theorized by philosophers of the commons in the Global South.[51] The idea of the environment being intertwined with human societies can be an innovative way to combine sustainability, digital autonomy, and the organization of workers.

Media scholar Paola Ricaurte Quijano employs the symbol "La Pachamama" (the Earth Mother revered by Indigenous people of the Andes) to illustrate how digital decentralization actually has environmental implications: "Forms of labor organization coming from own initiatives oriented to rights defense constitute an alternative to the platform economy. From the digital world, is it possible to return to the land? The impulse to alternative forms of labor organization that generate decent working conditions from regional proposals based on distributed infrastructures offers an opportunity to deconcentrate our center of life from urban metropolises."[52]

Taxing data is another solution, more in line with the institutional frameworks of northern countries. The report *The Taxation of the Digital Sector*, issued in 2012 by the French Minister of Economy and Finance, did not simply represent bureaucratic fluff. In fact, it was a thought experiment that asked a fundamental question: how does a platform get established? Not "where are its offices and factories located?" and not "where are its headquarters?" For tax purposes, a permanent establishment is the place where a company's business is run, and where taxes can be collected.

For platforms, it might be more appropriate to ask: where is the digital workplace of each platform's user? According to the authors, this is where users perform "free labor" for the platforms.[53] It looks like the authors of this report only considered social media contributors. As we get closer to the end of this book, we know that not all user labor is free, but that a large portion of it is unpaid. Still, it all creates value for the platforms and establishes a tax base. Therefore, a South Korean app that is used in Taiwan can be taxed by the Taiwanese government. In the same way, all the tax collected from Chinese and American platforms, whose billions of workers and users are scattered around the world, could be tapped to enact ambitious redistributive policies—even global redistribution policies between the North and South.[54]

The Sword of Damocles

Friday, July 20, 2018, 11 a.m., and I'm at the Ministry of Mining and Metallurgy in La Paz, Bolivia. I have an appointment in a small and dimly lit office with former minister Luis Alberto Echazú, now retired. Sporting a white mustache and glasses, he greets me as I enter, rising from his large mahogany desk. We sit in two small leather armchairs.

His assistant brings in a couple of brochures about the Salar de Uyuni, which holds the largest lithium reserves in the country. The topic might not be the most exciting, but Uyuni is something out of a dream. From the capital, it's a day's drive by Jeep on rough roads to the world's biggest salt lake, in southern Bolivia. Rainwater covers the surface of the salar in February, causing it to reflect the sky like a mirror. During summer, the ground is cracked like an old elephant's skin, a maze of clumps of salt. There are mainly tourists there, so far. There's a small village, an airport, and a hotel entirely carved out of salt. However, the Salar now boasts another attraction: a lithium factory.

The former minister explains: "You can't imagine how coveted that lake is. By foreign companies. That's why our government nationalized it. Common resources can't be divided into parcels. Even if we divided it into four and authorized four different companies, they would still be able to access the lithium beneath as they drill right at the border with each other."

Foreign mining giants—German, Chinese, Japanese, American—all want a share of Uyuni to ensure the supply of lithium for tomorrow's artificial intelligence. "There are tensions," he says, "not just externally. We're developing a value distribution system for the lithium factory. There are strong miner unions. And peasant unions with long-standing rights. And of course the government. We have to pay these three every time we sell lithium."

My official meeting ends with a handshake from the former minister. "You should go to the Salar," he adds. "I can arrange a visit to the factory. Be sure to wear boots. The company will provide a hard hat for you."

As I exit the building, the Avenida Arce is filled with cars. I catch a "micro," a city bus driven by a man. His ten-year-old son sells tickets. Then I board the teleférico, the suspended cable car system, and I ascend to the final stop. I'm at an altitude of 13,000 feet, El Alto, the twin city looming over La Paz, sharp-edged and rusty, like a sword of Damocles.

Diego is waiting for me at a nearby restaurant. A translator in his thirties, he knows a place where we can grab a bite. During lunch, he tells me he uses "an app called Upwork." "It used to be that translations paid well," he recalls. "[I] used to translate handbooks, user manuals . . . but now there are only small tasks. Maybe you won't translate the whole document, just a few pages." Taking out his smartphone, he says, "Look at this. It's a task: taking hundreds of images with English text, in the street, at the supermarket. . . . If you accept the task, you have to go out, take a bunch of photos, come back home, upload them onto the computer, and then write next to each of them what is written. It's insane; it takes an insane amount of time to complete this task, and it's paid peanuts. And, above all, it's not even a translator's job. I should be there with a dictionary, focusing on a text. I shouldn't go for strolls in the city, look for billboards or shop signs, then return home and become a transcriber. . . . Anyway, the situation is getting worse. Always worse. I sense something coming."

I feel it, too. I got a sore throat in El Alto; the wind was blowing and the sky suddenly darkened with black clouds on the horizon. I may have caught a cold. Or perhaps it's the political situation that creates this strange tension. There are rumors about a coup. It might be soon, maybe before the presidential election. Or it could be this conversation with Diego, the feeling that this life is slipping away, shrinking, disappearing.

Stepping out of the restaurant, I take a deep breath. My throat is getting worse. I might be incubating something. To get back to my apartment in La Paz, I have a vast descent in front of me.

The Gordian Knot of Remuneration

Neither trade unions nor the free market provide a satisfactory solution to the problem of payment for digital labor. The former overlooks the problem of invisible platform work; the latter contributes to the under-valuation of users' contributions by advocating the resale of data on a piecemeal basis.[55] I believe we must take a new approach entirely and instead favor a *universal digital income.*

Before I begin, I need to point out something important. My support for universal basic income has nothing to do with my belief that our society is witnessing the end of work. As I tried to show, work isn't ending, it is growing. My stance isn't influenced, either, by Silicon Valley libertarians who argue that basic income should compensate for job loss caused by automation. The way the epigones of Musk or Zuckerberg frame this policy is very different from mine. I don't see robots replacing workers; I see other, lower-paid workers replacing them. So universal basic income is not a way to compensate for unemployment, rather, it's an essential tool to fight exploitation.

The debates that have arisen around the idea of providing all members of a community with a basic income, and the divergent interpretations of this idea, as reflected in its competing names (universal income, participatory income, dignity income, life wage), have, at the same time, popularized and weakened popular support for the concept. Public misunderstanding has hindered any common political action around it. The confusion can also be caused by other factors, such as the difficulty of generalizing the results of universal basic income experiments. Initially concentrated in North America, they are now being conducted in several European countries, Brazil, and South Korea as well. Experiments can go wrong or be counterproductive, like in Finland and Italy. Finland experimented with giving UBI to 2000 unemployed people in 2017–2019. Results were inconclusive.[56] A major goal of the program was to reignite employment, which this measure may not be the right fit for. Universal basic income has been co-opted by a right-wing government in Italy, which cynically used it to stir up discontent about social policies.[57] Misunderstandings don't just happen with words.

Basic income has mostly been considered a remedial measure, often paired with other ineffective policies such as taxes on robots to discourage technological advancements that destroy jobs.[58] But this point of view means that UBI does not address the sources of the employment insecurity and demobilization of workers. It simply confines itself to tempering their most extreme effects. Another, complementary, approach is to consider this income as a sort of annuity. In a world where automation will allegedly supplant human labor, it represents the value produced by the automated machines.[59] This approach is also flawed, as it does not register the fact that, as we have seen, automation is, and will invariably remain, driven by human labor.

Unlike these half-measures, a universal *digital* income would be an individual's main revenue, and not a supplement to sums received from other sources. It would therefore not replace or compete with welfare assistance and *would be paid regardless of other social benefits*. The economists Jean-Marie Monnier and Carlo Vercellone[60] have defended the principle of this income in the name of the structural exploitation of information within today's "cognitive capitalism," and they have examined how it could be financed. Under platform capitalism, it is essential to extend the concept of productive work to inconspicuous digital activities (including the training of artificial intelligence) to obtain full recognition of users' contribution to the digital economy's value chain.

Universal digital income could be financed partly by the tax on platforms mentioned above. Governments can earn revenues by taxing platforms and AI vendors based on data generated by their users' digital labor, which is used for sales, online advertising, or machine learning model training. Citizens will be able to receive cash payments derived from corporate tax revenues. As a complementary method, the sum could be paid in the form of primary goods or services (on the model of social property),[61] but on condition that it is conceived as the "institution of a common" rather than as a state provision. Here, the idea refers to the collective data sharing initiative discussed above, where data is governed by a communal institution rather than a government. Disbursement of universal digital income by authorities cannot be based on tracking and surveillance of every human action online, nor should it be proportional to the quantity of data and information a platform extracts from a user. On the contrary, platforms would return what workers produced back to the commons. Jean-Marie Monnier and Carlo Vercellone argue that, "from this perspective, in the spirit of the mutualist tradition . . . the resources assembled to [finance this system] could be put into a common fund," which would be directly managed by the users, in line with the rules of democratic representation.[62]

As a matter of fact, this project has the feel of a moonshot. But why should moonshots be the exclusive domain of big tech platforms? The implications of this policy could be profound. The prospect of having to pay a collective contribution to the citizens of a state, would discourage platforms from subjugating users, disincentivize them from commodifying labor, and stop them from enclosing data. If platforms had to account and pay for digital labor, including inconspicuous and invisible labor, their business models would fall apart. Value capture would become unprofitable or even nonviable.

Moreover, if digital platforms had to move to non-predatory business models, the specter of automation would no longer be necessary to discipline the workforce. In tune with their original political theology, platforms could once again refocus on their threefold mission. Eliminate property, emancipate labor, and establish commons—but with a revised formulation: to substitute private property for social property, to replace subjugated labor with self-determination, and to trade data and labor enclosures for truly common infrastructures.

ACKNOWLEDGMENTS

This book is the result of almost ten years of research. Without the colleagues and friends I encountered along the way, it would not have been possible. There are many things for which I'm thankful. My research is communal in nature, and these ideas were developed most significantly as a result of a number of intellectual exchanges and stimulations. The list of all those who helped me will inevitably be incomplete. But I'll try my best.

The scholarship of Lilly Irani (UC San Diego), Payal Arora (Erasmus University Rotterdam), Trebor Scholz (The New School, New York City), Mary L. Gray (Microsoft Research), and Frank Pasquale (Cornell University) inspired this book. As a result of our conversations, I developed my own perspective on the various topics discussed in the previous chapters.

The ideas and research presented in this text were debated and refined in my seminar at the School for Advanced Studies in the Social Sciences (EHESS, Paris), "Studying Digital Cultures," which I facilitated from 2007 to 2021. By drawing out new themes and perspectives and working together on radically deconstructive approaches to technologies, the speakers and participants greatly enriched my understanding.

I am grateful for the opportunity to work with colleagues at the NEXA Center for Internet & Society, especially Juan Carlos De Martin, and those at the Weizenbaum Institute for the Networked Society, Berlin, in particular Milagros Miceli. During the last two years, it has been a pleasure to be a visiting professor at the Center for Artificial Intelligence and Machine Learning at TU Wien, led by Stefan Woltran, and at the Centre Internet et Société in Paris, co-founded by Melanie Dulong de Rosnay and Francesca Musiani.

Since the mid-2010s, my work has been closely associated with colleagues with whom I launched INDL, the International Network on Digital Labor, as well as with the students and researchers of my DiPLab team. Through these two initiatives, I have met and collaborated with extraordinary people. In no particular order, I'd like to thank them all: Julian

Posada (Yale University), Mark Graham (Oxford Internet Institute), Manolis Patiniotis (NKUA), Karen Gregory (University of Edinburgh), Rafael Grohmann (University of Toronto), Sarrah Kassem (University of Tübingen), Iraklis Vogiatsis (NKUA), Uma Rani (ILO), Gina Neff (Magdalene College, Cambridge), Ursula Huws (Hertfordshire Business School), José Luis Molina (Universitat Autònoma de Barcelona), Aida Ponce Del Castillo (ETUI), Vili Lehdonvirta (Aalto University), Nick Couldry (LSE), Valerio De Stefano (University of Ottawa), Ulises Ali Mejias (SUNY Oswego), Janine Berg (ILO), Myriam Raymond (Université d'Angers), Niels Van Doorn (University of Amsterdam), Marie Lechner (École Supérieure d'art et de design d'Orléans), Ulrich Laitenberger (Tilburg University), Adam Badger (Newcastle University), Matheus Viana Braz (Minas Gerais State University), Alessandro Delfanti (University of Toronto), Jack Linchuan Qiu (Nanyang Technological University), Jen Schradie (Sciences Po Paris), Johan Lindquist (University of Stockholm), Jonathan Corpus Ong (University of Massachusetts Amherst), Jeremias Adams-Prassl (University of Oxford), Baptiste Delmas (Paris 1 Panthéon-Sorbonne), Khantuta Muruchi (AGETIC Bolivia), Odile Chagny (IRES), Mohammad Amir Anwar (University of Edinburgh), Antonio Aloisi (IE Law School), Kylie Jarrett (Maynooth University), Gabriella Coleman (Harvard University), Otto Kässi (ETLA), Ivana Pais (Università Cattolica di Milano), and all my friends at the Fairwork project and the Tierra Común network.

I would like to extend my sincere gratitude to those who contributed to the English version of this book. Sarah Roberts, who honored me by writing the foreword, and all the others who assisted in the publication of this revised and updated version. My special thanks go out to Steve Corcoran (Parrhesia Philosophy School Berlin) and Fred Pailler (University of Luxembourg), who both reviewed the different permutations of this manuscript, in their respective languages, sometimes patiently, sometimes hurriedly.

As a final note, I express my cosmic love to Paola Tubaro. Without her, I would not be who I am. As a matter of fact, I fear I would not exist at all. The research we have conducted together has covered a wide range of topics, including urban riots, online mental health, privacy, and digital labor. Who knows, her wisdom and grace may even have knocked some sense into my head.

NOTES

Preface

1. Mary L. Gray and Siddharth Suri, *Ghost Work: How to Stop Silicon Valley from Building a New Global Underclass* (New York: Harper Business, 2019).

2. Kate Crawford, *Atlas of AI: Power, Politics, and the Planetary Costs of Artificial Intelligence* (New Haven, CT: Yale University Press, 2021).

3. Kylie Jarrett, *Digital Labor* (London: Wiley, 2022).

4. Sarah T. Roberts, *Behind the Screen: Content Moderation in the Shadows of Social Media* (New Haven, CT: Yale University Press, 2019).

5. James Muldoon, *Platform Socialism: How to Reclaim our Digital Future from Big Tech* (London: Pluto Press, 2022).

6. Olivia Erlanger and Luis Ortega Govela, *Garage* (Cambridge, MA: MIT Press, 2018).

Introduction

1. Nestor Maslej et al., "The AI Index 2023 Annual Report," AI Index Steering Committee, Institute for Human-Centered AI, Stanford University, Stanford, CA, April 2023, https://aiindex.stanford.edu/report/.

2. Folker Fröbel, Jürgen Heinrichs, and Otto Kreye, *The New International Division of Labour: Structural Unemployment in Industrialised Countries and Industrialisation in Developing Countries* (Cambridge, UK: Cambridge University Press, 1977).

3. Luke Munn, *Automation Is a Myth* (Stanford, CA, Stanford University Press, 2022).

4. Hamid R. Ekbia and Bonnie A. Nardi, *Heteromation, and Other Stories of Computing and Capitalism* (Cambridge, MA: MIT Press, 2017).

5. Astra Taylor, "The Automation Charade," *Logic(S) Magazine*, August 1, 2018, https://logicmag.io/failure/the-automation-charade/.

6. Jathan Sadowski, "Planetary Potemkin AI: The Humans Hidden inside Mechanical Minds," in *Digital Work in the Planetary Market*, ed. M. Graham and F. Ferrari (Cambridge, MA: MIT Press, 2022), 229–40.

7. At the end of the twentieth century, these two stances were typified by American economic theorist Jeremy Rifkin and by French philosopher Dominique Méda, respectively. Jeremy Rifkin, *The End of Work: The Decline of the Global Labor Force and the Dawn of the Post-Market Era* (New York: G. P. Putnam's Sons, 1995); Dominique Méda, *Le Travail: une valeur en voie de disparition?* (Paris: Aubier, 1995).

8. Justin M. Berg, Jane E. Dutton, and Amy Wrzesniewski, "Job Crafting and Meaningful Work," in *Purpose and Meaning in the Workplace*, ed. Bryan J. Dik, Zinta S. Byrne, and Michael F. Steger (American Psychological Association, 2013), 81–104.

9. Antonio Negri, "Archaeology and Project: The Mass Worker and the Social Worker," in *Revolution Retrieved: Writings on Marx, Keynes, Capitalist Crisis, and New Social Subjects (1967–83)* (London: Red Notes, 1988).

10. Yann Moulier-Boutang, *Cognitive Capitalism* (London: Wiley, 2012).

11. Nicholas Thoburn, "Autonomous Production? On Negri's 'New Synthesis,'" *Theory, Culture, and Society* 18, no. 5 (2021): 75–96.

12. An analysis of Marx's concept from a workerist standpoint can be found in Paolo Virno, "Notes on the 'General Intellect,'" in *Marxism Beyond Marxism*, ed. Saree Makdisi, Cesare Casarino, and Rebecca E. Karl (London: Routledge, 1996).

Chapter One

1. Stuart G. Shanker, "Wittgenstein versus Turing on the Nature of Church's Thesis," *Notre Dame Journal of Formal Logic* 28, no.4 (1987): 615–49, 616.

2. Alan M. Turing, "On Computable Numbers, with an Application to the Entscheidungsproblem," *Proceedings of the London Mathematical Society* s2-42, no. 1 (1937): 230–65.

3. Ludwig Wittgenstein, *Remarks on the Philosophy of Psychology* vol. 1, ed. G. E. M. Anscombe and G. H. von Wright, tr. G. E. M. Anscombe (Oxford: Blackwell, 1980 [1945]), § 1096. For an analysis of the meaning of this statement, see Juliet Floyd, "Wittgenstein's Diagonal Argument: A Variation on Cantor and Turing," in *Epistemology versus Ontology: Essays on the Philosophy and Foundations of Mathematics in Honour of Per Martin-Löf*, ed. Peter Dybjer et al. (New York: Springer Verlag, 2012), 25–44.

4. Andre Esteva et al., "Dermatologist-Level Classification of Skin Cancer with Deep Neural Networks," *Nature* 542 (2017): 115–18.

5. Judson Chambers Webb, *Mechanism, Mentalism, and Metamathematics: An Essay on Finitism* (Dordrecht: D. Reidel, 1980), 220.

6. Shanker, "Wittgenstein versus Turing," 634.

7. Clément Guillou and Alexandre Piquard, "Comment Jordan Bardella tente d'exploiter le thème de l'intelligence artificielle," *Le Monde*, October 2, 2023, https://lemonde.fr/politique/article/2023/10/02/comment-jordan-bardella-tente-d-exploiter-le-theme-de-l-intelligence-artificielle_6191920_823448.html.

8. Quintus Ennius, *Annales* Fragment 9, Diom.447K.

9. Thomas Mortimer, *Lectures on the Elements of Commerce, Politics, and Finance* (London: T. N. Longman and O. Rees, 1801), 72.

10. David Ricardo, *On the Principles of Political Economy and Taxation* (London: John Murray, 1821).

11. Andrew Ure, *The Philosophy of Manufactures* (London: Chas. Knight, 1835), 23.

12. Daniel Bell, *The Coming of Post-Industrial Society: A Venture in Social Forecasting* (New York: Basic Books, 1973).

13. Simon Nora and Alain Minc, *The Computerization of Society: A Report to the President of France* (Cambridge, MA: MIT Press, 1981).

14. Dominique Méda, *Le travail: une valeur en voie de disparition?* (Paris: Aubier, 1995).

15. James E. Bessen, "How Computer Automation Affects Occupations: Technology, Jobs, and Skills," 2016, *Boston University School of Law, Law and Economics Research Paper* no. 15-49.

16. Erik Brynjolfsson and Tom Mitchell, "What Can Machine Learning Do? Workforce Implications," *Science* 358, no. 6370 (2017): 1530-34.

17. Manuel Castells, *The Rise of the Network Society: The Information Age: Economy, Society, and Culture*, vol. 1 (Oxford: Blackwell Publishing, 2009), 272.

18. Aad Blok, "Introduction: Uncovering Labor in Information Revolutions, 1750-2000," *International Review of Social History* 48, S11 (2003): 1-11.

19. Maarten Goos and Alan Manning, "Lousy and Lovely Jobs: The Rising Polarization of Work in Britain," *The Review of Economics and Statistics* 89, no. 1 (2007): 118-33.

20. *World Employment and Social Outlook 2023: The Value of Essential Work* (Geneva: ILO, 2023), https://www.ilo.org/digitalguides/en-gb/story/weso2023-key-workers#home.

21. Tyna Eloundou et al., "GPTs are GPTs: An Early Look at the Labor Market Impact Potential of Large Language Models," arXiv.org, March 17, 2023, https://arxiv.org/abs/2303.10130.

22. Carl Benedikt Frey and Michael A. Osborne, "The Future of Employment: How Susceptible Are Jobs to Computerisation?," *Technological Forecasting and Social Change* 114 (2017): 254-80.

23. Attilio Di Battista et al., *The Future of Jobs Report 2023* (Geneva: World Economic Forum, 2023), https://www.weforum.org/reports/the-future-of-jobs-report-2023.

24. Leopold Till Alexander, Vesselina S. Ratcheva, and Saadia Zahidi, "The Future of Jobs Report 2018," *World Economic Forum*, vol. 2, 2018.

25. Ross Gruetzemacher, David Paradice, and Kang Bok Lee, "Forecasting Extreme Labor Displacement: A Survey of AI Practitioners," *Technological Forecasting and Social Change* 161 (2020): 120323.

26. Stuart Armstrong and Kaj Sotala, "How We're Predicting AI—or Failing To," in *Beyond Artificial Intelligence: Topics in Intelligent Engineering and Informatics*, vol. 9, ed. Jan Romportl, Eva Zackova, and Jozef Kelemen (Springer Cham, 2015), https://doi.org/10.1007/978-3-319-09668-1_2.

27. Bureau of Labor Statistics, "Productivity and Costs Second Quarter 2017—Revised," September 7, 2017, https://www.bls.gov/news.release/pdf/prod2.pdf.

28. Bureau of Labor Statistics, "Productivity and Costs."

29. Shawn Sprague, "The U.S. Productivity Slowdown: An Economy-Wide and Industry-Level Analysis," *Monthly Labor Review* 144 (April 2021), https://www.bls.gov/opub/mlr/2021/article/the-us-productivity-slowdown-the-economy-wide-and-industry-level-analysis.htm.

30. Georg Erber, Ulrich Fritsche, and Patrick Christian Harms, "The Global Productivity Slowdown: Diagnosis, Causes and Remedies," *Intereconomics* 52, no. 1 (2017): 45-50.

31. Dean Baker, "Badly Confused Economics: The Debate on Automation," *The Hankyoreh*, English Edition, February 5, 2017, http://english.hani.co.kr/arti/english_edition/e_editorial/781397.html.

32. Georg Graetz and Guy Michaels, "Robots at Work," IZA Discussion Papers, no. 8938, Institute for the Study of Labor (IZA), 2015, http://EconPapers.repec.org/RePEc:iza:izadps:dp8938.

33. International Federation of Robotics (IFR), "The Impact of Robots on Productivity, Employment, and Jobs" (position paper, Frankfurt, Germany, April 2018).

34. IFR, "Impact of Robots on Productivity, Employment, and Jobs: World Robotics Service Robots" (position paper, Frankfurt, Germany, December 2022).

35. Maria Savona et al., "The Design of Digital Automation Technologies: Implications for the Future of Work," *EconPol Forum* 23, no. 5 (2022): 4–10.

36. André Leroi-Gourhan, *Gesture and Speech* (Cambridge, MA: MIT Press, 1993), 248.

37. Melanie Arntz, Terry Gregory, and Ulrich Zierahn, "The Risk of Automation for Jobs in OECD Countries: A Comparative Analysis," *OECD Social, Employment and Migration Working Papers*, May 14, 2016, https://doi.org/10.1787/5jlz9h56dvq7-en.

38. David H. Autor, "Why Are There Still So Many Jobs? The History and Future of Workplace Automation," *Journal of Economic Perspectives* 29, no. 3 (2015): 3–30 (5).

39. James Bessen, "Toil and Technology," *Finance and Development* 52, no. 1 (2015): 16–19.

40. Michael Palm, "The Cost of Paying, or Three Histories of Swiping," in *Digital Labour and Prosumer Capitalism: The US Matrix*, ed. Olivier Frayssé and Mathieu O'Neil (London: Palgrave Macmillan, 2015), 51–65.

41. Marie-Anne Dujarier, "The Three Sociological Types of Consumer Work," *Journal of Consumer Culture* 16, no. 2 (2016): 555–71.

42. Ursula Huws, *The Making of a Cybertariat: Virtual Work in a Real World* (New York: Monthly Review Press, 2003), 19.

43. Olivier Frayssé, "Work and Labor as Metonymy and Metaphor," *triple C* 12, no. 2 (2014): 468–85.

44. Jérôme Porta and Alexandra Bidet, "Le travail à l'épreuve du numérique—Regards disciplinaires croisés, droit/sociologie," *Revue de Droit du Travail*, no. 6 (2016): 333.

45. Gilbert Simondon, *On the Mode of Existence of Technical Objects* (Minneapolis, MN: Univocal Publishing, 2017); André Leroi-Gourhan, *Milieu et techniques* (Paris: Albin Michel, 1973).

46. Mary L. Gray, "Your Job Is About to Get 'Taskified,'" *Los Angeles Times*, January 8, 2016, http://www.latimes.com/opinion/op-ed/la-oe-0110-digital-turk-work-20160110-story.html.

47. Robert Weideman, "2018 Predictions: Five Ways AI Will Make You Love Customer Service This Year," Nuance, January 5, 2018, https://whatsnext.nuance.com/customer-engagement/2018-customer-service-and-ai/.

48. Cade Metz, "Facebook's Human-Powered Assistant May Just Supercharge AI," *Wired*, August 26, 2015, https://www.wired.com/2015/08/how-facebook-m-works/.

49. I will discuss this in chap. 8.

50. Ray Kurzweil, *The Singularity Is Near: When Humans Transcend Biology* (New York: Penguin Publishing Group, 2005).

51. Walter Benjamin, *Illuminations* (New York: Schocken Books, 1968).

52. Nikos Smyrnaios, *Internet Oligopoly: The Corporate Takeover of Our Digital World* (Bingley: Emerald Publishing, 2018).

53. Dominic Rushe, "Apple and Google Settle Antitrust Lawsuit over Hiring Collu-

sion Charges," *The Guardian*, April 24, 2014, https://www.theguardian.com/technology/2014/apr/24/apple-google-settle-antitrust-lawsuit-hiring-collusion; Alyssa Stringer, "A Comprehensive List of 2023 Tech Layoffs," *TechCrunch*, June 5, 2023, https://techcrunch.com/2023/06/05/tech-industry-layoffs-2023/.

54. François Vatin, *Le Travail et ses valeurs* (Paris: Albin Michel, 2008), 160–62.

55. Franco Berardi (Bifo), *Contro il lavoro* (Milan: La Libreria, 1970), 111.

Chapter Two

1. This expression was popularized by Clayton Christensen, *The Innovator's Dilemma: When New Technologies Cause Great Firms to Fail* (Cambridge, MA: Harvard Business Review Press, 1997).

2. The former head of the multinational advertising and public relations company Publicis, Maurice Lévy, reputedly coined this term. Cf. Adam Thomson, "Maurice Lévy Tries to Pick Up Publicis after Failed Deal with Omnicom," *Financial Times*, December 14, 2014.

3. Henri Verdier and Nicolas Colin, *L'Âge de la multitude: Entreprendre et gouverner après la révolution numérique* (Paris: Armand Colin, 2012).

4. Arbeitskreis Industrie 4.0, "Deutschlands Zukunft als Produktionsstandort sichern: Umsetzungsempfehlungen für das Zukunftsprojekt Industrie 4.0, Abschlussbericht," April 2013, https://www.bmbf.de/de/zukunftsprojekt-industrie-4-0-848.html.

5. *Le Numérique déroutant*, BpiFrance Le Lab, 2015, https://www.bpifrance-lelab.fr/Analyses-Reflexions/Les-Travaux-du-Lab/Le-numerique-deroutant.

6. Nick Srnicek, *Platform Capitalism* (Cambridge, MA: Polity Press, 2017).

7. Antonio A. Casilli and Julian Posada, "The Platformization of Labor and Society," in *Society and the Internet: How Networks of Information and Communication Are Changing Our Lives*, 2nd ed., ed. Mark Graham and William H. Dutton (Oxford, UK: Oxford University Press, 2019), 293–306.

8. Jean-Charles Rochet and Jean Tirole, "Platform Competition in Two-Sided Markets," *Journal of the European Economic Association* 1, no. 4 (2003): 990–1029.

9. Annabelle Gawer and Michael A. Cusumano, *Platform Leadership: How Intel, Microsoft, and Cisco Drive Industry Innovation* (Boston: Harvard Business School Press, 2002); Thomas R. Eisenmann, Geoffrey Parker, and Marshall Van Alstyne, "Opening Platforms: How, When and Why?," in *Platforms, Markets and Innovation*, ed. Annabelle Gawer (Cheltenham: Edward Elgar, 2009), 131–62.

10. Jane I. Guyer, *Legacies, Logics, Logistics: Essays in the Anthropology of the Platform Economy* (Chicago: University of Chicago Press, 2016), 114.

11. David S. Evans, Andrei Hagiu, and Richard Schmalensee, *Invisible Engines: How Software Platforms Drive Innovation and Transform Industries* (Cambridge, MA: MIT Press, 2006).

12. Carliss Y. Baldwin and C. Jason Woodard, "The Architecture of Platforms: A Unified View," in *Platforms, Markets and Innovation*, 19–44.

13. John L. Hennessy and David A. Patterson, *Computer Architecture: A Quantitative Approach* (New York: Elsevier, 1990).

14. Tarleton L. Gillespie, "The Platform Metaphor, Revisited," Humboldt Insti-

tut für Internet und Gesellschaft, August 24, 2017, https://www.hiig.de/en/blog/the
-platform-metaphor-revisited/.

15. Tarleton L. Gillespie, "The Politics of 'Platforms,'" *New Media & Society* 12, no. 3 (2010): 364.

16. Hensleigh Wedgwood, *Dictionary of English Etymology*, vol. 2 (London: Trübner & Co., 1862), 525; Douglas Harper, "Platform," *Online Etymology Dictionary*, 2000, http://www.etymonline.com/index.php?term=platform.

17. Referring to the region of Kent, the translator mentions the "fertilitie of the soile" as "the onelye platforme of Englande" (chap. 35). Stephen Batman, *Batman vppon Bartholome his Booke De Proprietatibus Rerum, Newly Corrected, Enlarged and Amended* (London: Thomas East, 1582).

18. Robert Stephens, *Letters of Sr. Francis Bacon, Written during the Reign of King James* (London: Benjamin Tooke, 1702), 287.

19. Congregational Churches in Massachusetts, *The Cambridge Platform of Church Discipline* (Cambridge, MA: Synod, 1648).

20. Congregational Churches in England, *Savoy Declaration of Faith and Order* (London: John Field, 1658).

21. Gerrard Winstanley, *The Law of Freedom in a Platform*, printed by J. M. for the author (London, 1652).

22. Winston Churchill, *Divi Britannici: Being a Remark upon the Lives of all the Kings of this Isle from the Year of the World 2855, Unto the Year of Grace 1660* (London: Thomas Roycroft, 1675 [1660]), 356.

23. Tim O'Reilly, "Government as a Platform," *Innovations: Technology, Governance, Globalization* 6, no. 1 (2011): 13.

24. O'Reilly, "Government as a Platform," 33.

25. William Lazonick, "Marketization, Globalization, Financialization: The Fragility of the US Economy in an Era of Global Change," *2010 BHC Meeting*, Athens, Georgia, March 27, 2010, http://citeseerx.ist.psu.edu/viewdoc/download?doi=10.1.1.628.1345&rep=rep1&type=pdf.

26. William Lazonick, "The New Economy Business Model and the Crisis of US Capitalism," *Capitalism and Society* 4, no. 2 (2009): 26.

27. Lu Wang and Mark Whitehouse, "How Buybacks Came to Drive the Stock Market," *Bloomberg*, June 8, 2023, https://www.bloomberg.com/news/articles/2023-06-08/how-stock-buybacks-came-to-drive-the-stock-market-quicktake.

28. Blanche Segrestin and Armand Hatchuel, *Refonder l'entreprise* (Paris: Seuil/La République des idées, 2012), 63.

29. Ronald H. Coase, "The Nature of the Firm," *Economica* 4 (1937): 387.

30. Oliver E. Williamson, "The Vertical Integration of Production: Market Failure Considerations," *American Economic Review* 61, no. 2 (1971): 112-23.

31. Frank Pasquale, *The Black Box Society: The Secret Algorithms That Control Money and Information* (Cambridge, MA: Harvard University Press, 2015).

32. David S. Evans, *Platform Economics: Essays on Multi-Sided Businesses* (Competition Policy International, 2011), https://ssrn.com/abstract=1974020.

33. Kevin Kelly, "'I'll Pay You to Read My Book,'" *The Technium*, June 1, 2012, http://kk.org/thetechnium/ill-pay-you-to/.

34. Lisa Nakamura, "'Words with Friends': Socially Networked Reading on Goodreads," *PMLA* 128, no. 1 (2013): 238-43.

35. Amrit Tiwana, *Platform Ecosystems: Aligning Architecture, Governance, and Strategy* (San Francisco, Morgan Kaufmann, 2013), 80.

36. Arne L. Kalleberg, Jeremy Reynolds, and Peter V. Marsden, "Externalizing Employment: Flexible Staffing Arrangements in US Organizations," *Social Science Research* 32, no. 4 (December 1, 2003): 525–52, https://doi.org/10.1016/s0049-089x(03)00013-9.

37. Christophe Benavent, *Plateformes. Sites collaboratifs, marketplaces, réseaux sociaux* (Limoges: Fyp Éditions, 2016).

38. Michael L. Tushman and Richard R. Nelson, "Technology, Organizations and Innovation—Introduction," *Administrative Science Quarterly* 35, no. 1 (1990): 1–8.

39. Claudia U. Ciborra, "The Platform Organization: Recombining Strategies, Structures, and Surprises," *Organization Science* 7, no. 2 (1996): 103–18.

40. Cliff Bowman and Véronique Ambrosini, "Value Creation versus Value Capture: Towards a Coherent Definition of Value in Strategy," *British Journal of Management* 11, no. 1 (2000): 1–15.

41. Michael Ryall, "The New Dynamics of Competition," *Harvard Business Review* 91, no. 6 (2013): 80–87.

42. David J. Teece and Greg Linden, "Business Models, Value Capture, and the Digital Enterprise," *Journal of Organization Design* 6, no. 8 (2017), https://jorgdesign.springeropen.com/articles/10.1186/s41469-017-0018-x.

43. Garth Johnston, "Comcast Site Teams with Facebook," *Broadcasting & Cable* 137, no. 7 (2007): 16.

44. Nicole Cohen, "The Valorization of Surveillance: Towards a Political Economy of Facebook," *Democratic Communiqué* 22, no. 1 (2008): 5–22.

45. Juliette Garside, "Twitter Puts Trillions of Tweets Up for Sale to Data Miners," *The Guardian*, March 18, 2015, https://www.theguardian.com/technology/2015/mar/18/twitter-puts-trillions-tweets-for-sale-data-miners.

46. Kathleen Kuehn and Michael S. Daubs, "The Holy Trail: Rethinking 'Value' in Google's Ubiquitous Mapping Project," *MEDIANZ: Media Studies Journal of Aotearoa New Zealand* 16, no. 1 (2017), https://doi.org/10.11157/medianz-vol16iss1id199.

47. Dave Gooden, "How Airbnb Became a Billion-Dollar Company," *Dave Gooden Blog*, May 31, 2011, http://davegooden.com/2011/05/how-airbnb-became-a-billion-dollar-company/; Matt Rosoff, "Airbnb Farmed Craigslist to Grow Its Listings, Says Competitor," *Business Insider*, May 31, 2011, http://www.businessinsider.com/us/airbnb-harvested-craigslist-to-grow-its-listings-says-competitor-2011-5/.

48. Natt Garun, "Airbnb Will Now Use Foursquare Photos in Its City Guides," *The Verge*, December 7, 2016, https://www.theverge.com/2016/12/7/13869010/airbnb-buys-foursquare-photos-city-guides.

49. Hector Yee and Bar Ifrach, "Aerosolve: Machine Learning for Humans," *Medium*, June 4, 2015, https://medium.com/airbnb-engineering/aerosolve-machine-learning-for-humans-55efcf602665.

50. Niels Van Doorn and Adam Badger, "Dual Value Production as Key to the Gig Economy Puzzle," in *Platform Economy Puzzles: A Multidisciplinary Perspective on Gig Work*, ed. Jeroen Meijerink, Giedo Jansen, and Victoria Daskalova (Cheltenham, UK: Edgar Elgar, 2021), 123–39.

51. William H. Davidow, *The Virtual Corporation: Structuring and Revitalizing the Corporation for the 21st Century* (New York: HarperBusiness, 1992).

52. Raymond E. Miles et al., "Organizing in the Knowledge Age: Anticipating the Cellular Form," *Academy of Management Executive* 11, no. 4 (December 1997): 7-20.

53. Ikujiro Nonaka and Hirotaka Takeuchi, *The Knowledge Creating Company: How Japanese Companies Create the Dynamics of Innovation* (New York: Oxford University Press, 1995).

54. Gunnar Hedlund, "A Model of Knowledge Management and the N-Form Corporation," *Strategic Management Journal* 15, no. S2 (1994): 73-90.

55. Robert W. Zmud, "The Designing Organization in the Netcentric Economy," Netcentricity Symposium, Decision and Information Technologies, R. H. Smith Business School, University of Maryland, March 30-31, 2001, http://citeseerx.ist.psu.edu /viewdoc/download?doi=10.1.1.200.3892&rep=rep1&type=pdf.

56. Jathan Sadowski, "Alarmed by Admiral's Data Grab? Wait until Insurers Can See the Contents of Your Fridge," *The Guardian*, November 2, 2016, https://www .theguardian.com/technology/2016/nov/02/admiral-face-book-data-insurers -internet-of-things; Elisa Braün, "Le BHV aspire les données de ses clients, mais il est loin d'être le seul," *Le Figaro*, August 3, 2017, http://www.lefigaro.fr/secteur/high-tech /2017/08/02/32001-20170802ARTFIG00264-le-bhv-aspire-les-donnees-de-ses -clients-mais-il-est-loin-d-etre-le-seul.php.

57. Feargus O'Sullivan, "A German App for Free Transit Rides Is Too Popular for Its Own Good," *Bloomberg*, November 18, 2016, https://www.bloomberg.com /news/articles/2016-11-18/germany-s-welectgo-trades-ad-watching-for-free-transit -rides.

58. Julien Lausson, "La SNCF va installer le Wi-Fi dans 300 rames TGV d'ici mi-2017," *Numerama*, April 13, 2016, http://www.numerama.com/tech/162890-la-sncf-va -installer-le-wi-fi-dans-300-rames-tgv-dici-mi-2017.html.

59. Aurélie Barbaux, "La SNCF va permettre l'accès à Internet dans tous les trains... et vendre ses données," *L'Usine Digitale*, February 10, 2015, https://www.usine-digitale .fr/article/la-sncf-va-permettre-l-acces-a-inter-net-dans-tous-les-trains-et-vendre-ses -donnees.N312497.

60. Olivier Razemon, "Avec #TGVPop, la SNCF se rallie à l'uberisation' de la mobilité," *Le Monde*, June 18, 2015, http://www.lemonde.fr/entreprises/article/2015/06 /18/avec-tgvpop-la-sncf-se-rallie-a-l-uberisation-de-la-mobilite_4656471_1656994 .html.

61. Efrat Nechushtai, "Could Digital Platforms Capture the Media through Infrastructure?," *Journalism* 19, no. 8 (2018): 1043-58.

62. Michael Levine and Jennifer L. Forrence, "Regulatory Capture, Public Interest, and the Public Agenda: Toward a Synthesis," *Journal of Law, Economics & Organization* 6, Special Papers from the Organization of Political Institutions Conference (1990): 167-98.

63. Fabian Muniesa, "A Flank Movement in the Understanding of Valuation," *Sociological Review* 59, no. s2 (2011): 24-38.

64. Michel Callon, *Markets in the Making: Rethinking Competition, Goods, and Innovation* (New York: Zero Books, 2021).

65. Callon, *Markets in the Making*.

66. Claes-Fredrik Helgesson and Fabian Muniesa, "For What It's Worth: An Introduction to Valuation Studies," *Valuation Studies* 1, no. 1 (2013): 1-10.

67. Helgesson and Muniesa, "For What It's Worth," 183.

Chapter Three

1. Antonio Aloisi and Valerio De Stefano, *Your Boss Is an Algorithm. Artificial Intelligence, Platform Work and Labour* (London: Bloomsbury, 2022).

2. Juliet Schor, *After the Gig: How the Sharing Economy Got Hijacked and How to Win It Back* (Oakland: University of California Press, 2021).

3. Yochai Benkler and Helen Nissenbaum, "Commons-Based Peer Production and Virtue," *The Journal of Political Philosophy* 4, no. 14 (2006): 394–419.

4. Valerio De Stefano, "The Rise of the 'Just-in-Time Workforce': On-Demand Work, Crowdwork and Labor Protection in the 'Gig-Economy,'" *Conditions of Work and Employment Series* 71 (2016), http://www.ilo.org/wcmsp5/groups/public/---ed_protect/---protrav/---travail/documents/publication/wcms_443267.pdf.

5. Antonio Aloisi, "Commoditized Workers: Case Study Research on Labor Law Issues Arising from a Set of 'On-Demand/Gig Economy' Platforms," *Comparative Labor Law & Policy Journal* 37, no. 3 (2016): 37–38.

6. "Gig Work, Online Selling and Home Sharing," Pew Research Center: Internet & Technology, 2016, http://www.pewinternet.org/2016/11/17/gig-work-online-selling-and-home-sharing/; Monica Anderson et al., "The State of Gig Work in 2021," Pew Research Center, December 8, 2021, https://www.pewresearch.org/internet/2021/12/08/the-state-of-gig-work-in-2021/.

7. Andrew Garin et al., "Interactive Research Brief: The Evolution of Platform Gig Work, 2012–2021," Becker Friedman Institute for Economics at the University of Chicago, 2023, https://bfi.uchicago.edu/insight/research-summary/interactive-research-brief-the-evolution-of-platform-gig-work-2012-2021/.

8. OECD, "Measuring Platform Mediated Workers," *OECD Digital Economy Papers*, no. 282 (Paris: OECD Publishing, 2019), https://doi.org/10.1787/170a14d9-en.

9. Diana Farrell, Fiona Greig, and Amar Hamoudi, "The Online Platform Economy in 2018: Drivers, Workers, Sellers and Lessors," JPMorgan Chase Institute (September 2018), https://www.jpmorganchase.com/content/dam/jpmc/jpmorgan-chase-and-co/institute/pdf/institute-ope-2018.pdf.

10. Mohammad Amir Anwar, Jack Ong'Iro Odeo, and Elly Otieno, "'There Is No Future in It': Pandemic and Ride-Hailing Hustle in Africa," *International Labour Review* 162, no. 1 (2023): 23–44.

11. Maya Kosoff, "Two Workers Are Suing a Cleaning Startup Called Handy over Alleged Labor Violations," *Business Insider*, November 12, 2014, http://www.businessinsider.fr/us/handy-cleaning-lawsuit-2014-11/.

12. Hilary Osborne, "Uber Loses Right to Classify UK Drivers as Self-Employed," *The Guardian*, October 28, 2016, https://www.theguardian.com/technology/2016/oct/28/uber-uk-tribunal-self-employed-status; "L'Urssaf poursuit Uber pour requalifier ses chauffeurs en salaries," *Le Monde*, May 15, 2016, http://www.lemonde.fr/economie-francaise/article/2016/05/17/l-urssaf-poursuit-uber-pour-requalifier-ses-chauffeurs-en-salaries_4920825_1656968.html; David Streitfeld, "Uber Drivers Win Preliminary Class Action Status in Labor Case," *New York Times*, July 12, 2017, https://www.nytimes.com/2017/07/12/business/uber-drivers-class-action.html.

13. Mariagrazia Lamannis, "Collective Bargaining in the Platform Economy," *ETUI, The European Trade Union Institute*, February 16, 2023, https://www.etui.org/publications/collective-bargaining-platform-economy.

14. Maxime Cornet, Mandie Joulin, and Antonio Casilli, "Platform-Mediated Labor in Europe," *SWIRL—Slash Workers and Industrial ReLations*, VP/2018/0004/0041 WP1 report, https://www.swirlproject.eu/wp-content/uploads/2021/04/WP1_Platform _SWIRL-1.pdf.

15. Adam Satariano, "In a First, Uber Agrees to Classify British Drivers as 'Workers,'" *New York Times*, March 16, 2021, https://www.nytimes.com/2021/03/16/technology /uber-uk-drivers-worker-status.html.

16. Jean-Pierre Chauchard, "L'apparition de nouvelles formes d'emploi: L'exemple de l'ubérisation," in *Travail et protection sociale: de nouvelles articulations?*, ed. Michel Borgetto et al. (Paris: LGDJ, 2017), 73-88.

17. US Bureau of Labor Statistics, "Labor Force Characteristics (CPS): Contingent and alternative employment arrangements," Labor Force Statistics from the Current Population Survey, 2023, https://www.bls.gov/cps/lfcharacteristics.htm#contingent.

18. Eurostat, "Part-Time Employment and Temporary Contracts—Annual Data," Data Browser, 2023, https://ec.europa.eu/eurostat/databrowser/view/lfsi_pt_a/default /table?lang=en.

19. These estimates, though, often conflate on-demand on-location workers with freelancers that work remotely. PPMI, "Study to Support the Impact Assessment of an EU Initiative on Improving Working Conditions in Platform Work," *European Commission*, 2021, https://ec.europa.eu/social/main.jsp?catId=738&langId=en&pubId =8428&furtherPubs=yes.

20. Caroline Bruckner and Jonathan B. Forman, "Women, Retirement, and the Growing Gig Economy Workforce," *Georgia State University Law Review* 38, no. 2 (2022): 269.

21. Lawrence F. Katz and Alan B. Krueger, "The Rise and Nature of Alternative Work Arrangements in the United States, 1995-2015," working paper no. 22667, National Bureau of Economic Research, September 2016, http://dataspace.princeton.edu/jspui /bitstream/88435/dsp01zs25xb933/3/603.pdf.

22. "Non-standard Employment around the World: Understanding Challenges, Shaping Prospects," Genève, ILO, November 2016, http://www.ilo.org/global/publications /books/WCMS_534326/lang--en/index.htm.

23. US Government Accountability Office, "Nonstandard and Contracted: Work Arrangements: Data from the 2020 Annual Business Survey and Analysis of 2021 10-K Filings," GAO-23-106212, October 20, 2022, https://www.gao.gov/assets/gao-23 -106212.pdf.

24. Manjari Mahato, Nitish Kumar, and Lalatendu Kesari Jena, "Re-Thinking Gig Economy in Conventional Workforce Post-COVID-19: A Blended Approach for Upholding Fair Balance," *Journal of Work-Applied Management* 13, no. 2 (2021): 261-76.

25. *Women and Men in the Informal Economy: A Statistical Update* (Geneva: ILO, 2023).

26. Antonio A. Casilli et al., "From GAFAM to RUM: Platforms and Resourcefulness in the Global South," *Pouvoirs*, no. 185 (2023): 51-67.

27. Jonathan V. Hall and Alan B. Krueger, "An Analysis of the Labor Market for Uber's Driver-Partners in the United States," *ILR Review* 71, no. 3 (2018): 705-32.

28. Xiaotong Guo et al., "Understanding Multi-Homing and Switching by Platform Drivers," *Transportation Research Part C: Emerging Technologies* 154 (2023): 104233.

29. Henry Ross, "Ridesharing's House of Cards: O'Connor v. Uber Technologies,

Inc. and the Viability of Uber's Labor Model in Washington," *Washington Law Review* 90, no. 3 (2015): 1431–69.

30. Alex Bitter and Nancy Luna, "Delivery Drivers Who Make Their Living on Apps Like Instacart, DoorDash, and Grubhub Say They're Being Booted with Little Warning or Recourse," *Business Insider*, October 1, 2023, https://www.businessinsider.com /how-gig-workers-are-fighting-against-sudden-account-deactivations-2023-9?r=US &IR=T.

31. Gemma Newlands, "Algorithmic Surveillance in the Gig Economy: The Organization of Work through Lefebvrian Conceived Space," *Organization Studies* 42, no. 5 (2021): 719–37.

32. Ioulia Bessa et al., "A Global Analysis of Worker Protest in Digital Labor Platforms," working paper 70 (Geneva: ILO, 2022), https://www.ilo.org/global/publications /working-papers/WCMS_849215/lang--en/index.htm.

33. "Etsy Employees Call for Health Insurance for Sellers, Standardized Benefits, More Transparency," Freelancers Union, August 15, 2017, https://blog.freelancersunion .org/2017/08/15/etsy-employees-call-for-health-in-surance-for-sellers-standardized -benefits-more-transparency/.

34. Nicole Fallert, "Thousands of Etsy Sellers Strike over Company's Fee and Policy Changes," *USA Today*, April 4, 2022, https://www.usatoday.com/story/money/shopping /2022/04/11/etsy-sellers-on-strike/7276344001/.

35. Shweta Routh, "Urban Company Promises to Enhance Partners' Earnings after Women Employees Claim Exploitation," *The Logical Indian*, October 11, 2021, https:// thelogicalindian.com/trending/urban-company-protest-31162#google_vignette.

36. Paolo Marinaro and Katherine Maich, "Food Delivery Workers Shaping the Future of Work: #NiUnRepartidorMenos," Center for Global Workers' Rights, School of Labor and Employment Relations at the Pennsylvania State University, 2022, https://ler.la.psu.edu/wp-content/uploads/sites/4/2022/09/Food-Delivery-Workers -Report.pdf.

37. Pierre Duquesne, "L'esclavage moderne, livré à domicile," *L'Humanité*, April 21, 2016, https://www.humanite.fr/lesclavage-moderne-livre-domicile-605275; Ben Wray, "Jugarse la vida como rider en Bilbao por cuatro duros y sin amparo legal," *El Salto Diario*, December 4, 2023, https://www.elsaltodiario.com/repartidores/riders -glovo-ley-bilbao-salud-laboral-accidentes.

38. Ursula Huws, *The Making of a Cybertariat: Virtual Work in a Real World* (New York: Monthly Review Press, 2003), 19.

39. Willem Pieter De Groen and Ilaria Maselli, "The Impact of the Collaborative Economy on the Labour Market," *CEPS Special Report* 138, June 2016, https://www.ceps .eu/system/files/SR138Collaborative-Economy_0.pdf.

40. Eliza McCullough et al., "Prop 22 Depresses Wages and Deepens Inequities for California Workers," *National Equity Atlas*, September 21, 2022, https:// nationalequityatlas.org/prop22-paystudy#wagefloor.

41. Daniel Szomoru, "The Economic Case for Uber in France," *Uber Under the Hood: Insights and Updates from the Uber Public Policy Team*, March 3, 2016, https://medium .com/uber-under-the-hood/the-economic-case-for-uber-in-france-1530aa95365e.

42. "How Much Are People Making from the Sharing Economy?," *Earnest Blog*, June 13, 2017, https://www.earnest.com/blog/sharing-economy-income-data/.

43. Katie J. Wells, Kafui Attoh, and Declan Cullen, "The Uber Workplace in DC," Kal-

manovitz Initiative for Labor and the Working Poor, Georgetown University, 2019, https://lwp.georgetown.edu/wp-content/uploads/sites/319/uploads/Uber-Workplace.pdf.

44. Noam Scheiber, "Uber Drivers and Others in the Gig Economy Take a Stand," *New York Times*, February 2, 2016, https://www.nytimes.com/2016/02/03/business/uber-drivers-and-others-in-the-gig-economy-take-a-stand.html.

45. "Council Unanimously Adopts First-of-Its-Kind Legislation to Give Drivers a Voice on the Job," press release, Seattle City Council, December 14, 2015, https://www.seattle.gov/council/issues/giving-drivers-a-voice.

46. Natalie Tuck, "Deliveroo Loses Appeal over Dutch Pension Payments," *European Pensions*, November 30, 2023, https://www.europeanpensions.net/ep/Deliveroo-loses-appeal-over-Dutch-pension-payments.php.

47. Natasha Lomas, "Italy's DPA Fines Glovo-Owned Foodinho $3M, Orders Changes to Algorithmic Management Of Riders," *TechCrunch*, July 6, 2021, https://techcrunch.com/2021/07/06/italys-dpa-fines-glovo-owned-foodinho-3m-orders-changes-to-algorithmic-management-of-riders/.

48. Natasha Lomas, "Drivers in Europe Net Big Data Rights Win against Uber and Ola," *TechCrunch*, April 5, 2023, https://techcrunch.com/2023/04/05/uber-ola-gdpr-worker-data-access-rights-appeal/.

49. "Uber Fined €10 Million," press release, Autoriteit Persoonsgegevens, January 31, 2024, https://www.autoriteitpersoonsgegevens.nl/en/current/uber-fined-eu10-million-for-infringement-of-privacy-regulations.

50. Marion Schmid-Drüner, *The Situation of Workers in the Collaborative Economy*, EPRS: European Parliamentary Research Service, Belgium, 2016, https://doi.org/20.500.12592/bgp6qt.

51. Yanbo Ge et al., "Racial and Gender Discrimination in Transportation Network Companies," working paper no. 22776, National Bureau of Economic Research, October 2016, http://www.nber.org/papers/w22776.

52. Benjamin Edelman, Michael Luca, and Dan Svirsky, "Racial discrimination in the sharing economy: Evidence from a field experiment," *American Economic Journal: Applied Economics* 9, no. 2 (2017): 1-22.

53. Michael Luca, Elizaveta Pronkina, and Michelangelo Rossi, "Scapegoating and Discrimination in Times of Crisis: Evidence from Airbnb," working paper no. 30344, National Bureau of Economic Research, August 1, 2022, https://www.nber.org/papers/w30344.

54. Veena Dubal, "On Algorithmic Wage Discrimination," *Columbia Law Review* 123, no. 7 (2023): 1929-92.

55. Zephyr Teachout, "Algorithmic Personalized Wages," *Politics & Society* 51, no. 3 (2023): 436-58.

56. Christophe Degryse, "Digitalisation of the Economy and Its Impact on Labour Markets," *ETUI, The European Trade Union Institute*, July 14, 2022, https://www.etui.org/publications/working-papers/digitalisation-of-the-economy-and-its-impact-on-labour-markets.

57. Arun Sundararajan, *The Sharing Economy: The End of Employment and the Rise of Crowd-Based Capitalism* (Cambridge, MA: MIT Press, 2016).

58. Min Kyung Lee et al., "Working with Machines: The Impact of Algorithmic and Data-Driven Management on Human Workers," in *Proceedings of the 33rd Annual ACM Conference on Human Factors in Computing Systems*, Seoul, Republic of Korea, Associ-

ation for Computing Machinery, 2015, 1603–12; Alex Rosenblat, *Uberland: How Algorithms Are Rewriting the Rules of Work* (Oakland: University of California Press, 2018).

59. Alex Rosenblat and Luke Stark, "Algorithmic Labor and Information Asymmetries: A Case Study of Uber's Drivers," *International Journal of Communication* 10, no. 1 (2016): 3758–84.

60. Alex Rosenblat, "How Can Wage Theft Emerge in App-Mediated Work?" *The Rideshare Guy*, August 10, 2016, https://therideshareguy.com/how-can-wage-theft -emerge-in-app-mediated-work/.

61. Jay Cassano, "How Uber Profits Even While Its Drivers Aren't Earning Money," *Vice*, February 2, 2016, https://motherboard.vice.com/en_us/article/wnxd84/how-uber -profits-even-while-its-drivers-arent-earning-money; Camille Alloing, *La E-réputation: Médiation, calcul, émotion* (Paris: CNRS Éditions, 2016).

62. Didier Fouarge et al., "Gebruikers Uber app, gemaakte trips, verdiensten en tevredenheid," Research Center for Education and the Labor Market (ROA) of Maastricht University, 2023, https://roa.nl/sites/roa/files/roa_r_2023_1_uber_rapportage.pdf.

63. Thi Nguyen, "ETA Phone Home: How Uber Engineers an Efficient Route," *Uber Blog*, November 3, 2015, https://www.uber.com/blog/engineering-routing-engine/.

64. Sean M. Smith, Roxanna Edwards, and Hao C. Duong, "Unemployment Rises in 2020, as the Country Battles the COVID-19 Pandemic," *Monthly Labor Review*, June 2021, https://www.bls.gov/opub/mlr/2021/article/unemployment-rises-in-2020-as -the-country-battles-the-covid-19-pandemic.htm.

65. Fairwork, "The Gig Economy and Covid-19: Fairwork Report on Platform Policies," Oxford, UK, April 2020, https://fair.work/en/fw/publications/the-gig-economy- and-covid-19/.

66. *World Employment and Social Outlook 2023: The Value of Essential Work* (Geneva: ILO, 2023), https://www.ilo.org/digitalguides/en-gb/story/weso2023-key-workers #home.

67. Deliverance Milano, "Il decalogo dei rider," *Effimera*, December 14, 2018, https://effimera.org/decalogo-dei-rider-deliverance-milano/.

68. Claudio Agosti et al., "Exercising Workers Rights in Algorithmic Management Systems: Lessons Learned from the Glovo-Foodinho Digital Labour Platform Case," *Social Science Research Network*, January 1, 2023, https://doi.org/10.2139/ssrn.4606803.

69. Liat Clark, "Uber Denies Researchers' 'Phantom Cars' Map Claim," *Wired*, July 28, 2015, http://www.wired.co.uk/news/archive/2015-07/28/uber-cars-always-in -real-time.

70. CJUE, Uber Systems SpainSL C-434/15, December 20, 2017.

71. Hungyu Lin, Travis Kalanick, and Emily Wang, "System and Method for Providing Dynamic Supply Positioning for On-Demand Services," Patent US20140011522A1, January 9, 2014.

72. Jonathan Hall, Cory Kendrick, and Chris Nosko, "The Effects of Uber's Surge Pricing: A Case Study," *Uber Blog*, September 1, 2015, https://www.uber.com/blog /research/the-effects-of-ubers-surge-pricing-a-case-study/.

73. "Introducing the Uber API," *Uber Blog*, August 20, 2014, https://www.uber.com /blog/uber-api/.

74. Andreas Hein et al., "The Emergence of Native Multi-Sided Platforms and Their Influence on Incumbents," *Electronic Markets* 29, no. 4 (May 10, 2019): 631–47.

75. Carolyn Elerding and Roopika Risam, "Introduction: A Gathering of Feminist

Perspectives on Digital Labor," *First Monday*, March 1, 2018, https://doi.org/10.5210/fm.v23i3.8278.

76. Maureen Dowd, "Driving Uber Mad," *New York Times*, May 23, 2015, https://www.nytimes.com/2015/05/24/opinion/Sunday/maureen-dowd-driving-uber-mad.html.

77. Joseph-Albert Kuuire, "Uber Lowers Service Fees for Its Drivers To 20%," *Tech Labari*, January 27, 2022, https://techlabari.com/uber-lowers-service-fees-for-its-drivers-to-20/.

78. IHS Markit, "Fuel for Thought: Waiting for Autonomy," *IHS Markit* (blog), October 25, 2023, https://www.spglobal.com/mobility/en/research-analysis/fuel-for-thought-waiting-for-autonomy.html.

79. Brad Templeton, "Cruise 'Recalls' Robotaxis after Crash, but the Recall Is the Wrong Mechanism," *Forbes*, September 14, 2022, https://www.forbes.com/sites/bradtempleton/2022/09/14/cruise-recalls-robotaxis-after-crash-but-the-recall-is-the-wrong-mechanism/?sh=764128f04a2b.

80. Jordan Crook, "Uber Confirms It's Testing Self-Driving Cars in Pittsburgh," *TechCrunch*, May 19, 2016, http://techcrunch.com/2016/05/19/uber-confirms-its-testing-self-driving-cars-in-pittsburgh/.

81. Evelyn Cheng, "China's Capital City Is Letting the Public Take Fully Driverless Robotaxis—And Has Bigger Rollout Plans, Startup Pony.Ai Says," CNBC, September 19, 2023, https://www.cnbc.com/2023/09/19/chinas-capital-city-beijing-has-big-plans-for-robotaxis-ponyai-says.html.

82. Krystal Hu, "Uber in Talks to Sell Its ATG Self-Driving Unit to Aurora," Reuters, November 14, 2020, https://www.reuters.com/article/idUSKBN27T30N/.

83. Andrew J. Hawkins, "The False Promises of Tesla's Full Self-Driving," *The Verge*, August 23, 2023, https://www.theverge.com/2023/8/23/23837598/tesla-elon-musk-self-driving-false-promises-land-of-the-giants.

84. Ryan Felton, "Cruise's Driverless-Car Permits Suspended in California by DMV," *Wall Street Journal*, October 24, 2023, https://www.wsj.com/business/autos/cruise-driverless-car-permits-suspended-in-california-by-dmv-db0f2c8e?mod=e2tw.

85. David Ingram, "SFPD AV Interaction Guidelines," 2022, https://www.documentcloud.org/documents/23731977-sfpd-av-interaction-guidlines.

86. Max Chafkin, "Uber's First Self-Driving Fleet Arrives in Pittsburgh This Month," *Bloomberg*, August 18, 2016, https://www.bloomberg.com/news/features/2016-08-18/uber-s-first-self-driving-fleet-arrives-in-pittsburgh-this-month-is06r7on.

87. Mengdi Chu et al., "Work with AI and Work for AI: Autonomous Vehicle Safety Drivers' Lived Experiences," in *Proceedings of the 2023 CHI Conference on Human Factors in Computing Systems*, Hamburg, Germany, Association for Computing Machinery, 2023, Article 753, 1-16.

88. Sam Levin, "Video Released of Uber Self-Driving Crash That Killed Woman in Arizona," *The Guardian*, March 22, 2018, https://www.theguardian.com/technology/2018/mar/22/video-released-of-uber-self-driving-crash-that-killed-woman-in-arizona.

89. Laurel Wamsley, "Uber Not Criminally Liable in Death of Woman Hit by Self-Driving Car, Prosecutor Says," *NPR*, March 6, 2019, https://www.npr.org/2019/03/06/700801945/uber-not-criminally-liable-in-death-of-woman-hit-by-self-driving-car-says-prosec.

90. Lauren Smiley, "'I'm the Operator': The Aftermath of a Self-Driving Tragedy," *Wired*, March 8, 2022, https://www.wired.com/story/uber-self-driving-car-fatal-crash/.

91. Mack DeGeurin, "Human Pleads Guilty in First Ever Self-Driving Pedestrian Death," *Gizmodo*, July 31, 2023, https://gizmodo.com/uber-self-driving-car-death-rafaela-vasquez-guilty-az-1850691829.

92. Scott Le Vinea, Alirez Zolfagharib. and John Polak, "Autonomous Cars: The Tension between Occupant Experience and Intersection Capacity," *Transportation Research Part C: Emerging Technologies* 52, no. 1 (2015): 1–14.

93. "On the Road to Fully Self-Driving," Waymo Safety Report, October 12, 2017, https://storage.googleapis.com/sdc-prod/v1/safety-report/waymo-safety-report-2017-10.pdf.

94. Cara Bloom et al., "Self-Driving Cars and Data Collection: Privacy Perceptions of Networked Autonomous Vehicles," in *Proceedings of the Thirteenth Symposium on Usable Privacy and Security (SOUPS '17)* (Santa Clara, CA: USENIX Association, 2017), 357–75.

95. "Waymo Open Dataset," Waymo, 2023, https://waymo.com/open/.

96. Mark Harris, "The Radical Scope of Tesla's Data Hoard," IEEE Spectrum, August 3, 2022, https://spectrum.ieee.org/tesla-autopilot-data-scope.

97. Antuan Goodwin, "Aurora Partners with Amazon Web Services on Autonomous Vehicle Development," *CNET*, December 1, 2021, https://www.cnet.com/roadshow/news/aurora-drive-aws-autonomous-vehicle-development/.

98. Uber Advanced Technologies Group, "Our Road to Self-Driving Vehicles: Uber ATG," YouTube, October 6, 2017, https://youtu.be/27OuOCeZmwI.

99. Anthony Levandowski, "Self-Driving Cars," Distinguished Innovator Lecture, University of California, Berkeley, November 12, 2013.

Chapter Four

1. David Durward, Ivo Blohm, and Jan Marco Leimeister, "Principal Forms of Crowdsourcing and Crowd Work," in *The Digital Economy and The Single Market: Employment Prospects and Working Conditions in Europe*, ed. Werner Wobbe, Elva Bova, and Catalin Dragomirescu-Gaina (Brussels: Foundation for European Progressive Studies, 2016), 39–55.

2. Fairwork, "Fairwork Cloudwork Ratings 2023: Work in the Planetary Labour Market," Oxford, UK, July 2023, https://fair.work/wp-content/uploads/sites/17/2023/07/Fairwork-Cloudwork-Ratings-2023-Red.pdf.

3. Milagros Miceli and Julian Posada, "The Data-Production Dispositif," *Proceedings of the ACM on Human–Computer Interaction* 6, CSCW2 (2022): art. 460.

4. For a general history of human-based computation, see David Alan Grier, *When Computers Were Human* (Princeton University Press, 2007).

5. Edgar Allan Poe, "Maelzel's Chess Player" (Text-02), *Southern Literary Messenger* 2 (April 1836): 318–26; 318, col. 1; 321, col. 2; 322, col. 2, for the ensuing citations.

6. Walter Benjamin, *Illuminations* (New York: Schocken Books, 1968).

7. Birgitta Bergvall-Kåreborn and Debra Howcroft, "Amazon Mechanical Turk and the Commodification of Labor," *New Technology, Work and Employment* 29, no. 3 (2014): 213–23.

8. Jeff Bezos, "Opening Keynote and Keynote Interview," MIT World Special Events and Lectures, September 27, 2006, https://techtv.mit.edu/videos/16180-opening-keynote-and-keynote-interview-with-jeff-bezos.

9. Mechanical Turk Tracker, http://demographics.mturk-tracker.com.

10. Jonathan Robinson et al., "Tapped Out or Barely Tapped? Recommendations for How to Harness the Vast and Largely Unused Potential of the Mechanical Turk Participant Pool," *PLOS ONE* 14, no. 12 (2019): e0226394; Aaron J. Moss et al., "Demographic Stability on Mechanical Turk Despite COVID-19," *Trends in Cognitive Sciences* 24, no. 9 (2020):678-80.

11. Mechanical Turk Tracker.

12. Ellie Pavlick, Matt Post, Ann Irvine, Dmitry Kachaev, and Chris Calli-son-Burch, "The Language Demographics of Amazon Mechanical Turk," *Transactions of the Association for Computational Linguistics* 2, no 1 (2014): 79–92.

13. Ned Rossiter, *Software, Infrastructure, Labor: A Media Theory of Logistical Nightmares* (New York: Routledge, 2016).

14. Turkernation, "The Reasons Why Amazon Mechanical Turk No Longer Accepts International Turkers," *Tips For Requesters On Mechanical Turk*, January 17, 2013, http://turkrequesters.blogspot.com/2013/01/the-reasons-why-amazon-mechanical-turk.html.

15. Ryan Kennedy et al., "How Venezuela's Economic Crisis Is Undermining Social Science Research—About Everything," *Washington Post*, November 7, 2018, https://www.washingtonpost.com/news/monkey-cage/wp/2018/11/07/how-the-venezuelan-economic-crisis-is-undermining-social-science-research-about-everything-not-just-venezuela/; Margaret A. Webb and June P. Tangney, "Too Good to Be True: Bots and Bad Data From Mechanical Turk," *Perspectives on Psychological Science*, November 7, 2022, 174569162211200, https://doi.org/10.1177/17456916221120027.

16. Mechanical Turk Tracker. Almost all studies classify Turkers according to binary gender classifications. Initially, MTurk was thought to be particularly useful for sampling hard-to-reach populations, including LGBTQ people. In spite of this, when a team of researchers planned to recruit more than 2000 transgender individuals via the platform, they were forced to abandon the idea because of unreliable and invalid data. To my knowledge, the only survey measuring the presence of transgender and gender queer people (in a non-representative sample) was conducted in 2017 and found that they account for 0.6 percent and 0.9 percent of the user base, respectively. Nicholas A. Smith et al., "A Convenient Solution: Using MTurk to Sample From Hard-To-Reach Populations," *Industrial and Organizational Psychology* 8, no. 2 (2015): 220–28; Logan S. Casey et al., "Intertemporal Differences Among MTurk Workers: Time-Based Sample Variations and Implications for Online Data Collection," *SAGE Open* 7, no. 2 (2017): 215824401771277; Carola Binder, "Time-of-Day and Day-of-Week Variations in Amazon Mechanical Turk Survey Responses," *Journal of Macroeconomics* 71 (2022): 103378; C. Blair Burnette et al., "Concerns and Recommendations for Using Amazon MTurk for Eating Disorder Research," *International Journal of Eating Disorders* 55, no. 2 (2022): 263–72.

17. Panagiotis G. Ipeirotis, "MTurk Demographics," *GitHub*, 2023, https://github.com/ipeirotis/mturk_demographics.

18. Panagiotis G. Ipeirotis, "Demographics of Mechanical Turk (April 2015 edition)," *Behind The Enemy Lines*, April 6, 2015, http://www.behind-the-enemy-lines.com/2015/04/demographics-of-mechanical-turk-now.html.

19. Djellel Eddine Difallah et al., "The Dynamics of Micro-Task Crowdsourcing—The Case of Amazon Mechanical Turk," *Proceedings of the 24th International Conference on World Wide Web*, Geneva, International World Wide Web Conferences Steering Committee, 2015, 238-47.

20. Paul Hitlin, "Research in the Crowdsourcing Age, a Case Study," Pew Research Center: Internet & Technology, July 11, 2016, http://www.pewinternet.org/2016/07/11/research-in-the-crowdsourcing-age-a-case-study/.

21. John Joseph Horton and Lydia B. Chilton, "The Labor Economics of Paid Crowdsourcing," in *Proceedings of the 11th ACM Conference on Electronic Commerce* (Cambridge, MA: Association for Computing Machinery, 2010), 209-18.

22. Kotaro Hara et al., "A Data-Driven Analysis of Workers' Earnings on Amazon Mechanical Turk," in *Proceedings of the 2018 CHI Conference on Human Factors in Computing Systems*, Montreal QC, Canada, Association for Computing Machinery, 2018, Paper 449, 1-14.

23. Abi Adams-Prassl et al., "The Gender Wage Gap in an Online Labor Market: The Cost of Interruptions," *The Review of Economics and Statistics* (2023): 1-23.

24. Leib Litman and Jonathan Robinson, *Conducting Online Research on Amazon Mechanical Turk and Beyond* (Thousand Oaks, CA: SAGE, 2021).

25. Caitlin Harrington, "ChatGPT Is Reshaping Crowd Work," *Wired*, July 7, 2023, https://www.wired.com/story/chatgpt-is-reshaping-crowd-work/.

26. Paola Tubaro, "Learners in the Loop: Hidden Human Skills in Machine Intelligence," *Sociologia del Lavoro*, 163 (2022): 110-29.

27. Fabion Kauker, Kayan Hau, and John Iannello, "An Exploration of Crowdwork, Machine Learning and Experts for Extracting Information from Data," in *Human Interface and the Management of Information: Interaction, Visualization, and Analytics*, ed. Sakae Yamamoto and Hirohiko Mori, Conference Proceedings of 20th International Conference, HIMI 2018, Las Vegas, NV, USA July 15-20, 2018 (Springer, 2018), 643-57.

28. "Announcing the Sentiment App on Mechanical Turk," *Amazon Mechanical Turk* (blog), October 1, 2012, https://blog.mturk.com/announcing-the-sentiment-app-on-mechanical-turk-245a80c2ca57.

29. Ryan Grim and Akela Lacy, "Pete Buttigieg's Campaign Used Notoriously Low-Paying Gig-Work Platform for Polling," *The Intercept*, January 16, 2020, https://theintercept.com/2020/01/16/pete-buttigieg-amazon-mechanical-turk-gig-workers/; Mattathias Schwartz, "Facebook Failed to Protect 30 Million Users from Having Their Data Harvested by Trump Campaign Affiliate," *The Intercept*, March 30, 2017, https://theintercept.com/2017/03/30/facebook-failed-to-protect-30-million-users-from-having-their-data-harvested-by-trump-campaign-affiliate/.

30. Oscar Schwartz, "Untold History of AI: How Amazon's Mechanical Turkers Got Squeezed Inside the Machine," *IEEE Spectrum*, April 2019, https://spectrum.ieee.org/untold-history-of-ai-mechanical-turk-revisited-tktkt.

31. Panagiotis G. Ipeirotis, "Analyzing the Amazon Mechanical Turk Marketplace," *XRDS: Crossroads* 17, no. 2 (2010): 16-21.

32. Casey et al., "Intertemporal Differences."

33. "Amazon Mechanical Turk Pricing," Amazon Mechanical Turk, 2018, https://requester.mturk.com/pricing.

34. Danielle N. Shapiro, Jesse Chandler, and Pam A. Mueller, "Using Mechanical Turk to Study Clinical Populations," *Clinical Psychological Science* 1, no. 2 (2013): 213-

20; Patrick W. Corrigan et al., "The Public Stigma of Mental Illness Means a Difference between You and Me," *Psychiatry Research* 226, no. 1 (2015): 186–91; Adam J. Vanhove, Andrew Miller, and Peter Harms, "Unemployed and Turking: Is Amazon Mechanical Turk a Viable Source for Unemployment Research?," *Academy of Management Proceedings*, no. 1 (2018):17538.

35. Janine Berg, *Income Security in the On-Demand Economy: Findings and Policy Lessons from a Survey of Crowdworkers* (Geneva: ILO, 2016), Conditions of Work and Employment Series, no. 74,11.

36. "FAQs: My Work Was Rejected, What Can I Do?," Amazon Mechanical Turk, 2018, https://www.mturk.com/mturk/help?helpPage=worker#work_rejected.

37. Lilly C. Irani and M. Six Silberman, "Turkopticon: Interrupting Worker Invisibility in Amazon Mechanical Turk," in *Proceedings of the SIGCHI Conference on Human Factors in Computing Systems*, Paris, France, Association for Computing Machinery, 2013, 611–20.

38. "Participation Agreement," Amazon Mechanical Turk, March 25, 2020, https://www.mturk.com/participation-agreement.

39. "Amazon Mechanical Turk Pricing," Amazon Mechanical Turk, 2018, https://requester.mturk.com/pricing.

40. "Dear Jeff Bezos," WeAreDynamo, 2014, http://www.wearedynamo.org/dearjeffbezos.

41. Miranda Katz, "Amazon's Turker Crowd Has Had Enough," *Wired*, August 23, 2017, https://www.wired.com/story/amazons-turker-crowd-has-had-enough.

42. Stanford Crowd Research Collective, "Daemo: A Self-Governed Crowdsourcing Marketplace," UIST '15 Adjunct: *Proceedings of the 28th Annual ACM Symposium on User Interface Software & Technology* (New York, 2015), 101–2.

43. Erin Binney, "What Are Employers' Ethical Obligations to Gig Workers?," SHRM, 2019, https://www.shrm.org/resourcesandtools/hr-topics/technology/pages/what-are-employers-ethical-obligations-to-gig-workers.aspx.

44. "Crowd Code of Ethics," Appen, 2022, https://c.connect.appen.com/crowd-code-of-ethics/.

45. "Ensuring Fairness and Integrity in Crowdworking," Crowdsourcing Code of Conduct, 2023, https://crowdsourcing-code.com.

46. "Data Enrichment Sourcing Guidelines," Partnership on AI, 2022, https://partnershiponai.org/wp-content/uploads/2022/11/data-enrichment-guidelines.pdf.

47. Patrick Lucas Austin, "Amazon Drone Delivery Was Supposed to Start By 2018. Here's What Happened Instead," *TIME*, November 2, 2021, https://time.com/6093371/amazon-drone-delivery-service/; Spencer Soper and Matt Day, "Amazon Abandons Home Delivery Robot Tests in Latest Cost Cuts," *Bloomberg*, October 6, 2022, https://www.bloomberg.com/news/articles/2022-10-06/amazon-abandons-autonomous-home-delivery-robot-in-latest-cut.

48. Luis von Ahn and Laura Dabbish, "Labeling Images with a Computer Game," in *Proceedings of the SIGCHI Conference on Human Factors in Computing Systems*, Vienna, Austria, Association for Computing Machinery, 2004, 319–26.

49. Yaoli Mao et al., "How Data Scientists Work Together with Domain Experts in Scientific Collaborations," *Proceedings of the ACM on Human–Computer Interaction* 3, no. GROUP (December 5, 2019): 1–23, https://doi.org/10.1145/3361118.

50. Aniket Kittur et al., "The Future of Crowd Work," *CSCW '13: Proceedings of the*

2013 Conference on Computer Supported Cooperative Work, New York, Association for Computing Machinery, 2013, 1301–18.

51. Rion Snow et al., "Cheap and Fast—But Is It Good? Evaluating Non-Expert Annotations for Natural Language Tasks," in *EMNLP '08: Proceedings of the Conference on Empirical Methods in Natural Language Processing*, Stroudsburg, Association for Computational Linguistics, 2008, 254–63.

52. Ian McGraw et al., "Collecting Voices from the Cloud," *Proceedings of the Language Resources and Evaluation*, 2010.

53. Yotam Gingold, Ariel Shamir, and Daniel Cohen-Or, "Micro Perceptual Human Computation for Visual Tasks," *ACM Transactions on Graphics* 31, no. 5 (2012): art. no. 119.

54. Greg Little et al., "TurKit: Human Computation Algorithms on Mechanical Turk," *UIST '10: Proceedings of the 23rd Annual ACM Symposium on User Interface Software and Technology*, New York, Association for Computing Machinery, 2010, 57–66.

55. Scale AI, 2023, https://www.scale.com/.

56. Gokul Yenduri et al., "Generative Pre-trained Transformer: A Comprehensive Review on Enabling Technologies, Potential Applications, Emerging Challenges, and Future Directions," arXiv.org, May 11, 2023, https://doi.org/10.48550/arxiv.2305.10435.

57. "Introducing Amazon SageMaker Ground Truth—Build Highly Accurate Training Datasets Using Machine Learning," AWS, November 2018, https://aws.amazon.com/fr/about-aws/whats-new/2018/11/introducing-amazon-sagemaker-groundtruth/.

58. These examples are discussed in the following publications, which I coauthored with other members of my research team at DiPLab: Matheus Viana Braz, Paola Tubaro, and Antonio A. Casilli, "Microwork in Brazil: Who Are the Workers behind Artificial Intelligence?," 2023, https://hal.science/hal-04140411; Paola Tubaro and Antonio A. Casilli, "Human Listeners and Virtual Assistants: Privacy and Labor Arbitrage in the Production of Smart Technologies," in *Digital Work in the Planetary Market*, ed. Mark Graham and Fabian Ferrari (Cambridge, MA: MIT Press, 2022), 175–90; Clément Le Ludec, Maxime Cornet, and Antonio A. Casilli, "The Problem with Annotation: Human Labour and Outsourcing between France and Madagascar," *Big Data & Society* 10, no. 2 (July 1, 2023), https://journals.sagepub.com/doi/10.1177/20539517231188723.

59. Cf. Paola Tubaro, Antonio A. Casilli, and Marion Coville, "The Trainer, the Verifier, the Imitator: Three Ways in Which Human Platform Workers Support Artificial Intelligence," *Big Data & Society* 7, no. 1 (2020), https://doi.org/10.1177/2053951720919776.

60. Venky Harinarayan, Anand Rajaraman, and Anand Ranganathan, "Hybrid Machine/Human Computing Arrangement," United States Patent US7197459B1, October 12, 2001, https://patentimages.storage.googleapis.com/1c/a5/6e/1704a00b34ed7d/US7197459.pdf.

61. Matt McGee, "Yes, Bing Has Human Search Quality Raters & Here's How They Judge Web Pages," Search Engine Land, August 15, 2012, https://searchengineland.com/bing-search-quality-rating-guidelines-130592.

62. "While most tasks have a very obvious connection to advertising, sometimes the connection might not be as obvious—for example, you might be asked to judge whether two terms mean the same thing, find something on a map, or identify different parts of a sentence. Often this data is used to make specific kinds of improvements or to test dif-

ferent parts of the ads program. We won't always be able to tell you what the task is for, but it's always something we consider important." Rafael Sánchez-Vallejo Bey, "Ads Evaluation Rater Hub," Scribd, 2016, https://www.scribd.com/document/310542844/Ads-Evaluation-Rater-Hub.

63. Paola Tubaro et al., "Where Does AI Come From? A Global Case Study across Europe, Africa, and Latin America," *New Political Economy* (forthcoming).

64. Jess Casey, "Job Losses at IT Firm Linked to Apple," *The Irish Examiner*, August 21, 2019, https://www.irishexaminer.com/news/arid-30945604.html.

65. "MightyAI," CrunchBase, 2023, https://www.crunchbase.com/organization/spare5.

66. Janine Berg, Martine Humblet, and Sergei Soares, *Working from Home: From Invisibility to Decent Work* (Geneva: ILO, 2021).

67. Jonathan I. Dingel and Brent Neiman, "How Many Jobs Can Be Done at Home?," *Journal of Public Economics* 189 (2020): 104235.

68. See Paola Tubaro and Antonio A. Casilli, "Who Bears the Burden of a Pandemic? COVID-19 and the Transfer of Risk to Digital Platform Workers," *American Behavioral Scientist*, January 15, 2022, 000276422110660, https://doi.org/10.1177/00027642211066027.

69. Siou Chew Kuek et al., *The Global Opportunity in Online Outsourcing* (Washington, DC: The World Bank Group, 2015), http://documents.worldbank.org/curated/en/138371468000900555/The-global-opportunity-in-online-outsourcing; Namita Datta et al., "Working without Borders: The Promise and Peril of Online Gig Work" (Washington, DC: World Bank, 2023), https://openknowledge.worldbank.org/entities/publication/ebc4a7e2-85c6-467b-8713-e2d77e954c6c.

70. "Explore Online Labour Markets with the Online Labour Observatory," Oxford Internet Institute, 2023, https://onlinelabourobservatory.org.

71. Paola Tubaro, Clément Le Ludec, and Antonio A. Casilli, "Counting 'Micro-Workers': Societal and Methodological Challenges around New Forms of Labour," *Work Organisation, Labour & Globalisation* 14, no. 1 (2020): 67–82.

72. "Appen 2020 Annual Report," 2021, https://appen2020.qreports.com.au/.

73. clickworker GmbH, "Your Virtual Workforce. On Demand. Worldwide," 2021, https://cdn.clickworker.com/wp-content/uploads/2021/04/Company-Presentation-2021.pdf.

74. "Freelance Forward," Upwork Research Institute, 2023, https://www.upwork.com/research/freelance-forward-2023-research-report.

75. Robinson et al., "Tapped Out?"

76. This estimate is based on the data provided in Otto Kässi, Vili Lehdonvirta, and Fabian Stephany, "How Many Online Workers Are There in the World? A Data-Driven Assessment," *Open Research Europe* 1, no. 53 (2021), https://open-research-europe.ec.europa.eu/articles/1-53/v4.

77. Vili Lehdonvirta and Mirko Ernkvist, "Converting the Virtual Economy into Development Potential: Knowledge Map of the Virtual Economy," Washington, *infoDev/World Bank*, 2011, https://openknowledge.worldbank.org/handle/10986/27361.

78. Saori Imaizumi and Indhira Santos, "Online Outsourcing: A Global Job Opportunity for Everyone?," *Let's Talk Development*, The World Bank, December 1, 2016, https://blogs.worldbank.org/developmenttalk/online-outsourcing-global-job-opportunity-everyone.

79. Vili Lehdonvirta, "Do Platforms Connect Clients Directly to Providers? The New Network Patterns of Digital Work," The Digital Inequality Group, Oxford Internet Institute, University of Oxford, August 20, 2015, https://www.oii.ox.ac.uk/news-events /do-platforms-connect-clients-directly-to-providers-the-new-network-patterns-of -digital-work/.

80. Mark Graham, Isis Hjorth, and Vili Lehdonvirta, "Digital Labour and Development: Impacts of Global Digital Labour Platforms and the Gig Economy on Worker Livelihoods," *Transfer: European Review of Labour and Research* 23 (2017): 135-62.

81. Vili Lehdonvirta, Helena Barnard, and Mark Graham, "Online Labour Markets— Leveling the Playing Field for International Service Markets?," paper presented at IPP2014: Crowdsourcing for Politics and Policy Conference, Oxford Internet Institute, 2014; Vili Lehdonvirta, "Online Platforms, Diversity and Fragmentation," paper presented at Digital Transformations of Work Conference, Oxford Internet Institute, Oxford, March 10, 2016.

82. Gang Wang et al., "Serf and Turf: Crowdturfing for Fun and Profit," *WWW '12: Proceedings of the 21st International Conference on World Wide Web*, New York, Association for Computing Machinery, 2012, 679-88.

83. Feng Zhu, Weiru Chen, and Shirley Sun, "ZBJ: Building a Global Outsourcing Platform for Knowledge Workers," Harvard Business School Case 618-044 (Boston: Harvard Business School Publishing, January 2018; revised October 2019), https://www .hbs.edu/faculty/Pages/item.aspx?num=53948.

84. Vili Lehdonvirta et al., "Online Labour Markets and the Persistence of Personal Networks: Evidence from Workers in Southeast Asia," paper presented at the conference of the American Sociological Association 2015, Chicago, August 2015.

85. Cited in Lehdonvirta, "Do Platforms Connect Clients?"

86. Ming Yin et al., "The Communication Network within the Crowd," *WWW '16: Proceedings of the 25th International Conference on World Wide Web*, Montreal, Quebec, 2016, 1293-303.

87. David Martin et al., "Turking in a Global Labour Market," *Computer Supported Cooperative Work (CSCW)* 25, no. 1 (2016): 39-77.

88. Sara Constance Kingsley, Mary L. Gray, and Siddharth Suri, "Accounting for Market Frictions and Power Asymmetries in Online Labor Markets," *Policy & Internet* 7, no. 4 (2015): 383-400.

89. Tobias Hoßfeld, Matthias Hirth, and Phuoc Tran-Gia, "Modeling of Crowdsourcing Platforms and Granularity of Work Organization in Future Internet," *ITC '11: 23rd International Teletraffic Congress Pages*, San Francisco, 2011, 142-49; Graham, Hjorth, and Lehdonvirta, "Digital Labour."

90. Siou Chew Kuek et al., "Feasibility Study: Microwork for the Palestinian Territories," *Country Management Unit for the Palestinian Territories* (MNC04)/Information Communication Technologies Unit (TWICT), The World Bank, February 2013, http:// siteresources.worldbank.org/INTWESTBANKGAZA/Resources/Finalstudy.pdf; Alison Gillwald, Onkokame Mothobi, and Aude Schoentgen, "What Is the State of Microwork in Africa? A View from Seven Countries," *Research ICT Africa Policy Paper Series 5: After Access—Assessing digital inequality in Africa*, 2017, https://researchictafrica .net/wp/wp-content/uploads/2017/12/Microwork_WorkingPaper_FINAL.pdf; Mrunal Gawade, Rajan Vaish, Mercy Nduta Waihumbu, and James Davis, "Exploring Employment Opportunities through Microtasks via Cybercafes," *Proceedings of the 2012 IEEE*

Global Humanitarian Technology Conference, Washington, IEEE Computer Society, 2012, 77-82.

91. Antonio A. Casilli, "'End-to-End' Ethical Artificial Intelligence: Taking into Account the Social and Natural Environments of Automation," in *Ethical AI*, ed. Aida Ponce Del Castillo (Brussels: ETUI, 2024), 91-100.

92. *World Employment and Social Outlook 2023: The Value of Essential Work* (Geneva: ILO, 2023), https://www.ilo.org/digitalguides/en-gb/story/weso2023-key-workers #home.

93. ILO, "ILO Modelled Estimates Database" (Self-employed, total, % of total employment), *ILOSTAT*, January 2021, https://data.worldbank.org/indicator/SL.EMP .SELF.ZS; OECD, Self-Employment Rate (Indicator), 2023, https://data.oecd.org/emp /self-employment-rate.htm#indicator-chart.

94. *World Employment and Social Outlook 2016: Trends for Youth* (Geneva: ILO, 2016).

95. Ulrich Laitenberger et al., "Unemployment and Online Labor: Evidence from Microtasking," *MIS Quarterly* 47, no. 2 (2023): 771-802.

96. *World Social Protection Report 2020-22* (Geneva: ILO, 2021).

97. "Social Security Programs throughout the World: Africa, 2017," Woodlawn, MD, *The United States Social Security Administration*, 2017, https://www.ssa.gov/policy/docs /progdesc/ssptw/2016-2017/africa/index.html; "Social Security Programs Throughout the World: Asia and the Pacific, 2016," *The United States Social Security Administration*, 2017, https://www.ssa.gov/policy/docs/progdesc/ssptw/2016-2017/asia/index .html.

98. Mark Graham et al., "Digital Labour and Development: New Knowledge Economies or Digital Sweatshops," paper at the conference *Digital Transformations of Work*, Oxford, Oxford Internet Institute, 2016.

99. The Rockefeller Foundation, "Evaluation of Impact: The Rockefeller Foundation's Digital Jobs Africa Initiative," April 14, 2020, https://www.rockefellerfoundation .org/report/evaluation-impact-rockefeller-foundations-digital-jobs-africa-initiative/.

100. The Rockefeller Foundation, "Job Creation through Building the Field of Impact Sourcing," April 13, 2020, https://www.rockefellerfoundation.org/report/job -creation-through-building-the-field-of-impact-sourcing/.

101. Dave Lee, "Why Big Tech Pays Poor Kenyans to Teach Self-Driving Cars," BBC News, November 3, 2018, https://www.bbc.com/news/technology-46055595.

102. Billy Perrigo, "Inside Facebook's African Sweatshop," *TIME*, February 14, 2022, https://time.com/6147458/facebook-africa-content-moderation-employee -treatment/.

103. Billy Perrigo, "Exclusive: OpenAI Used Kenyan Workers on Less than $2 per Hour to Make ChatGPT Less Toxic," *TIME*, January 18, 2023, https://time.com /6247678/openai-chatgpt-kenya-workers/.

104. "The Facebook-Sama Layoffs: Union Busting Disguised as Redundancy in Nairobi," Foxglove, March 24, 2023, https://www.foxglove.org.uk/2023/03/24/facebook -sama-layoffs-redundancy-nairobi/.

105. James Muldoon et al., "The Poverty of Ethical AI: Impact Sourcing and AI Supply Chains," *AI and Society*, 2023, https://link.springer.com/article/10.1007/s00146 -023-01824-9.

106. "Isahit - Ethical On-demand Workforce Platform for Your Digital Tasks," Isahit, 2023, http://www.isahit.com.

107. Isabelle Mashola, "La transformation digitale pour réconcilier business et impact social en Afrique," *La Tribune Afrique*, September 10, 2017, https://afrique.latribune.fr/think-tank/tribunes/2017-09-10/la-transformation-digitale-pour-reconcilier-business-et-impact-social-en-afrique-749685.html.

108. "Isahit: Edwige Working for Isahit in Abidjan, Ivory Coast," YouTube, December 5, 2016, https://youtu.be/hS9EXxdFJVc.

109. "Isahit: Betty Working for Isahit in Congo, Pointe Noire," YouTube, December 22, 2016, https://youtu.be/PKixsyRSBaA.

110. "Isahit—hiteuse du mois de juillet 2017, Diane Guessan, Côte d'Ivoire," YouTube, July 28, 2017, https://youtu.be/sQFlYDtLcjE.

111. Cf. documentary by Henri Poulain, "Invisibles. Les travailleurs du clic," *France Télévisions*, 2022, https://www.france.tv/slash/invisibles/3302449-invisibles-les-travailleurs-du-clic-version-longue-2022.html.

112. Fareesa Malik, Brian Nicholson, and Sharon Morgan, "Towards a Taxonomy and Critique of Impact Sourcing," paper given at the *8th International Conference in Critical Management Studies*, Manchester, 2013.

113. Gary Gereffi, John Humphrey, and Timothy Sturgeon, "The Governance of Global Value Chains," *Review of International Political Economy* 12, no. 1 (2005): 78–104.

114. William Milberg and Deborah Winkler, *Outsourcing Economics: Global Value Chains in Capitalist Development* (Cambridge, UK: Cambridge University Press, 2013).

115. Tatiana López et al., "Digital Value Chain Restructuring and Labour Process Transformations in the Fast-Fashion Sector: Evidence from the Value Chains of Zara & H&M," *Global Networks-A Journal of Transnational Affairs* 22, no. 4 (2021): 684–700.

116. Uday M. Apte and Richard O. Mason, "Global Disaggregation of Information-Intensive Services," *Management Science* 41, no. 7 (1995): 1250–62.

117. Hamid Reza Motahari-Nezhad, Bryan Stephenson, and Sharad Singhal, "Outsourcing Business to Cloud Computing Services: Opportunities and Challenges," Semantic Scholar, 2009, https://api.semanticscholar.org/CorpusID:167735615.

118. See next chapter for a discussion of the notion of "doubly free" platforms.

119. "FAQ," Crowdsource Help, 2023, https://support.google.com/crowdsource/answer/7070475?hl=en.

120. Google LLC, "Crowdsource," Google Play, 2023, https://play.google.com/store/apps/details?id=com.google.android.apps.village.boond.

121. Ignacio Garcia, "Beyond Translation Memory: Computers and the Professional Translator," *Journal of Specialized Translation*, no. 12 (2009): 199–214.

122. Paul Baker and Amanda Potts, "'Why Do White People Have Thin Lips?' Google and the Perpetuation of Stereotypes via Auto-Complete Search Forms," *Critical Discourse Studies* 10, no. 2 (2013): 189.

123. "Quick, Draw!," Quick Draw with Google, 2023, https://quickdraw.withgoogle.com/#.

124. Ian Goodfellow et al., "Generative Adversarial Nets," in *Advances in Neural Information Processing Systems 27 (NIPS 2014)*, 2014, 2672–80.

125. Raul Fernandez-Fernandez et al., "Quick, Stat!: A Statistical Analysis of the Quick, Draw! Dataset," arXiv.org, October 23, 2019, https://arxiv.org/abs/1907.06417.

126. David Ha and Douglas Eck, "A Neural Representation of Sketch Drawings," arXiv.org, May 19, 2017, https://arxiv.org/pdf/1704.03477.pdf.

127. Paško Bilić, "Search Algorithms, Hidden Labour and Information Control," *Big Data & Society* 3, no. 1 (2016): 1–9.

128. Antonio A. Casilli, "Digital labor: Travail, technologies et conflictualités," in *Qu'est-ce que le digital labor?* (Paris: INA, 2016), 22.

129. "reCAPTCHA: Easy on Humans, Hard on Bots," 2020, https://www.google.com/recaptcha/intro/?zbcode=inc5000.

130. Luis von Ahn et al., "reCAPTCHA: Human-Based Character Recognition via Web Security Measures," *Science* 321, no. 5895 (2008): 1465–68.

131. von Ahn et al., "reCAPTCHA."

Chapter Five

1. Lev Grossman, "You—Yes You—Are TIME's Person of the Year," *TIME*, December 13, 2006, http://www.time.com/time/magazine/article/0,9171,1569514,00.html.

2. David Weinberger et al., *Cluetrain Manifesto: The End of Business as Usual* (New York: Basic Books, 2000).

3. Alice Marwick and Danah Boyd, "To See and Be Seen: Celebrity Practice on Twitter," *Convergence* 1, no. 7 (2011): 125–37.

4. Alvin Toffler, *Powershift: Knowledge, Wealth, and Violence at the Edge of the 21st Century* (New York: Bantam, 1990).

5. Axel Bruns, "From Prosumer to Produser: Understanding User-Led Content Creation," paper presented at the conference "Transforming Audiences," London, 2009.

6. George Ritzer and Nathan Jurgenson, "Production, Consumption, Prosumption: The Nature of Capitalism in the Age of the Digital 'Prosumer,'" *Journal of Consumer Culture* 10, no. 1 (2010): 13–36.

7. Mike Featherstone, *Consumer Culture and Postmodernism* (London: Sage, 1991).

8. Colin Campbell, "The Craft Consumer Culture: Craft and Consumption in a Postmodern Society," *Journal of Consumer Culture* 5, no. 1 (2005): 23–42.

9. Carol Hayes, "Generative Artificial Intelligence and Copyright: Both Sides of the Black Box," *Social Science Research Network*, January 1, 2023, https://doi.org/10.2139/ssrn.4517799.

10. Lucia Vesnić-Alujević and Maria Francesca Murru, "Digital Audiences' Disempowerment: Participation or Free Labour," *Participations: Journal of Audience & Reception Studies* 13, no. 1 (2016): 422–30.

11. Henry Jenkins, *Convergence Culture: Where Old and New Media Collide* (New York: NYU Press, 2008), 259.

12. Christian Fuchs, *Social Media: A Critical Introduction* (London: Sage, 2013).

13. Dominique Cardon, "Internet par gros temps," in *Qu'est-ce que le digital labor?* (Paris: INA, 2016), 41–79.

14. Raymond Williams, "Means of Communication as Means of Production," in *Prilozi: Drustvenost Komunikacije* (Zagreb, 1978), cited in Williams, *Problems in Materialism and Culture: Selected Essays* (Verso Books, 1997).

15. Sut Jhally and Bill Livant, "Watching as Working: The Valorization of Audience Consciousness," *Journal of Communication* 36, no. 3 (1986): 124–43.

16. Tiziana Terranova, "Free Labor: Producing Culture for the Digital Economy," *Social Text* 18, no. 2 (2000): 33.

17. Mark Andrejevic, "Personal Data: Blind Spot of the 'Affective Law of Value'?," *The Information Society* 31, no. 1 (2015): 5-12.

18. PJ Rey, "Alienation, Exploitation, and Social Media," *American Behavioral Scientist* 56, no. 4 (2012): 399-420 (415).

19. Sacha Wunsch-Vincent and Graham Vickery, "Participative Web: User-Created Content," report presented to the Working Party on the Information Economy, December 2006, http://www.oecd.org/dataoecd/57/14/38393115.pdf.

20. Mark Andrejevic, "Social Network Exploitation," in *A Networked Self: Identity, Community, and Culture on Social Network Sites*, ed. Zizi Papachaissi (New York: Routledge, 2011), 86.

21. "Terms of Service," Facebook, January 12, 2024, https://www.facebook.com/legal/terms.

22. Christian Fuchs and Sebastian Sevignani, "What Is Digital Labour? What Is Digital Work? What's Their Difference? And Why Do These Questions Matter for Understanding Social Media?," *tripleC* 11, no. 2 (2013): 237-93.

23. Fuchs and Sevignani, "What Is Digital Labour?," 287.

24. Trebor Scholz, "Introduction: Why Digital Labor Matters Now?," in *Digital Labor: The Internet as Playground and Factory*, ed. Trebor Scholz (New York: Routledge, 2012), 2.

25. Scholz, "Why Digital Labor Matters Now?," 4.

26. Robert Gehl, *Reverse Engineering Social Media: Software, Culture, and Political Economy in New Media Capitalism* (Philadelphia: Temple University Press, 2014).

27. Adam Arvidsson and Elanor Colleoni, "Value in Informational Capitalism and On the Internet," *Information Society* 28, no. 3 (2012): 135-50.

28. Paolo Virno, "Quelques notes à propos du general intellect," *Futur antérieur*, no. 10 (1992): 45-53.

29. Howard Rheingold, *Smart Mobs: The Next Social Revolution* (New York: Basic Books, 2002).

30. Yochai Benkler, *The Wealth of Networks: How Social Production Transforms Markets and Freedom* (New Haven, CT: Yale University Press, 2006).

31. Clay Shirky, *Cognitive Surplus: Creativity and Generosity in a Connected Age* (New York: Penguin Press, 2010).

32. Patrice Flichy, *Le Sacre de l'amateur: Sociologie des passions ordinaires à l'ère numérique* (Paris: Seuil/La République des idées, 2010), 90.

33. Charles Leadbeater and Paul Miller, *The Pro-Am Revolution: How Enthusiasts Are Changing Our Economy and Society* (London: Demos, 2004).

34. Patrice Flichy, "Le digital labor, un amateurisme heureux ou un travail qui s'ignore?," *INA Global*, January 21, 2016, https://www.inaglobal.fr/numerique/article/le-digital-labor-un-amateurisme-heureux-ou-un-travail-qui-s-ignore-8737.

35. Shirky, *Cognitive Surplus*.

36. Fuchs and Sevignani, "What Is Digital Labour?"

37. Xavier Ochoa and Erik Duval, "Relevance Ranking Metrics for Learning Objects," *IEEE Transactions on Learning Technologies* 1, no. 1 (2008): 34-48; Jakob Nielsen, "Participation Inequality: Encouraging More Users to Contribute," *Jakob Nielsen's Alertbox*, Newsletter no. 9, 2006.

38. Charlene Li, "Social Technographics: Mapping Participation in Activities Forms the Foundation of a Social Strategy," Forrester Report, April 19, 2007, https://www.forrester.com/go?docid=42057.

39. Jennifer Gerson, Anke C. Plagnol, and Philip J. Corr, "Passive and Active Facebook Use Measure (PAUM): Validation and Relationship to the Reinforcement Sensitivity Theory," *Personality and Individual Differences* 117 (2017): 81–90.

40. Sara Rosenthal and Kathleen Mckeown, "Detecting Influencers in Multiple Online Genres," *ACM Transactions on Internet Technology* 17, no. 2 (2017): art. no. 12; Lana El Sanyoura and Ashton Anderson, "Quantifying the Creator Economy: A Large-Scale Analysis of Patreon," *Proceedings of the International AAAI Conference on Web and Social Media* 16, no. 1 (2022): 829–40.

41. José Van Dijck, "Users Like You? Theorizing Agency in User-Generated Content," *Media Culture Society* 31, no. 1 (2009): 41–58.

42. Van Dijck, "Users Like You?," 49–50.

43. Leland Tabares, "Professional Amateurs: Asian American Content Creators in YouTube's Digital Economy," *Journal of Asian American Studies* 22, no. 3 (2019): 387–417.

44. Zoe Haylock, "Pornhub Just Deleted Most of Its Content," *Vulture*, December 14, 2020, https://www.vulture.com/2020/12/pornhub-deletes-all-unverified -content-millions-of-videos.html.

45. John Paul Stadler, "Pornographic Altruism, or, How to Have Porn in a Pandemic," *Synoptique: An Online Journal of Film and Moving Image Studies* 9, no. 2 (2021): 201–16, https://www.synoptique.ca/issue-9-2.

46. Daniel Laurin, "Subscription Intimacy: Amateurism, Authenticity and Emotional Labour in Direct-to-Consumer Gay Pornography," *AG About Gender: Rethinking Gender and Agency in Pornography: Producers, Consumers, Workers, and Contexts* 8, no. 16 (2019): 61–79.

47. Jirasaya Uttarapong et al., "Social Support in Digital Patronage: OnlyFans Adult Content Creators as an Online Community," in *Extended Abstracts of the 2022 CHI Conference on Human Factors in Computing Systems*, New Orleans, LA, Association for Computing Machinery, 2022, Article 333, 1–7.

48. Gwyn Easterbrook-Smith, "OnlyFans as Gig-Economy Work: A Nexus of Precarity and Stigma," *Porn Studies* 10, no. 3 (2023): 252–67.

49. Bonnie Ruberg, "Doing It for Free: Digital Labour and the Fantasy of Amateur Online Pornography," *Porn Studies* 3, no. 2 (2016): 147–59.

50. Susanna Paasonen, "Labors of Love: Netporn, Web 2.0 and the Meanings of Amateurism," *New Media & Society* 12, no. 8 (2010): 1297–312.

51. Kylie Jarrett, "Showing Off Your Best Assets: Rethinking Commodification in the Online Creator Economy," *Sociologia del Lavoro* 163 (2022): 90–109.

52. Detlev Zwick, Samuel K. Bonsu, and Aron Darmody, "Putting Consumers to Work: 'Co-Creation' and New Marketing Govern-Mentality," *Journal of Consumer Culture* 8, no. 2 (2008): 163–96; John Banks and Sal Humphreys, "The Labour of User Co-Creators: Emergent Social Network Markets?," *Convergence* 14, no. 4 (2008): 412.

53. Claire Q. Bi and Libo Liu, "From Fame to Fortune: Influencer Entrepreneurship in a Digital Age," *Proceedings of PACIS* (Pacific Asia Conference on Information Systems), 2022, https://ir.canterbury.ac.nz/server/api/core/bitstreams/8e9a8f48-5acd -4b0f-817f-3cd379741710/content.

54. Lauren Rouse and Anastasia Salter, "Cosplay on Demand? Instagram, OnlyFans, and the Gendered Fantrepreneur," *Social Media + Society* 7, no. 3 (2021).

55. Fabian Hoose and Sophie Rosenbohm, "Self-Representation as Platform Work: Stories about Working as Social Media Content Creators," *Convergence*, July 15, 2023, https://doi.org/10.1177/13548565231185863.

56. Lisa Margonelli, "Inside AOL's 'Cyber-Sweatshop,'" *Wired*, October 1, 1999, https://www.wired.com/1999/10/volunteers/.

57. Eric Goldman, "Court Says Yelp Reviewers Aren't Employees," *Forbes*, August 17, 2015, https://www.forbes.com/sites/ericgoldman/2015/08/17/court-says-yelp-reviewers-arent-employees/.

58. Mavee, "Valve Sunsetting the Steam Translation Server by the End of March," Reddit, 2021, https://www.reddit.com/r/Steam/comments/sz9544/valve_sunsetting_the_steam_translation_server_by/.

59. Rebecca Taylor and Jane Parry, "As CEO Exits, Reddit Finds to Its Cost That Even Unpaid Workers Can Go on Strike," *The Conversation*, July 13, 2015, https://theconversation.com/as-ceo-exits-reddit-finds-to-its-cost-that-even-unpaid-workers-can-go-on-strike-44530.

60. Richard Miskella, "The Reddit Strike and the Outer Edges of Work," Lewis Silkin, July 12, 2023, https://www.lewissilkin.com/en/insights/the-reddit-strike-and-the-outer-edges-of-work.

61. Dave Lee, "Reddit's Ellen Pao Resigns after Community's Criticism," BBC News, July 10, 2015, https://www.bbc.com/news/technology-33490814.

62. Julia Guinamard, "RGPD et GAFAM: la Quadrature du Net accuse la CNIL," *Siècle Digital*, May 27, 2021, https://siecledigital.fr/2021/05/27/quadrature-net-accuse-cnil/.

63. Ed Pilkington, "*Huffington Post* Bloggers Lose Legal Fight for AOL Millions," *The Guardian*, April 1, 2012, https://www.theguardian.com/media/2012/apr/01/huffington-post-bloggers-aol-millions.

64. Wendy Davis, "YouTube Sued by Zombie Go Boom over 'Brand-Safe' Ad Policies," *Media Post*, July 14, 2017, https://www.mediapost.com/publications/article/304353/youtube-sued-by-zombie-go-boom-over-brand-safe-a.html?edition=104242.

65. Doreen St. Felix, "Black Teens Are Breaking the Internet and Seeing None of the Profits," *Fader*, December 3, 2015, http://www.thefader.com/2015/12/03/on-fleek-peaches-monroee-meechie-viral-vines.

66. Taylor Lorenz, "Inside the Secret Meeting That Changed the Fate of Vine Forever," *Mic*, October 29, 2016, https://mic.com/articles/157977/inside-the-secret-meeting-that-changed-the-fate-of-vine-forever.

67. Mathew Ingram, "Twitter Rolls Out New Program to Share Revenue with Video Creators," *Fortune*, August 30, 2016, http://fortune.com/2016/08/30/twitter-video-revenue/.

68. Maya Kosoff, "The Inside Story of Vine's Demise," *Vanity Fair*, October 28, 2016, http://www.vanityfair.com/news/2016/10/what-happened-to-vine.

69. Antonio A. Casilli, "Le wikipédien, le chercheur et le vandale," in *Wikipédia, objet scientifique non identifié*, ed. Lionel Barbe, Louise Merzeau, and Valérie Schafer (Paris: Presses universitaires de Paris Ouest, 2015), 91–104.

70. FAIR, Facebook's deep learning research lab ("Fundamental AI Research") uses FastText, a library of text recognition and classification codes in multilingual vectors. These vectors have been trained on Wikipedia's multilingual articles, which serve as an example bank for classifying words and ensuring that Facebook AIs are "taught to read." See Piotr Bojanowski, Edouard Grave, Armand Joulin, and Tomas Mikolov, "Enriching Word Vectors with Subword Information," *Transactions of the Association for Computational Linguistics* 5, no. 1 (2017): 135–46.

71. Barry Schwartz, "Google's Freebase to Close after Migrating to Wikidata: Knowledge Graph Impact?," *Search Engine Roundtable*, December 17, 2014, https://www.seroundtable.com/google-freebase-wikidata-knowledge-graph-19591.html.

72. Jon Gertner, "Wikipedia's Moment of Truth," *New York Times*, July 18, 2023, https://www.nytimes.com/2023/07/18/magazine/wikipedia-ai-chatgpt.html.

73. Caitlin Dewey, "You Don't Know It, but You're Working for Facebook. For Free," *Washington Post*, July 22, 2015, https://www.washingtonpost.com/news/the-intersect/wp/2015/07/22/you-dont-know-it-but-youre-working-for-facebook-for-free/.

74. Hexatekin (pseud. Dorothy Howard), "Wikipedia: Thoughts on Wikipedia Editing and Digital Labor," Wikipedia, April 19, 2014, https://en.wikipedia.org/wiki/Wikipedia:Thoughts_on_Wikipedia_Editing_and_Digital_Labor.

75. Arwid Lund and Juhana Venäläinen, "Monetary Materialities of Peer-Produced Knowledge: The Case of Wikipedia and Its Tensions with Paid Labour," *tripleC* 14, no. 1 (2016): 78-98.

76. Rebecca Johinke, "Social Production as Authentic Assessment: Wikipedia, Digital Writing, and Hope Labour," *Studies in Higher Education* 45, no. 5 (2020): 1015-25.

77. José Van Dijck and Hector Postigo, "Emerging Sources of Labor on the Internet: The Case of America Online Volunteers," *International Review of Social History* 48, no. 1 (2003): 205-23.

78. Kathleen Kuehn and Thomas F. Corrigan, "Hope Labor: The Role of Employment Prospects in Online Social Production," *The Political Economy of Communication* 1, no. 1 (2013): 9-25.

79. Mark R. Johnson and Jamie Woodcock, "'It's Like the Gold Rush': The Lives and Careers of Professional Video Game Streamers on Twitch.tv," *Information, Communication & Society* 22, no. 3 (2017): 336-51.

80. Andrew Ross, "The Mental Labor Problem," *Social Text* 18, no. 2 (2000): 1-32.

81. Aaron Smith, "Searching for Work in the Digital Era," Pew Research Center: Internet & Technology, November 19, 2015, http://www.pewinternet.org/2015/11/19/searching-for-work-in-the-digital-era/.

82. Earnest Wheeler and Tawanna R. Dillahunt, "Navigating the Job Search as a Low-Resourced Job Seeker," in *Proceedings of the 2018 CHI Conference on Human Factors in Computing Systems*, Montreal QC, Canada, Association for Computing Machinery, 2018, Paper 48, 10.

83. Tawanna R Dillahunt et al., "Examining the Use of Online Platforms for Employment: A Survey of U.S. Job Seekers," in *Proceedings of the 2021 CHI Conference on Human Factors in Computing Systems*, Yokohama, Japan, Association for Computing Machinery, 2021, Article 562, 1-23.

84. Kyounghwa Yonnie Kim, "The Landscape of Keitai Shōsetsu: Mobile Phones as a Literary Medium among Japanese Youth," *Continuum Journal of Media & Cultural Studies* 26, no. 3 (2012): 475-85.

85. Gabriella Lukacs, "Dreamwork: Cell Phone Novelists, Labor, and Politics in Contemporary Japan," *Cultural Anthropology* 28, no. 1 (2013): 44-64.

86. Nana Mgbechikwere Nwachukwu, Jennafer Shae Roberts, and Laura N Montoya, "The Glamorization of Unpaid Labor: AI and Its Influencers," arXiv.org, September 16, 2023, https://arxiv.org/abs/2308.02399.

87. Li Li et al., "A Network Analysis Approach to the Relationship between Fear of Missing Out (FOMO), Smartphone Addiction, and Social Networking Site Use among

a Sample of Chinese University Students," *Computers in Human Behavior* 128 (2022): 107086; Users Patrycja Uram and Sebastian Skalski, "Still Logged In? The Link Between Facebook Addiction, FOMO, Self-Esteem, Life Satisfaction and Loneliness in Social Media Users," *Psychological Reports* 125, no. 1 (2022): 218–31.

88. Emaline Friedman, *Internet Addiction: A Critical Psychology of Users* (Routledge, 2020).

89. Laura Robinson et al., "Digital Inequalities 2.0: Legacy Inequalities in the Information Age," *First Monday* 25, no. 7 (2020), https://firstmonday.org/ojs/index.php/fm/article/view/10842; Laura Robinson et al., "Digital Inequalities 3.0: Emergent Inequalities in the Information Age," *First Monday* 25, no. 7 (2020), https://firstmonday.org/ojs/index.php/fm/article/view/10844.

90. Eszter Hargittai, "Digital Na(t)ives? Variation in Internet Skills and Uses among Members of the 'Net Generation,'" *Sociological Inquiry* 80, no. 1 (2010): 92–113.

91. Laura Robinson, "A Taste for the Necessary: A Bourdieuian Approach to Digital Inequality," *Information, Communication & Society* 12, no. 4 (2009): 488–507.

92. Monica Anderson, Michelle Faverio, and Jeffrey Gottfried, "Teens, Social Media and Technology 2023," Pew Research Center, December 2023, https://www.pewresearch.org/internet/2023/12/11/teens-social-media-and-technology-2023/.

93. Naomi Schaefer Riley, "America's Real Digital Divide," *New York Times*, February 11, 2018, https://www.nytimes.com/2018/02/11/opinion/america-digital-divide.html.

94. Sophie Jehel and Serge Proulx, "Le travail émotionnel des adolescents face au web affectif: L'exemple de la réception d'images violentes, sexuelles et haineuses," *Communiquer* 28 (2020), https://id.erudit.org/iderudit/1069905ar.

95. Danah Boyd, *It's Complicated: The Social Lives of Networked Teens* (New Haven, CT: Yale University Press, 2015).

96. Office of the Assistant Secretary for Health, "Surgeon General Issues New Advisory About Effects Social Media Use Has on Youth Mental Health," US Department of Health and Human Services, May 23, 2023, https://www.hhs.gov/about/news/2023/05/23/surgeon-general-issues-new-advisory-about-effects-social-media-use-has-youth-mental-health.html.

97. Jörg Matthes et al., "'Too Much to Handle': Impact of Mobile Social Networking Sites on Information Overload, Depressive Symptoms, and Well-Being," *Computers in Human Behavior* 105 (April 1, 2020): 106217, https://doi.org/10.1016/j.chb.2019.106217.

98. "Get More Rewards Points with Microsoft Edge," Bing, 2017, http://www.bing.com/explore/rewards-browse-and-earn.

99. Matt Burgess, "Microsoft Is Now 'Paying' UK Users to Ditch Google and Search with Bing," *Wired*, May 31, 2017, http://www.wired.co.uk/article/microsoft-rewards-uk-bing.

100. Casey Johnston, "Google Paying Users to Track 100% of Their Web Usage via Little Black Box," *ArsTechnica*, February 8, 2012, https://arstechnica.com/gadgets/2012/02/google-paying-users-to-track-100-of-their-web-usage-via-little-black-box/.

101. "Earn Bitcoin by Replying to Emails and Completing Tasks," Earn, https://earn.com/.

102. Klint Finley, "Facebook Is Blocking an Upstart Rival—But It's Complicated," *Wired*, November 11, 2015, https://www.wired.com/2015/11/facebook-banning-tsu -rival-social-network/.

103. Andrejevic, "Personal Data."

104. Gehl, *Reverse Engineering Social Media*, 118.

105. Abraham Brown and Abigail Freeman, "Top-Earning TikTok-ers 2022: Charli And Dixie D'Amelio and Addison Rae Expand Fame—And Paydays," *Forbes*, January 7, 2022, https://www.forbes.com/sites/abrambrown/2022/01/07/top-earning-tiktokers -charli-dixie-damelio-addison-rae-bella-poarch-josh-richards/; Abraham Brown and Abigail Freeman, "The Highest-Paid YouTube Stars: MrBeast, Jake Paul and Markiplier Score Massive Paydays," *Forbes*, January 14, 2022, https://www.forbes.com/sites /abrambrown/2022/01/14/the-highest-paid-youtube-stars-mrbeast-jake-paul-and -markiplier-score-massive-paydays/.

106. Claire Wardle, Sam Dubberley, and Pete Brown, "Amateur Footage: A Global Study of User-Generated Content in TV and Online News Output," New York, Columbia University School of Journalism, Tow Center for Digital Journalism, 2014, http://towcenter.org/wp-content/uploads/2014/05/Tow-Center-Amateur-Footage -A-Global-Study-of-User-Generated-Content-in-TV-and-Online-News-Output .pdf.

107. Jake Coyle, "In Hollywood Writers' Battle against AI, Humans Win (for Now)," Associated Press, September 27, 2023, https://apnews.com/article/hollywood-ai-strike -wga-artificial-intelligence-39ab72582c3a15f77510c9c30a45ffc8.

108. Michael M. Grynbaum and Ryan Mac, "The Times Sues OpenAI and Microsoft Over A.I. Use of Copyrighted Work," *New York Times*, December 27, 2023, https:// www.nytimes.com/2023/12/27/business/media/new-york-times-open-ai-microsoft -lawsuit.html.

109. Alexandra Alter and Elizabeth A. Harris, "Franzen, Grisham and Other Prominent Authors Sue OpenAI," *New York Times*, September 20, 2023, https://www.nytimes .com/2023/09/20/books/authors-openai-lawsuit-chatgpt-copyright.html.

110. Blake Brittain, "Pulitzer-Winning Authors Join OpenAI, Microsoft Copyright Lawsuit," Reuters, December 21, 2023, https://www.reuters.com/legal/pulitzer -winning-authors-join-openai-microsoft-copyright-lawsuit-2023-12-20/.

111. Emanuel Maiberg, "Whoa, Valve Just Monetized Mods," *Motherboard*, April 23, 2015, https://motherboard.vice.com/en_us/article/jp5n4p/whoa-valve-just-monetized -mods.

112. "Vimeo on Demand," Vimeo Help Center, 2024, https://help.vimeo.com/hc /en-us/sections/12397318066705-Vimeo-on-Demand.

113. Kurt Wagner, "How Facebook Is Using Your Photos in Ads," *Mashable*, January 5, 2013, https://mashable.com/2013/09/05/facebook-ads-photo/.

114. "About Us: Terms of Service," DeviantArt, 2023, https://www.deviantart.com /about/policy/service.

115. Steven Englehardt, "No Boundaries: Exfiltration of Personal Data by Session-Replay Scripts," *Freedom to Tinker*, November 15, 2017, https://freedom-to-tinker.com /2017/11/15/no-boundaries-exfiltration-of-personal-data-by-session-replay-scripts/.

116. Dominique Cardon, "Regarder les données," *Multitudes*, no. 49 (2012): 139.

117. "Personal Data: The Emergence of a New Asset Class," Geneva, World Economic Forum, 2013, http://www.weforum.org/reports/personal-data-emergence-new

-asset-class; Christopher Soghoian, *The Spies We Trust: Third Party Service Providers and Law Enforcement Surveillance*, PhD diss., Indiana University, Department of Computer Science, 2012.

118. Alessandro Acquisti, "The Economics of Personal Data and the Economics of Privacy," background paper, *OECD Roundtable on The Economics of Privacy and Personal Data* (Paris, December 1, 2010).

119. Olivia Solon, "How Much Data Did Facebook Have on One Man? 1,200 Pages of Data in 57 Categories," *Wired*, December 28, 2012, http://www.wired.co.uk/article /privacy-versus-facebook.

120. Judith Duportail, "I Asked Tinder for My Data. It Sent Me 800 Pages of My Deepest, Darkest Secrets," *The Guardian*, September 26, 2018, https://www.theguardian .com/technology/2017/sep/26/tinder-personal-data-dating-app-messages-hacked -sold.

121. Laurent Gille, *Aux sources de la valeur: Des biens et des liens* (Paris: L'Harmattan, 2006).

122. Andrejevic, "Social Network Exploitation."

123. Lukasz Olejnik, Tran Minh-Dung, and Claude Castelluccia, "Selling Off Privacy at Auction," paper presented at the Network and Distributed System Security (NDSS) Symposium, San Diego, February 23–26, 2014; Clark Boyd and Ximena Sanchez, "Breaking Down the Facebook Auction," *Journal of Digital & Social Media Marketing* 6, no. 2 (2018): 160–67.

124. Elitagm, "I Will Create 1100 FACEBOOK Likes or Followers," Fiverr, 2014, https://www.fiverr.com/elitagm/create-1100-facebook-likes-or-followers; jonathgb25, "I Will Do 1100 Real and Active Facebook Likes," Fiverr, 2016, https://www.fiverr.com /jonathgb25/do-1100-real-and-active-facebook-likes.

125. Courtney Kettmann, "Is a Facebook 'Like' Worth $174? Probably Not," *Wired*, July 1, 2013, https://www.wired.com/insights/2013/07/is-a-facebook-like-worth-174 -probably-not/.

126. Drake Bennett, "How Much Is a Tweet Worth?," *Bloomberg*, April 10, 2012, https://www.bloomberg.com/news/articles/2012-04-10/how-much-is-a-tweet -worth.

127. "About X Premium," Twitter, 2023, https://help.twitter.com/en/using-x/x -premium#tbpricing-bycountry.

128. Marc Graser, "How Much Is a Tweet Worth to Hollywood? $560 at the Box Office," *Variety*, February 27, 2015, http://variety.com/2015/digital/news/the -power-of-twitter-on-a-movies-box-office-and-why-its-not-what-you-might-expect -1201442361/.

129. Alexander R. Galloway, *Protocol: How Control Exists after Decentralization* (Cambridge, MA: MIT Press, 2004); Van Dijck, "Users Like You?"

130. Ignacio Siles et al., "The Mutual Domestication of Users and Algorithmic Recommendations on Netflix," *Communication, Culture & Critique* 12, no. 4 (2019): 499–518, https://doi.org/10.1093/ccc/tcz025.

131. Sergiu Viorel Chelaru, Claudia Orellana-Rodriguez and Ismail Sengor Altingovde, "Can Social Features Help Learning to Rank YouTube Videos?," *Proceedings of the Web Information Systems Engineering—WISE 2012* (Heidelberg: Springer, 2012), 552–66.

132. Annika Richterich, "'Karma, Precious Karma!' Karmawhoring on Reddit and the Front Page's Econometrisation," *Journal of Peer Production* 4 (2014), http://

peerproduction.net/issues/issue-4-value-and-currency/peer-reviewed-articles
/karma-precious-karma/.

133. Nick Hagar and Nicholas Diakopoulos, "Algorithmic Indifference: The Dearth of News Recommendations on TikTok," *New Media & Society*, August 30, 2023, https://doi.org/10.1177/14614448231192964.

134. Christian Esposito, Vincenzo Moscato, and Giancarlo Sperlì, "Trustworthiness Assessment of Users in Social Reviewing Systems," *IEEE Transactions on Systems, Man, and Cybernetics: Systems* 52, no. 1 (2022): 151–65.

135. Karim Nader and Min Kyung Lee, "Folk Theories and User Strategies on Dating Apps," in *Information for a Better World: Shaping the Global Future*, ed. Malte Smits, iConference 2022, *Lecture Notes in Computer Science* 13192 (2022): 445–58.

136. Sahin Cem Geyik, Stuart Ambler, and Krishnaram Kenthapadi, "Fairness-Aware Ranking in Search & Recommendation Systems with Application to LinkedIn Talent Search," in *Proceedings of the 25th ACM SIGKDD International Conference on Knowledge Discovery & Data Mining*, KDD '19, New York, Association for Computing Machinery, 2019, 2221–31.

137. Heloisa Sturm Wilkerson, Martin J. Riedl, and Kelsey N. Whipple, "Affective Affordances: Exploring Facebook Reactions as Emotional Responses to Hyperpartisan Political News," *Digital Journalism* 9, no. 8 (2021): 1040–61.

138. Danah Boyd, "Untangling Research and Practice: What Facebook's 'Emotional Contagion' Study Teaches Us," *Research Ethics* 12, no. 1 (2015): 4–13.

139. "Facebook Editor Community," Facebook, 2020, https://www.facebook.com/editorcommunity/.

140. Share Lab, "Immaterial Labour and Data Harvesting: Facebook Algorithmic Factory (1)," August 21, 2016, https://labs.rs/en/facebook-algorithmic-factory-immaterial-labour-and-data-harvesting/; Vladan Joler, "Facebook Algorithmic Factory," Ars Electronica, 2019, https://ars.electronica.art/outofthebox/en/facebook/.

141. "Solutions Explorer," Facebook Business, 2018, https://www.facebook.com/business/solutions-explorer/campaign_management/.

142. Edith Ramirez, Julie Brill, Maureen K. Ohlhausen, Joshua D. Wright, and Terrell McSweeny, "Data Brokers: A Call for Transparency and Accountability," Report of the Federal Trade Commission, 2014, https://www.ftc.gov/system/files/documents/reports/data-brokers-call-transparency-accountability-report-federal-trade-commission-may-2014/140527databrokerreport.pdf.

143. Yann LeCun, "Deep Learning," presentation at the conference USI, Paris, 2015, https://youtu.be/RgUcQceqC_Y.

144. Nicolas Jones, "Computer Science: The Learning Machines," *Nature* 505 (2014): 146–48.

145. Yann LeCun, "Qu'est-ce que l'intelligence artificielle?," Collège de France Annual Chair on Research on Artificial Intelligence, Computing and the Digital Sciences, 2015-2016, https://www.college-de-france.fr/media/yann-lecun/UPL4485925235409209505_Intelligence_Artificielle____Y._LeCun.pdf.

146. Alexia Tsotsis, "Facebook Scoops Up Face.com for $55–60M to Bolster Its Facial Recognition Tech (Updated)," *TechCrunch*, June 18, 2012, https://techcrunch.com/2012/06/18/facebook-scoops-up-face-com-for-100m-to-bolster-its-facial-recognition-tech/.

147. Josh Constine, "Facebook's Facial Recognition Now Finds Photos You're Untagged In," *TechCrunch*, December 19, 2017, https://techcrunch.com/2017/12/19/facebook-facial-recognition-photos/.

148. Peter Granitz, "Texas Sues Meta, Saying It Misused Facial Recognition Data," *NPR*, February 15, 2022, https://www.npr.org/2022/02/15/1080769555/texas-sues-meta-for-misusing-facial-recognition-data.

149. Jerome Pesenti, "An Update on Our Use of Face Recognition," Meta, November 2, 2021, https://about.fb.com/news/2021/11/update-on-use-of-face-recognition/.

150. Chris Burt, "Clearview AI Tops 40 Billion Reference Images in Facial Recognition Database," Biometric Update, November 24, 2023, https://www.biometricupdate.com/202311/clearview-ai-tops-40-billion-reference-images-in-facial-recognition-database.

151. Mikel Artetxe and Holger Schwenk, "Massively Multilingual Sentence Embeddings for Zero-Shot Cross-Lingual Transfer and Beyond," arXiv.org, September 25, 2019, https://arxiv.org/abs/1812.10464.

152. "AI Scheduling Assistants from X.ai, Now on Customers' Microsoft Office 365," press release, PR Newswire, January 23, 2018, http://news.sys-con.com/node/4224048.

153. "Chatbot Vendor Directory Released," press release, Hypergrid Business, October 27, 2011, http://www.hypergridbusiness.com/2011/10/chatbot-vendor-directory-released.

154. "Messenger," Facebook Business, 2017, https://www.facebook.com/business/products/messenger-for-business.

155. Mai-Hanh Nguyen, "AI Driven Adult Chatbots Are Being Used for Virtual Boyfriends and Girlfriends—And They Keep Getting Smarter and Sexier," *Business Insider*, November 29, 2017, http://www.businessinsider.fr/us/virtual-friend-dirty-talking-adult-chatbots-2017-11/.

156. "What Is ChatGPT?," OpenAI, 2023, https://help.openai.com/en/articles/6783457-what-is-chatgpt.

157. Jay M. Patel, "Introduction to Common Crawl Datasets," in *Getting Structured Data from the Internet* (Berkeley, CA: Apress, 2020), https://doi.org/10.1007/978-1-4842-6576-5_6.

158. Reece Rogers, "Facebook Trains Its AI on Your Data. Opting Out May Be Futile," *Wired*, September 7, 2023, https://www.wired.com/story/facebook-trains-ai-your-data-opt-out/.

159. Dave Lee, "Tay: Microsoft Issues Apology over Racist Chatbot Fiasco," *BBC News*, March 26, 2016, https://www.bbc.com/news/technology-35902104.

160. Gina Neff and Peter Nagy, "Talking to Bots: Symbiotic Agency and the Case of Tay," *International Journal of Communication* 10 (2016): 4915–31.

161. Cade Metz, "Facebook's Human-Powered Assistant May Just Supercharge AI," *Wired*, August 26, 2015, https://www.wired.com/2015/08/how-facebook-m-works/.

162. Nanette George, "People Make AI-Powered Virtual Assistants More Human," *CloudFactory Blog*, October 31, 2017, https://blog.cloudfactory.com/people-powered-ai-virtual-assistants.

163. Erin Griffith and Tom Simonite, "Facebook's Virtual Assistant M Is Dead. So Are Chatbots," *Wired*, January 8, 2018, https://www.wired.com/story/facebooks-virtual-assistant-m-is-dead-so-are-chatbots/.

164. ClickHappier, "Google-Affiliated Requesters Master List (Sergey Schmidt,

Project Endor, and Way Too Many More)," Forum MTurk Crowd, January 12, 2016, http://www.mturkcrowd.com/threads/google-affiliated-requesters-master-list-sergey -schmidt-project-endor-and-way-too-many-more.1958/.

165. "Why Amazon SageMaker," Amazon, 2024, https://aws.amazon.com /sagemaker/.

166. Michael Nunez, "Former Facebook Workers: We Routinely Suppressed Conservative News," Gizmodo, May 9, 2016, https://gizmodo.com/former-facebook-workers -we-routinely-suppressed-conser-1775461006.

167. Sam Thielman, "Facebook Fires Trending Team, and Algorithm without Humans Goes Crazy," The Guardian, August 29, 2016, https://www.theguardian.com /technology/2016/aug/29/facebook-fires-trending-topics-team-algorithm.

168. Richard Nieva, "Facebook Dumps Personalized Trending Topics after Backlash," CNET, January 25, 2017, https://www.cnet.com/news/facebook-changes-trending -topics-section-after-backlash/.

169. Adam Mosseri, "Addressing Hoaxes and Fake News," Facebook Newsroom, December 15, 2016, https://newsroom.fb.com/news/2016/12/news-feed-fyi -addressing-hoaxes-and-fake-news/.

170. Benjamin Mullin and Alexios Mantzarlis, "Facebook Has a Plan to Fight Fake News: Here's Where We Come In," Poynter, December 15, 2016, http://www .poynter.org/2016/facebook-has-a-plan-to-fight-fake-news-heres-where-we-come-in /442649/; Kaveh Waddell, "Should Facebook Buy Snopes?," The Atlantic, November 11, 2016, https://www.theatlantic.com/technology/archive/2016/11/should-facebook-buy -snopes/507359/; Andrew Anker, Sara Su, and Jeff Smith, "New Test to Provide Context about Articles," Facebook Newsroom, October 5, 2017, https://newsroom.fb.com /news/2017/10/news-feed-fyi-new-test-to-provide-context-about-articles/.

171. Morgane Tual, "Inside Facebook's Top Secret Moderation Center," World Crunch, December 7, 2018, https://worldcrunch.com/tech-science/inside-facebooks -top-secret-moderation-center-1; Adam Satariano and Mike Isaac, "How Facebook Relies on Accenture to Scrub Toxic Content," New York Times, October 28, 2021, https:// www.nytimes.com/2021/08/31/technology/facebook-accenture-content-moderation .html; Niamh McIntyre, Rosie Bradbury, and Billy Perrigo, "Behind TikTok's Boom: A Legion of Traumatised, $10-a-Day Content Moderators," The Bureau of Investigative Journalism, October 2022, https://www.thebureauinvestigates.com/stories/2022 -10-20/behind-tiktoks-boom-a-legion-of-traumatised-10-a-day-content-moderators; Billy Perrigo, "Inside Facebook's African Sweatshop," TIME, February 14, 2022, https:// time.com/6147458/facebook-africa-content-moderation-employee-treatment/.

172. Craig Silverman, "Here's Why It Doesn't Matter If People Trust Facebook's Fake News Label in the News Feed," BuzzFeed, September 14, 2017, https://www .buzzfeed.com/craigsilverman/its-all-about-the-data-and-algorithms.

173. Miranda Sissons, "Our Approach to Maintaining a Safe Online Environment in Countries at Risk," Meta, October 31, 2021, https://about.fb.com/news/2021/10 /approach-to-countries-at-risk/.

174. Zhaodi Chen and Dali L. Yang, "Governing Generation Z in China: Bilibili, Bidirectional Mediation, and Online Community Governance," The Information Society 39, no. 1 (2023): 1–16.

175. Sarah T. Roberts, "Commercial Content Moderation: Digital Laborers' Dirty Work," in The Intersectional Internet: Race, Sex, Class and Culture, ed. Brendesha M.

Tynes and Safiya Umoja Noble (New York: Peter Lang, 2016), 147-60; Sarah T. Roberts, *Behind the Screen: Content Moderation in the Shadows of Social Media* (New Haven, CT: Yale University Press, 2019).

176. Miriah Steiger et al., "The Psychological Well-Being of Content Moderators: The Emotional Labor of Commercial Moderation and Avenues for Improving Support," in *Proceedings of the 2021 CHI Conference on Human Factors in Computing Systems*, Yokohama, Japan, Association for Computing Machinery, 2021, Article 341, 1-14.

177. Feliz Solomon, "Microsoft Is Being Sued by Online Moderators, Who Say Watching Violent Images Gave Them PTSD," *TIME*, January 12, 2017, http://time.com /4632504/microsoft-ptsd-online-moderators-child-abuse-violence/; Olivia Solon, "Facebook Is Hiring Moderators. But Is the Job Too Gruesome to Handle?," *The Guardian*, May 4, 2017, https://www.theguardian.com/technology/2017/may/04/facebook -content-moderators-ptsd-psychological-dangers; Associated Press, "Facebook Content Moderators Call the Work They Do 'Torture.' Their Lawsuit May Ripple Worldwide," *EuroNews*, June 29, 2023, https://www.euronews.com/next/2023/06/29/facebook -content-moderators-call-the-work-they-do-torture-their-lawsuit-may-ripple-worldwi.

178. Sana Ahmad and Martin Krzywdzinski, "Moderating in Obscurity: How Indian Content Moderators Work in Global Content Moderation Value Chains," in *Digital Work in the Planetary Market*, ed. Mark Graham and Fabian Ferrari (Cambridge, MA: MIT Press, 2022), 77-95.

179. Sarah T. Roberts, "Digital Refuse: Canadian Garbage, Commercial Content Moderation and the Global Circulation of Social Media's Waste," *Media Studies Publications* (2016): art. no. 14, http://ir.lib.uwo.ca/commpub/14.

180. Nikos Smyrnaios and Emmanuel Marty, "Profession 'nettoyeur du net': De la modération des commentaires sur les sites d'information," *Réseaux*, no. 205 (2017): 57-90 (86).

181. Ciaran Cassidy and Adrian Chen, dirs., *The Moderators*, short film, *Field of Vision*, 2017, https://vimeo.com/213152344.

182. Roberts, *Behind the Screen*.

183. "Growing Our Trusted Flagger Program into YouTube Heroes," *YouTube Official Blog*, September 22, 2016, https://youtube.googleblog.com/2016/09/growing-our -trusted-flagger-program.html.

184. "Get Involved with YouTube Contributors," Google Support, 2018, https:// support.google.com/youtube/answer/7124236.

185. Davey Alba, "The Hidden Laborers Training AI to Keep Ads off Hateful You- Tube Videos," *Wired*, April 21, 2017, https://www.wired.com/2017/04/zerochaos -google-ads-quality-raters.

186. Jack Nicas, "YouTube Subjecting All 'Preferred' Content to Human Review," *Wall Street Journal*, January 16, 2018, https://www.wsj.com/articles/youtube-subjecting -all-preferred-content-to-human-review-1516143751.

187. Roberts, "Commercial Content."

188. "Get Paid to Write from Home," Writers Domain, 2018, https://www .writersdomain.net/.

189. "Quality Content When You Need It," The Content Authority, 2018, http:// thecontentauthority.com/.

190. TextBroker, for example, proudly advertised rates between 0.7 and 5 cents per word. MyAMS workers can achieve $13.20 an hour (equivalent to half the hourly rate

of private-sector workers in the US, $33.88). But these earnings could only be achieved by writing tens of thousands of words a day. Obviously, this figure does not take into account the pace of work and the volatility of tasks on these platforms. Furthermore, generative AI has undoubtedly disrupted the text mill industry. Cf. Chris Stokel-Walker, "True Confessions: I Wrote for an Internet Content Mill," *ArsTechnica*, July 13, 2015, http://arstechnica.com/information-technology/2015/07/inside-an-online-content -mill-or-writing-4156-words-a-day-just-to-earn-lunch-money/.

191. Bronwyn E. Howell and Petrus H. Potgieter, "AI-Generated Lemons: A Sour Outlook for Content Producers?," *32nd European Regional ITS Conference* (Madrid: International Telecommunications Society, 2023), http://hdl.handle.net/10419/277971.

192. Charles Arthur, "How Low-Paid Workers at 'Click Farms' Create Appearance of Online Popularity," *The Guardian*, August 2, 2013, https://www.theguardian.com /technology/2013/aug/02/click-farms-appearance-online-popularity.

193. Vili Lehdonvirta and Mirko Ernkvist, "Converting the Virtual Economy into Development Potential: Knowledge Map of the Virtual Economy," Washington, *infoDev/World Bank*, 2011, https://openknowledge.worldbank.org/handle/10986/27361.

194. Yuli Liu et al., "Pay Me and I'll Follow You: Detection of Crowdturfing following Activities in Microblog Environment," *Proceedings of the 25th International Joint Conference on Artificial Intelligence (IJCAI2016)*, Palo Alto, CA, AAAI Press/International Joint Conferences on Artificial Intelligence, 2016, 3789–96.

195. Liang Wu and Huan Liu, "Detecting Crowdturfing in Social Media," in *Encyclopedia of Social Network Analysis and Mining*, ed. Reda Alhajj and Jon Rokne (New York: Springer, 2016), 1–9.

196. Ming Lim, "Why Many Click Farm Jobs Should Be Understood as Digital Slavery," *The Conversation*, January 17, 2018, http://theconversation.com/why-many-click -farm-jobs-should-be-understood-as-digital-slavery-83530.

197. Johan Lindquist, "Illicit Economies of the Internet: Click Farming in Indonesia and Beyond," *Made in China Journal*, October–December 2019, https:// madeinchinajournal.com/2019/01/12/illicit-economies-of-the-internet-click-farming -in-indonesia-and-beyond/; Johan Lindquist, "Good Enough Imposters: The Market for Instagram Followers in Indonesia and Beyond," in *The Imposter as Social Theory*, ed. Steve Woolgar et al. (Bristol University Press, 2021), 269–92.

198. Associated Press, "Inside a Click Farm That Helps Fake Online Popularity," *New York Post*, June 13, 2017; Bradley Jolly, "The Bizarre 'Click Farm' of 10,000 Phones That Give Fake 'Likes' to Our Most-Loved Apps," *The Mirror*, May 15, 2017, https://www .mirror.co.uk/news/world-news/bizarre-click-farm-10000-phones-10419403.

199. Marti Motoyama, Damon McCoy, Kirill Levchenko, Stefan Savage, and Geoffrey M. Voelker, "Dirty Jobs: The Role of Freelance Labor in Web Service Abuse," *Proceedings of the 20th USENIX Conference on Security* (Berkeley: USENIX Association, 2011), 4–14.

200. Jonathan Marciano, "Fake Online Reviews Cost $152 Billion a Year," World Economic Forum, August 10, 2021, https://www.weforum.org/agenda/2021/08/fake -online-reviews-are-a-152-billion-problem-heres-how-to-silence-them/.

201. David Nevado-Catalán et al., "An Analysis of Fake Social Media Engagement Services," *Computers & Security* 124 (2023): 103013.

202. Wang et al., "Serf and Turf."

203. Matthias Hirth, Tobias Hoßfeld and Phuoc Tran-Gia, "Anatomy of a Crowdsourcing Platform—Using the Example of Microworkers.com," *Fifth International Con-*

ference on Innovative Mobile and Internet Services in Ubiquitous Computing, Seoul, IEEE, 2011, 322–29.

204. Kyumin Lee, Prithivi Tamilarasan, and James Caverlee, "Crowdturfers, Campaigns, and Social Media: Tracking and Revealing Crowdsourced Manipulation of Social Media," *Proceedings of the Seventh International AAAI Conference on Weblogs and Social Media*, 2013, 331–40.

205. Nicole Perlroth, "Fake Twitter Followers Become Multimillion-Dollar Business," *New York Times*, April 5, 2013, https://bits.blogs.nytimes.com/2013/04/05/fake-twitter-followers-becomes-multimillion-dollar-business/; Charles Arthur, "Facebook Spammers Make $200m Just Posting Links, Researchers Say," *The Guardian*, August 28, 2013, https://www.theguardian.com/technology/2013/aug/28/facebook-spam-202-million-italian-research.

206. Parmy Olson, "This App Is Cashing in on Giving the World Free Data," *Forbes*, July 29, 2015, https://www.forbes.com/sites/parmyolson/2015/07/29/jana-mobile-data-facebook-internet-org/.

207. Mark Bergen, "Verizon's Venture Arm Backs Mobile Startup Jana, AOL's Tim Armstrong Joins Advisory Board," *Vox*, February 18, 2016, https://www.vox.com/2016/2/18/11587948/verizons-venture-arm-backs-mobile-startup-jana-aols-tim-armstrong.

208. Michaela Lindenmayr and Jens Foerderer, "Purchased Popularity: Fake Follower Use by Firms and Investor Reactions," *SSRN*, 2023, https://papers.ssrn.com/sol3/papers.cfm?abstract_id=4598296.

209. Johan Lindquist, "'Follower Factories' in Indonesia and Beyond: Automation and Labor in a Transnational Market," in Graham and Ferrari, *Digital Work in the Planetary Market*, 59–75.

210. Jonathan Corpus Ong and Jason Vincent A. Cabañes, "When Disinformation Studies Meets Production Studies: Social Identities and Moral Justifications in the Political Trolling Industry," *International Journal of Communication* 13, no. 20 (2019), https://ijoc.org/index.php/ijoc/article/view/11417/2879.

211. Rafael Grohmann et al., "Platform Scams: Brazilian Workers' Experiences of Dishonest and Uncertain Algorithmic Management," *New Media & Society* 24, no. 7 (2022): 1611–31.

212. Sarah Khaled, Neamat El-Tazi, and Hoda M. O Mokhtar, "Detecting Fake Accounts on Social Media," *2018 IEEE International Conference on Big Data (Big Data)*, Seattle, WA, 2018, 3672–81.

213. Dave Lee, "Instagram Deletes Millions of Accounts in Spam Purge," *BBC News*, December 19, 2014, http://www.bbc.com/news/technology-30548463.

214. Karen Gullo and Peter Burrows, "Microsoft Sues over Online Advertising 'Click Fraud,'" *Bloomberg*, May 20, 2010, https://www.bloomberg.com/news/articles/2010-05-19/microsoft-sues-web-site-over-new-form-of-online-advertising-click-fraud-.

215. Stephanie Clifford, "Microsoft Sues Three in Click-Fraud Scheme," *New York Times*, June 15, 2009, http://www.nytimes.com/2009/06/16/business/media/16adco.html.

216. "Fake Likes Put Facebook in Hot Water with Advertisers," *Marketsmith*, February 21, 2014, https://www.marketsmithinc.com/fake-likes-put-facebook-hot-water-advertisers/; Jim Edwards, "Here Are the Sealed Court Papers on 'Invalid Clicks' Facebook Doesn't Want You to See," *Business Insider*, November 5, 2012, http://

www.businessinsider.fr/us/facebook-advertisers-invalid-click-class-action-lawsuit
-2012-10.

217. "Measure and Improve Business Page Promotions," Facebook Business, 2018, https://www.facebook.com/business/help/249469901849237.

218. Emiliano De Cristofaro et al., "Paying for Likes? Understanding Facebook like Fraud Using Honeypots," in *Proceedings of the 2014 Conference on Internet Measurement Conference*, Van Couver, BC, Association for Computing Machinery, 2014, 129–36; Rory Cellan-Jones, "Who 'Likes' My Virtual Bagels?," *BBC News*, July 13, 2012, http://www.bbc.com/news/technology-18819338; Veritasium, "Facebook Fraud," YouTube, February 10, 2014, https://youtu.be/oVfHeWTKjag; Marcela De Vivo, "Is Paying for Likes on Facebook Worth It?," *Search Engine Journal*, February 11, 2014, https://www.searchenginejournal.com/paying-likes-facebook-worth-case-study-search-engine-journal-exclusive/88111/; Shishir Nagaraja and Ryan Shah, "Clicktok: Click Fraud Detection Using Traffic Analysis," in *Proceedings of the 12th Conference on Security and Privacy in Wireless and Mobile Networks*, Miami, FL, Association for Computing Machinery, 2019, 105–116, https://doi.org/10.1145/3317549.3323407.

219. Prudhvi Ratna Badri Satya, *Fake Likers Detection on Facebook*, MS thesis, Utah State University, 2016.

220. Keith A. Quesenberry and Michael K. Coolsen, "What Makes Facebook Brand Posts Engaging? A Content Analysis of Facebook Brand Post Text That Increases Shares, Likes, and Comments to Influence Organic Viral Reach," *Journal of Current Issues & Research in Advertising* 40, no. 3 (2019): 229–44.

221. Julia Carrie Wong, "Facebook Overhauls News Feed in Favor of 'Meaningful Social Interactions,'" *The Guardian*, January 12, 2018, https://www.theguardian.com/technology/2018/jan/11/facebook-news-feed-algorithm-overhaul-mark-zuckerberg.

222. Marshall Manson, "Facebook Zero: Considering Life After the Demise of Organic Reach," Social@Ogilvy, March 5, 2014, https://www.techenet.com/wp-content/uploads/2014/03/Facebook-Zero-a-Social@Ogilvy-White-Paper.pdf; Divya Sharma, Biswatosh Saha, and Uttam K. Sarkar, "Affordance Lost, Affordance Regained, and Affordance Surrendered," in *Beyond Interpretivism? New Encounters with Technology and Organization, IFIP Advances in Information and Communication Technology*, ed. Lucas Introna et al. (Springer Cham, 2016), 73–89; Seb Joseph, "'Organic Reach on Facebook Is Dead': Advertisers Expect Price Hikes after Facebook's Feed Purge," *Digiday*, January 15, 2018, https://digiday.com/marketing/organic-reach-facebook-dead-advertisers-will-spend-reach-facebooks-feed-purge/.

223. Daniel Mochon et al., "What Are Likes Worth? A Facebook Page Field Experiment," *Journal of Marketing Research* 54, no. 2 (2017): 306–17.

224. "Learn the Difference between Organic, Paid and Post Reach," Facebook, 2023, https://www.facebook.com/help/285625061456389.

225. "Organic Reach on Facebook Is Decreasing," Facebook, n.d., https://www.facebook.com/business/news/Organic-Reach-on-Facebook (emphasis added).

226. Jonathan Stempel, "Facebook Sues ILikeAd, Alleges Ad Fraud," Reuters, December 12, 2019, https://www.reuters.com/article/uk-facebook-ilikead-lawsuit-idINKBN1Y92J3/; Chris Keall, "NZ Company Must Pay Facebook $800K in Fake 'Likes' Case," *New Zealand Herald*, October 3, 2019, https://www.nzherald.co.nz/business/nz-company-must-pay-facebook-800k-in-fake-likes-case/HR23KG4PGCXFEAJHRVOXK2JNOU/.

227. Theo Araujo et al., "From Purchasing Exposure to Fostering Engagement: Brand-Consumer Experiences in the Emerging Computational Advertising Landscape," *Journal of Advertising* 49, no. 4 (2020): 428–45.

228. Wallace Witkowski, "Facebook Accused of Fake Audience Numbers," Market Watch, September 7, 2017, https://www.marketwatch.com/story/facebook-accused-of -fake-audience-numbers-2017-09-06.

229. Lucy Handley, "Procter & Gamble Chief Marketer Slams 'Crappy Media Supply Chain,' Urges Marketers to Act," CNBS, January 31, 2017, https://www.cnbc.com /2017/01/31/procter-gamble-chief-marketer-slams-crappy-media-supply-chain.html.

230. Suzanne Vranica and Jack Marshall, "Facebook Overestimated Key Video Metric for Two Years," *Wall Street Journal*, September 22, 2016, https://www.wsj.com /articles/facebook-overestimated-key-video-metric-for-two-years-1474586951.

Chapter Six

1. Marie-Anne Dujarier, "The Activity of the Consumer: Strengthening, Transforming or Contesting Capitalism?," *American Quarterly* 56, no. 3 (2015): 460–71.

2. Marie-Anne Dujarier, "The Three Sociological Types of Consumer Work," *Journal of Consumer Culture* 16, no. 2 (2016): 555–71.

3. Ursula Huws, "The Future of Work: Neither Utopias nor Dystopias but New Fields of Accumulation and Struggle," Transform! Europe, February 21, 2017, https://www .transform-network.net/en/publications/yearbook/overview/article/yearbook-2017 /the-future-of-work-neither-utopias-nor-dystopias-but-new-fields-of-accumulation -and-struggle/.

4. Huws, "The Future of Work."

5. Robert M. Bauer and Thomas Gegenhuber, "Crowdsourcing: Global Search and the Twisted Roles of Consumers and Producers," *Organization* 22, no. 5 (2015): 661–81.

6. Antonio Aloisi and Elena Gramano, "Workers without Workplaces and Unions without Unity: Non-Standard Forms of Employment and Collective Rights," *Bulletin of Comparative Labour Relations* 107 (2019): 37–57.

7. Tabea Lakemann and Jann Lay, "Digital Platforms in Africa: the 'Uberisation' of Informal Work" (GIGA Focus Afrika, 7), Hamburg: GIGA German Institute of Global and Area Studies—Leibniz-Institut für Globale und Regionale Studien, Institut für Afrika-Studien, 2019, https://nbn-resolving.org/urn:nbn:de:0168-ssoar-65910-4.

8. Salonie Muralidhara Hiriyur, "Informal Workers Harnessing the Power of Digital Platforms in India," in *Social Contracts and Informal Workers in the Global South*, ed. Laura Alfers, Martha Chen, and Sophie Plagerson (Edgar Elgar, 2022), 169–88.

9. Aditi Surie, "On-Demand Platforms and Pricing: How Platforms Can Impact the Informal Urban Economy, Evidence from Bengaluru, India," *Work Organisation, Labour & Globalisation* 14, no. 1 (2020): 83–100.

10. Feminist authors have explored the continuities between domestic labor and platform labor in recent issues of international journals, for example in the Special Issue on "Gender and Digital Labor" of *First Monday* 23, no. 3–5 (2018), or in "Digital Labour," in the *Feminist Review* 122 (2019).

11. Kylie Jarrett, *Feminism, Labour and Digital Media: The Digital Housewife* (London: Routledge, 2015).

12. Roopika Risam, "Diversity and Work and Digital Carework in Higher Educa-

tion," *First Monday* 23, no. 3–5 (2018), http://firstmonday.org/ojs/index.php/fm/article/view/8241/6651.

13. Morgan G. Ames et al., "Making Love in the Network Closet: The Benefits and Work of Family Videochat," in *Proceedings of the 2010 ACM Conference on Computer Supported Cooperative Work*, Savannah, GA, Association for Computing Machinery, 2010, 145–54; Marije Nouwen et al., "Communication Between Grandparents and Young Grandchildren over Distance: Establishing Contact with Constitutive Nonhumans," *New Media & Society*, July 12, 2023, https://doi.org/10.1177/14614448231183703.

14. Amanda Menking and Ingrid Erickson, "The Heart Work of Wikipedia: Gendered, Emotional Labor in the World's Largest Online Encyclopedia," in *Proceedings of the 33rd Annual ACM Conference on Human Factors in Computing Systems*, Seoul, Republic of Korea, Association for Computing Machinery, 2015, 207–10.

15. Mariarosa Dalla Costa and Selma James, *The Power of Women and the Subversion of Community* (Bristol: Falling Wall Press, 1972); Ruth Schwartz Cowan, *More Work For Mother: The Ironies of Household Technology from the Open Hearth To The Microwave* (New York: Basic Books, 1985), 210; Kylie Jarrett, "The Relevance of 'Women's Work': Social Repro-Duction and Immaterial Labor in Digital Media," *Television & New Media* 15, no. 1 (2014): 14–29.

16. Carlo Perrotta, Neil Selwyn, and Carrie A. Ewin, "Artificial Intelligence and the Affective Labour of Understanding: The Intimate Moderation of a Language Model," *New Media & Society*, February 2, 2022, 146144482210752, https://doi.org/10.1177/14614448221075296.

17. Brooke Erin Duffy, "Amateur, Autonomous, Collaborative: Myths of Aspiring Female Cultural Producers in Web 2.0," *Critical Studies in Media Communication* 23, no. 1 (2015): 48–64.

18. Giovanna Franca Dalla Costa, *The Work of Love: Unpaid Housework, Poverty and Sexual Violence at the Dawn of the 21st Century* (Autonomedia, 2008), 42.

19. Kylie Jarrett, "Labour of Love: An Archaeology of Affect as Power in E-Commerce," *Journal of Sociology* 34, no. 4 (2003): 335–51.

20. Laurel Ptak, *Wages for Facebook*, web installation, 2014, http://wagesforfacebook.com/.

21. Silvia Federici, *Wages against Housework* (Bristol: Power of Women Collective and Falling Wall Press, 1975), 2.

22. Erik Brynjolfsson, Seon Tae Kim, and Joo Hee Oh, "The Attention Economy: Measuring the Value of Free Goods on the Internet," *Information Systems Research*, August 31, 2023, https://pubsonline.informs.org/doi/10.1287/isre.2021.0153.

23. Herbert A. Simon, Karl W. Deutsch, and Emilio Q. Daddario, "Designing Organizations for an Information-Rich World," in *Computers, Communication, and the Public Interest*, ed. Martin Greenberger (Baltimore, MD: Johns Hopkins University Press, 1971), 37–52.

24. Dallas W. Smythe, "Communications: Blindspot of Western Marxism," *Canadian Journal of Political and Social Theory* 1, no. 3 (1977): 1–27.

25. Dallas W. Smythe, "Rejoinder to Graham Murdock," *Canadian Journal of Political and Social Theory* 2, no. 2 (1978): 120–27.

26. Brice Nixon, "Recovering Audience Labor from Audience Commodity Theory: Advertising as Capitalizing on the Work of Signification," in *Explorations in Critical Studies of Advertising*, ed. James F. Hamilton, Robert Bodle, and Ezequiel Korin (New York: Routledge, 2017), 42–53.

27. Sut Jhally and Bill Livant, "Watching as Working: The Valorization of Audience Consciousness," *Journal of Communication* 36, no. 3 (1986): 124–43.

28. Eran Fisher, "Class Struggles in the Digital Frontier: Audience Labour Theory and Social Media Users," *Information, Communication & Society* 18, no. 9 (2015): 1108–22.

29. Mark Andrejevic, "The Work of Being Watched: Interactive Media and the Exploitation of Self-Disclosure," *Critical Studies in Media Communication* 19, no. 2 (2002): 230–48.

30. Christian Fuchs, "The Digital Labour Theory of Value and Karl Marx in the Age of Facebook, YouTube, Twitter, and Weibo," in *Reconsidering Value and Labour in the Digital Age*, ed. Eran Fisher and Christian Fuchs (Basingstoke: Palgrave Macmillan, 2015), 26–41; Christian Fuchs, "Dallas Smythe and Digital Labor," in *Routledge Companion to Labor and Media*, ed. Richard Maxwell (New York: Routledge, 2016), 51–62; Ippolita, *Tecnologie del dominio: Lessico minimo per l'autodifesa digitale* (Milan: Meltemi Editore, 2017).

31. Theodor W. Adorno, "Free Time," in *Critical Models. Interventions and Catchwords* (1969; repr. New York: Columbia University Press, 2005), 167–75.

32. Julian Kücklich, "Precarious Playbour: Modders and the Digital Games Industry," *Fibreculture Journal* 25, no. 1 (2005), http://five.fibreculturejournal.org/fcj-025 -precarious-playbour-modders-and-the-digital-games-industry/.

33. Alan Chorney, "Taking the Game Out of Gamification," *Dalhousie Journal of Interdisciplinary Management* 8, no. 1 (2012), https://ojs.library.dal.ca/djim/article/view /2012vol8Chorney.

34. Mathieu Cocq, "Constitution and exploitation du capital communautaire: Le travail des streamers sur la plateforme Twitch," *La Nouvelle Revue du travail* 13 (2018), http://journals.openedition.org/nrt/3911.

35. Niels Justesen et al., "Deep Learning for Video Game Playing," *IEEE Transactions on Games* 12, no. 1 (2020): 1–20.

36. Max Jaderberg et al., "Human-Level Performance in 3D Multiplayer Games with Population-Based Reinforcement Learning," *Science* 364, no. 6443 (2019): 859–65.

37. Joseph Suarez et al., "Neural MMO: A Massively Multiagent Game Environment for Training and Evaluating Intelligent Agents," arXiv.org, March 2, 2019, https://arxiv .org/abs/1903.00784.

38. Kücklich, "Precarious Playbour."

39. Greig de Peuter and Chris J. Young, "Contested Formations of Digital Game Labor," *Television & New Media* 20, no. 8 (2019): 747–55.

40. Mark R. Johnson and Jamie Woodcock, "'It's Like the Gold Rush': The Lives and Careers of Professional Video Game Streamers on Twitch.tv," *Information, Communication & Society* 22, no. 3 (2017): 336–51.

41. Anne-Marie Schleiner, *The Player's Power to Change the Game: Ludic Mutation* (Amsterdam University Press, 2017).

42. Dayana Hristova et al., "The Social Media Game? How Gamification Shapes Our Social Media Engagement," in *The Digital Gaming Handbook*, ed. Roberto Dillon (CRC Press, 2020), 64–92.

43. Marianna Sigala, "The Application and Impact of Gamification Funware on Trip Planning and Experiences: The Case of TripAdvisor's Funware," *Electronic Markets* 25, no. 3 (2015): 189–209.

44. Laurentiu Catalin Stanculescu et al., "Work and Play: An Experiment in Enter-

prise Gamification," in *Proceedings of the 19th ACM Conference on Computer-Supported Cooperative Work & Social Computing* (San Francisco, CA: Association for Computing Machinery, 2016), 346–58.

45. Ayoung Suh et al., "Gamification in the Workplace: The Central Role of the Aesthetic Experience," *Journal of Management Information Systems* 34, no. 1 (2017): 268–305; Florin Oprescu, Christian Jones, and Mary Katsikitis, "I Play at Work—Ten Principles for Transforming Work Processes through Gamification," *Frontiers in Psychology* 5, no. 14 (2014), https://www.frontiersin.org/articles/10.3389/fpsyg.2014.00014/full; Marie-Pierre Feuvrier, "Bonheur et travail, oxymore ou piste de management stratégique de l'entreprise?," *Management & Avenir* 2, no. 68 (2014): 164–82.

46. Luc Boltanski and Ève Chiapello, *The New Spirit of Capitalism* (Verso, 2018).

47. Ethan R. Mollick and Nancy Rothbard, "Mandatory Fun: Consent, Gamification and the Impact of Games at Work," *The Wharton School Research Paper Series*, 2014, https://papers.ssrn.com/sol3/papers.cfm?abstract_id=2277103.

48. Ian Bogost, "Exploitationware," in *Rhetoric/Composition/Play through Video Games*, ed. Richard Colby and Matthew S. S. Johnson (New York: Palgrave Macmillan, 2013), 139–47.

49. Ritu Agarwal and Elena Karahanna, "Time Flies When You're Having Fun: Cognitive Absorption and Beliefs about Information Technology Usage," *MIS Quarterly* 24, no. 4 (2000): 665–94.

50. Jonathan Crary, *24/7: Late Capitalism and the Ends of Sleep* (Verso, 2014), 17.

51. Peter Kafka, "Amazon? HBO? Netflix Thinks Its Real Competitor Is . . . Sleep," CNBC, April 17, 2017, https://www.cnbc.com/2017/04/17/amazon-hbo-netflix-thinks-its-real-competitor-is--sleep.html.

52. Crary, 24/7, 53.

53. Maurizio Lazzarato, "Immaterial Labor," in *Radical Thought in Italy: A Potential Politics*, ed. Paolo Virno and Michael Hard (University of Minnesota Press, 2006), 133–47.

54. Paolo Virno, "General Intellect," in *Lessico Postfordista*, ed. Adelino Zanini and Ubaldo Fadini (Milan: Feltrinelli, 2001), 146–51.

55. Gilles Deleuze and Félix Guattari, *A Thousand Plateaus: Capitalism and Schizophrenia* (University of Minnesota Press, 1987).

56. Lazzarato developed the concept of machinic enslavement featured in *A Thousand Plateaus*. Machines enslave people if they enclose them into a "machinic assemblage" where they have no agency and no awareness of what they're doing. The author uses driving a car as an example: "When we drive, we activate subjectivity and a multiplicity of partial consciousnesses connected to the car's technological mechanisms. There is no 'individuated subject' that says, 'you must push this button, you must press this pedal.' If one knows how to drive, one acts without thinking about it, without engaging reflexive consciousness, without speaking or representing what one does. We are guided by the car's machinic assemblage. Our actions and subjective components (memory, attention, perception, etc.) are 'automatized,' a part of the machinic, hydraulic, electronic, etc., apparatuses, constituting, like mechanical (nonhuman) components, parts of the assemblage." Maurizio Lazzarato, *Signs and Machines: Capitalism and the Production of Subjectivity* (Cambridge, MA: MIT Press, 2014), 89.

57. Deleuze and Guattari, *A Thousand Plateaus*, 492.

58. Michel Volle, "De la main-d'œuvre au cerveau d'œuvre," in *Qu'est-ce qu'un ré-*

gime de travail réellement humain?, ed. Pierre Musso and Alain Supiot (Paris: Hermann, 2018), 341–55.

59. Alfred Sohn-Rethel, *Intellectual and Manual Labor: A Critique of Epistemology* (Basingstoke, UK: Macmillan, 1978).

60. Barry Wellman, Milena Gulia, and Marilyn Tremaine, "Net Surfers Don't Ride Alone: Virtual Communities as Communities," in *Networks in The Global Village*, ed. Barry Wellman (Routledge, 2020), 331–66; Michele White, *The Body and the Screen: Theories of Internet Spectatorship* (Cambridge, MA: MIT Press, 2006).

61. Susan Leigh Star and Anselm Strauss, "Layers of Silence, Arenas of Voice: The Ecology of Visible and Invisible Work," *Computer Supported Cooperative Work (CSCW)* 8, no. 1–2 (1999): 9–30.

62. Star and Strauss, "Layers of Silence."

63. Star and Strauss, "Layers of Silence."

64. Jérôme Denis, "Data and Its Invisible Work," presented at Science + Technology by Other Means—4S/EASST Conference, Barcelona, Spain, August 2016, ffhalshs-01364311v2f, https://shs.hal.science/halshs-01364311/document.

65. Ivan Illich, "Shadow Work," *Philosophica* 26, no. 2 (1980): 7.

66. Illich, "Shadow Work," 8.

67. Jethro Lieberman, *The Tyranny of the Experts: How Professionals and Specialists Are Closing the Open Society* (New York: Walker, 1970).

68. Lieberman, *The Tyranny of the Experts*, 90.

69. Mary L. Gray and Siddharth Suri, *Ghost Work: How to Stop Silicon Valley from Building a New Global Underclass* (New York: Harper Business, 2019).

70. Ian Bogost, *The Geek's Chihuahua: Living with Apple* (Minneapolis: University of Minnesota Press, 2015).

71. Jane Thier, "Tim Cook Called Remote Work 'The Mother of All Experiments.' Now Apple Is Cracking Down on Employees Who Don't Come in 3 Days a Week, Report Says," *Fortune*, March 24, 2023, https://fortune.com/2023/03/24/remote-work-3-days-apple-discipline-terminates-tracks-tim-cook/; Kali Hays and Ashley Stewart, "Meta Stops Offering Remote Work New Job Postings," *Business Insider*, March 31, 2023, https://www.businessinsider.com/meta-stops-offering-remote-work-new-job-postings-zuckerberg-2023-3; Gabriela Riccardi, "Google Has Officially Changed Its Mind about Remote Work," *Yahoo!Finance*, June 8, 2023, https://finance.yahoo.com/news/google-officially-changed-mind-remote-204500513.html.

72. Annesha Enam, Joshua Auld, and Taha H. Rashidi, "Do People Spend Travel Time the Way They Think They Would? A Comparative Study of Generic and Trip-Specific Travel Time Allocation Using Hybrid Multiple Discrete Continuous (MDC) Framework," *Transportation Letters* (March 8, 2023): 1–12, https://doi.org/10.1080/19427867.2023.2186609; Lukas Hartwig, Astrid Gühnemann, and Reinhard Hössinger, "Decomposing Mode-Specific Values of Travel Time Savings with Respect to Different Levels of Travel-Based Multitasking: A Revealed Preference Study," *Travel Behaviour and Society* 34 (2024): 100700.

73. Hancheng Cao et al., "Large Scale Analysis of Multitasking Behavior During Remote Meetings," in *Proceedings of the 2021 CHI Conference on Human Factors in Computing Systems*, Yokohama, Japan, Association for Computing Machinery, 2021, Article 448, 1–13.

74. Bogost, *The Geek's Chihuahua*, 48.

75. Bogost, *The Geek's Chihuahua*.

76. Bogost, *The Geek's Chihuahua*, 50–51.

Chapter Seven

1. Jean-Pierre Durand, *La Chaîne invisible: Travailler aujourd'hui: flux tendu et servitude volontaire* (Paris: Seuil, 2004).

2. "Starbursting," *The Economist*, March 24, 2011, https://www.economist.com /business/2011/03/24/starbursting.

3. Amanda Hoover, "Tech Layoffs Are Feeding a New Startup Surge," *Wired*, February 22, 2023, https://www.wired.com/story/tech-layoffs-are-feeding-a-new-startup -surge/.

4. I borrow this expression from André Gorz, *Critique of Economic Reason* (London: Verso, 1989). The author defines "work for oneself" not as becoming self-employed, but rather as pertaining to activities that individuals perform directly to satisfy their own needs.

5. Stanley Aronowitz and William Difazio, *The Jobless Future: Sci-Tech and the Dogma of Work* (Minneapolis: University of Minnesota Press, 1995); Stanley Aronowitz and Jonathan Cutler, *Post-Work: The Wages of Cybernation* (New York: Routledge, 1998).

6. ILO, "Employment by Economic Class, Globally and in Regions," *Key Indicators of the Labour Market (KILM)*, 2013, http://www.ilo.org/global/about-the-ilo/multimedia /maps-and-charts/WCMS_2320--/lang--fr/index.htm.

7. David Autor and Elisabeth Reynolds, "The Nature of Work after the COVID Crisis: Too Few Low-Wage Jobs," Brookings Institution, July 2020, https://policycommons .net/artifacts/4135903/the-nature-of-work-after-the-covid-crisis/4943803/.

8. "The Impact of COVID-19 on Employment and Jobs," OECD, 2020, https://www .oecd.org/employment/covid-19.htm.

9. Ramon Gomez-Salvador and Michel Soudan, "The US Labour Market after the COVID-19 Recession," *European Central Bank Occasional Paper Series*, no. 298 (2022), https://www.ecb.europa.eu/pub/pdf/scpops/ecb.op298-f3f39e0b4f.en.pdf; John Hurley, Dragos Adăscăliței, and Elisa Staffa, "Recovery from COVID-19: The Changing Structure of Employment in the EU," *Eurofound*, Publications Office of the European Union, 2022, https://www.eurofound.europa.eu/en/publications/2021/recovery-covid -19-changing-structure-employment-eu.

10. Marc Bacchetta, Ekkehard Ernst, and Juana P. Bustamante, *Globalization and Informal Jobs in Developing Countries* (Geneva: ILO and WTO, 2009), http://www.ilo .org/wcmsp5/groups/public/@dgreports/@dcomm/documents/publication/wcms _115087.pdf.

11. Zygmunt Bauman, *Work, Consumerism and the New Poor* (Buckingham, PA: Open University Press, 1998); Ulrich Beck, *The Brave New World of Work* (Cambridge, MA: Polity Press, 2000).

12. Richard Sennett, *The Corrosion of Character: The Personal Consequences of Work in the New Capitalism* (New York: Norton, 1998).

13. Tim Strangleman, "The Nostalgia for Permanence at Work? The End of Work and Its Commentators," *The Sociological Review* 55, no. 1 (2007): 81–103 (85).

14. Gina Neff, *Venture Labor: Work and the Burden of Risk in Innovative Industries* (Cambridge, MA: MIT Press, 2012); Antonio A. Casilli, "How Venture Labor Sheds

Light on the Digital Platform Economy," *International Journal of Communication* 17 (2017): 2067-70.

15. Anne-Laure Fayard, "Notes on the Meaning of Work: Labor, Work, and Action in the 21st Century," *Journal of Management Inquiry* 30, no. 2 (2021): 207-20.

16. Alexandrea J. Ravenelle, *Hustle and Gig: Struggling and Surviving in the Sharing Economy* (University of California Press, 2019).

17. "Do the Hustle! Empowerment from Side-Hustles and Its Effects on Full-time Work Performance," *Academy of Management Journal* 64 (2020): 235-64.

18. Maarten Keune, "Flexicurity: A Contested Concept at the Core of the European Labour Market Debate," *Intereconomics* 43, no. 2 (2008): 92-98.

19. Lilly Irani, "The Cultural Work of Microwork," *New Media & Society* 17, no. 5 (2015): 720-39.

20. Neff, *Venture Labor*, 10.

21. Bernard Gazier, *Tous "sublimes": Vers un nouveau plein emploi* (Paris: Flammarion, 2003).

22. Robert Castel, *From Manual Workers to Wage Laborers: Transformation of the Social Question* (Transaction Publishers, 2003).

23. Pekka Himanen, *The Hacker Ethic and the Spirit of the Information Age* (New York: Random House, 1999).

24. Michel Lallement, *L'Âge du faire: Hacking, travail, anarchie* (Paris: Seuil, 2015), 156.

25. Fred Turner, *From Counterculture to Cyberculture: Stewart Brand, The Whole Earth Network, and the Rise of Digital Utopianism* (Chicago: University of Chicago Press, 2006).

26. Michael J. Piore, "Dualism in the Labor Market: A Response to Uncertainty and Flux. The Case of France," *Revue économique* 29, no. 1 (1978): 26-48.

27. Michael Reich, David M. Gordon, and Richard C. Edwards, "Dual Labor Markets: A Theory of Labor Market Segmentation," *American Economic Review* 63, no. 2 (1973): 361.

28. Paola Tubaro et al., "Hidden Inequalities: The Gendered Labour of Women on Micro-Tasking Platforms," *Internet Policy Review* 11, no. 1 (2022), https://policyreview .info/articles/analysis/hidden-inequalities-gendered-labour-women-micro-tasking -platforms.

29. Bernard Friot, "Employment and the Wage: Relationships on the Move," in *The Wage under Attack: Employment Policies in Europe*, ed. Bernard Friot and Bernadette Clasquin (Peter Lang, 2013), 47-70.

30. Michel Lallement, "Work and the Challenge of Autonomy," *Social Science Information* 54, no. 2 (2015): 229-48.

31. Lallement, "Work and the Challenge of Autonomy."

32. Isabelle Berrebi-Hoffmann et al., "Hyper-Flexibility in the IT Sector: Myth or Reality?," in *Globalization and Precarious Forms of Production and Employment: Challenges for Workers and Unions*, ed. Carole Thornley, Steve Jefferys, and Béatrice Appay (Cheltenham: Edward Elgar, 2010), 138.

33. Isabelle Berrebi-Hoffmann, "Les consultants et informaticiens: Un modèle d'encadrement de professionnels à l'échelle industrielle," *Revue française de gestion* 32, no. 168-169 (2006): 157-76.

34. Mark Chen et al., "Evaluating Large Language Models Trained on Code," arXiv .org, July 7, 2021, https://arxiv.org/abs/2107.03374.

35. Reed Albergotti and Louise Matsakis, "OpenAI Has Hired an Army of Contractors to Make Basic Coding Obsolete," *Semafor*, January 27, 2023, https://www.semafor .com/article/01/27/2023/openai-has-hired-an-army-of-contractors-to-make-basic -coding-obsolete.

36. Alain Cottereau, "Droit et bon droit: Un droit des ouvriers instauré, puis évincé par le droit du travail (France, XIXe siècle)," *Annales: Histoire, sciences sociales* 57, no. 6 (2022): 1521–57.

37. Claude Didry, *L'Institution du travail: Droit et salariat dans l'histoire* (Paris: La Dispute, 2016), 12.

38. "Drive With Uber: Be Your Own Boss," Uber, 2024, https://www.uber.com/sk /en/drive/.

39. See the London Employment Tribunal's decision of October 28, 2016, which reclassified two drivers from self-employed to workers. In the USA, the Superior Court of the County of San Francisco in California held that drivers for the ride-sharing company Uber were employees, not contractors (see Uber Technologies Inc. v. Barbara Berwick). In France, URSSAF Île-de-France brought proceedings against Uber for concealed work.

40. Cottereau, "Droit et bon droit," 1551.

41. Vili Lehdonvirta et al., "Online Labour Markets and the Persistence of Personal Networks: Evidence from Workers in Southeast Asia," paper presented at the conference of the American Sociological Association 2015, Chicago, August 2015.

42. Antonio A. Casilli et al., "From GAFAM to RUM: Platforms and Resourcefulness in the Global South," *Pouvoirs* 185 (2023): 51–67.

43. Alice Littlefield and Larry T. Reynolds, "The Putting-Out System: Transitional Form or Recurrent Feature of Capitalist Production?," *The Social Science Journal* 27, no. 4 (1990): 359–72; Elizabeth Prugl and Eileen Boris, *Homeworkers in Global Perspective Invisible No More* (Routledge, 1996); Prabin Baishya, "The Putting Out System in Ancient India," *Social Scientist* 25, no. 7/8 (1997): 51–56.

44. David Martín et al., "Turking in a Global Labour Market," *Computer Supported Cooperative Work* 25, no. 1 (2016): 39–77.

45. Ali Alkhatib, Michael S. Bernstein, and Margaret Levi, "Examining Crowd Work and Gig Work through the Historical Lens of Piecework," in *Proceedings of the 2017 CHI Conference on Human Factors in Computing Systems*, Denver, CO, Association for Computing Machinery, 2017, 4599–616; Veena Dubal, "The Time Politics of Home-Based Digital Piecework," *Center for Ethics Journal: Perspectives on Ethics*, Symposium Issue "The Future of Work in the Age of Automation and AI," 50, July 12, 2020, https:// c4ejournal.net/2020/07/04/v-b-dubal-the-time-politics-of-home-based-digital -piecework-2020-c4ej-xxx/.

46. Kiah Hawker and Nicholas Carah, "Snapchat's Augmented Reality Brand Culture: Sponsored Filters and Lenses as Digital Piecework," *Continuum: Journal of Media & Cultural Studies* 35, no. 1 (October 9, 2020): 12–29, https://doi.org/10.1080 /10304312.2020.1827370.

47. Renato Rodrigues da Silva, *The Anglo-Saxon Elite: Northumbrian Society in the Long Eighth Century* (Amsterdam University Press, 2021).

48. Alain Supiot, "Les nouveaux visages de la subordination," *Droit social*, no. 2 (2000): 131–45.

49. "Work That Fits around Your Life," Deliveroo, 2024, https://riders.deliveroo.fr/en/apply.

50. "MTurk Now Supports Amazon SNS Notifications," *Amazon Mechanical Turk Blog*, October 5, 2017, https://blog.mturk.com/mturk-now-supports-amazon-sns-notifications-b4847a6589e6.

51. Miriam A. Cherry and Antonio Aloisi, "'Dependent Contractors' in the Gig Economy: A Comparative Approach," *American University Law Review* 33, no. 3 (2016): 635–89.

52. Lilly Irani, "Difference and Dependence among Digital Workers: The Case of Amazon Mechanical Turk," *South Atlantic Quarterly* 114, no. 1 (2015): 225–34.

53. Sébastien Appiotti, "The Injunction to Share Photographs: A Form of Participation That Benefits the Public or the Institution?," *Hybrid*, no. 8, http://journals.openedition.org/hybrid/1975.

54. I borrow this formulation from the ethnologist Olivier D'Hont, *Techniques et savoirs des communautés rurales: Approche ethnographique du développement* (Paris: Karthala, 2005), while adapting it to the context of digital platforms. D'Hont used it to refer to relations between members of rural communities and intermediaries on the market of suppliers of agricultural materials.

55. BJ Fogg, "A Behavior Model for Persuasive Design," *Persuasive '09: Proceedings of the 4th International Conference on Persuasive Technology*, New York, Association for Computing Machinery, 2009, art. 40.

56. Maurizio Ferraris, *Mobilitazione totale* (Laterza, 2015).

57. Christian Licoppe, "Pragmatique de la notification," *Tracés*, no. 16 (2009): 77–98.

58. Danah Boyd, "Privacy and Security: The Politics of 'Real Names,'" *Communications of the ACM* 55, no. 8 (2012): 29–31; Oliver L. Haimson and Anna Lauren Hoffmann, "Constructing and Enforcing 'Authentic' Identity Online: Facebook, Real Names, and Non-Normative Identities" *First Monday* 21, no. 6 (2016), https://firstmonday.org/ojs/index.php/fm/article/view/6791.

59. Saadi Lahlou, "Attracteurs cognitifs et travail de bureau," *Intellectica* 30, no. 1 (2000): 75–113.

60. Caroline Datchary, *La Dispersion au travail* (Toulouse: Octarès, 2011).

61. Laurent Thevenot, "L'action en plan," *Sociologie du travail* 37, no. 3 (1995): 411–34.

62. Alessandro Delfanti, *The Warehouse: Workers and Robots at Amazon* (London: Pluto Press, 2021).

63. Paul M. Leonardi, Marleen Huysman, and Charles Steinfield, "Enterprise Social Media: Definition, History, and Prospects for the Study of Social Technologies in Organizations," *Journal of Computer-Mediated Communication* 19, no. 1 (October 1, 2013): 1–19.

64. Castel, *From Manual Workers to Wage Laborers*.

65. Cottereau, "Droit et bon droit."

66. David Golumbia, "Marxism and Open Access in the Humanities: Turning Academic Labor against Itself," *Workplace*, no. 28 (2016): 74–114.

67. Sarah T. Roberts, *Behind the Screen: Content Moderation in the Shadows of Social Media* (New Haven, CT: Yale University Press, 2019).

68. MicroSourcing, "Virtual Captives," 2016, https://www.micro-sourcing.com/virtual-captives.asp.

69. Erran Carmel, Mary Lacity, and Joseph W. Rottman, "Impact Sourcing: Employing Prison Inmates to Perform Digitally-Enabled Business Services," *Communications of the Association for Information Systems* 34, no. 1 (2014): art. no. 51.

70. Tuukka Lehtiniemi and Minna Ruckenstein, "Prisoners Training AI Ghosts, Humans and Values in Data Labour," in *Everyday Automation: Experiencing and Anticipating Emerging Technologies*, ed. Sarah Pink et al. (Abingdon, UK: Routledge, 2022), 184–96; Morgan Meaker, "These Prisoners Are Training AI," *Wired*, September 11, 2023, https://www.wired.com/story/prisoners-training-ai-finland/.

71. Kevin Rothrock, "There's a Prison in Russia Making Viral YouTube Videos," *GlobalVoices*, March 4, 2016, https://globalvoices.org/2016/03/04/theres-a-prison-in-russia-making-viral-youtube-videos/.

72. Clare Chambers-Jones, "Virtual World Financial Crime: Legally Flawed," *Law and Financial Markets Review* 7, no. 1 (2013): 48–56; Nick Dyer-Witheford, *Games of Empire: Global Capitalism and Video Games* (Minneapolis: University of Minnesota Press, 2009).

73. Nick Dyer-Witheford, *Games of Empire: Global Capitalism and Video Games* (Minneapolis: University of Minnesota Press, 2009).

74. Shannon Sims, "The End of American Prison Visits: Jails End Face-to-Face Contact—And Families Suffer," *The Guardian*, December 9, 2017, https://www.theguardian.com/us-news/2017/dec/09/skype-for-jailed-video-calls-prisons-replace-in-person-visits.

75. Dave Maass, "The Hidden Cost of JPay's Prison Email Service," Electronic Frontier Foundation, May 5, 2015, https://www.eff.org/deeplinks/2015/05/hidden-cost-jpays-prison-email-system.

76. Bennett Cyphers and Karen Gullo, "Inside the Invasive, Secretive 'Bossware' Tracking Workers," Electronic Frontier Foundation, June 30, 2020, https://www.eff.org/deeplinks/2020/06/inside-invasive-secretive-bossware-tracking-workers.

77. David Kravets, "Worker Fired for Disabling GPS App That Tracked Her 24 Hours a Day," *Ars Technica*, May 11, 2015, http://arstechnica.com/tech-policy/2015/05/worker-fired-for-disabling-gps-app-that-tracked-her-24-hours-a-day.

78. Kirstie Ball, "Workplace Surveillance: An Overview," *Labor History* 51, no. 1 (2010): 87–106.

79. Barbara Ehrenreich, *Nickel and Dimed: On (Not) Getting by in America* (New York: Metropolitan Books, 2001).

80. Samuel Warren and Louis Brandeis, "The Right to Privacy," *Harvard Law Review* 4, no. 5 (1890): 193–220; John Deigh, "Privacidad, democracia e Internet," in *Internet y el Futuro de la Democracia*, ed. Serge Champeau and Daniel Innerarity (Barcelona: Paidós, 2012), 119–32.

81. Judith Donath, "Signals in Social Supernets," *Journal of Computer-Mediated Communication* 13, no. 1 (2007): 231–51.

82. Helen Nissenbaum, *Privacy in Context: Technology, Policy, and the Integrity of Social Life* (Redwood City, CA: Stanford University Press, 2009).

83. Paola Tubaro, Antonio A. Casilli, and Yasaman Sarabi, *Against the Hypothesis of the End of Privacy* (Springer Cham, 2014), https://doi.org/10.1007/978-3-319-02456-1.

84. Ifeoma Ajunwa, Kate Crawford, and Jason Schultz, "Limitless Worker Surveillance," *California Law Review* 105, no. 3 (2017): 101–42.

85. Julie E. Cohen, "The Surveillance-Innovation Complex: The Irony of the Participatory Turn," in *The Participatory Condition in The Digital Age*, ed. Darin Barney (Minneapolis: University of Minnesota Press, 2016), 207-26.

86. Shoshana Zuboff, *The Age of Surveillance Capitalism: The Fight for a Human Future at the New Frontier of Power* (New York: Public Affairs, 2018).

87. Alex Rosenblat, Tamara Kneese, and Danah Boyd, "Workplace Surveillance," *Open Society Foundations' Future of Work Commissioned Research Papers*, October 2014, https://datasociety.net/pubs/fow/WorkplaceSurveillance.pdf.

88. Neff, *Venture Labor*.

89. Antonio A. Casilli, "Four Theses on Digital Mass Surveillance and the Negotiation of Privacy," presented at the 8th Annual Privacy Law Scholar Congress, Berkeley Center for Law & Technology, June 2015, Berkeley, CA, https://shs.hal.science/halshs-01147832.

90. Jamais Cascio, "Participatory Panopticon," *Institute for the Future, Ten-Year Forecast: Perspectives*, SR-1064 (2007), 20-27.

91. "Make Work Better," BetterWorks, 2023, https://www.betterworks.com.

92. Ajunwa, Crawford, and Schultz, "Limitless Worker Surveillance."

93. Alberto Romele et al., "Panopticism Is Not Enough: Social Media as Technologies of Voluntary Servitude," *Surveillance and Society* 15, no. 2 (2017): 204-21.

94. Gary T. Marx, "Soft Surveillance: The Growth of Mandatory Volunteerism in Collecting Personal Information—'Hey Buddy Can You Spare a DNA?,'" in *Surveillance and Security: Technological Politics and Power in Everyday Life*, ed. Torin Monahan (New York: Routledge, 2006), 37-56.

95. Valérie Sédallian, "Les conditions générales d'utilisation ont-elles une juridique?," *Documentaliste—Sciences de l' information* 49, no. 1 (2012): 16-19.

96. David Chau, "Uber Eats Imposes 'Unfair Contracts' and Ruins Deliveries, Restaurateurs Allege," ABC News, April 21, 2018, https://www.abc.net.au/news/2018-04-22/uber-eats-criticised-over-conditions-on-restaurant-owners/9662814.

97. Amazon Mechanical Turk, "Participation Agreement," March 25, 2020, https://www.mturk.com/worker/participation-agreement.

98. Howtank, "General Terms and Conditions in USA," February 10, 2020, https://www.howtank.com/conditions-generales#usa.

99. Valerio De Stefano, "The Rise of the 'Just-in-Time Workforce': On-Demand Work, Crowdwork and Labor Protection in the 'Gig-Economy,'" *Conditions of Work and Employment Series* 71 (2016), http://www.ilo.org/wcmsp5/groups/public/---ed_protect/---protrav/---travail/documents/publication/wcms_443267.pdf.

100. "TaskRabbit Global Terms of Service," TaskRabbit, February 19, 2021, https://www.taskrabbit.com/terms.

101. Zach Zhizhong Zhou and Kevin Zhu, "Platform Battle with Lock-In," *ICIS 2006 Proceedings*, AIS Electronic Library (AISeL), art. 20, 2006, 267-84, http://aisel.aisnet.org/cgi/viewcontent.cgi?article=1141&context=icis2006.

102. Marios Kokkodis and Panagiotis G. Ipeirotis, "Reputation Transferability in Online Labor Markets," *Management Science* 62, no. 6 (2015): 1687-706; Directorate for Financial and Entreprise Affairs Competition Committee, "Data Portability, Interoperability and Competition—Note by TUAC," OECD, June 9, 2021, https://one.oecd.org/document/DAF/COMP/WD(2021)35/en/pdf.

103. "The data subject shall have the right to receive the personal data concerning

him or her, which he or she has provided to a controller, in a structured, commonly used and machine-readable format and have the right to transmit those data to another controller without hindrance from the controller to which the personal data have been provided." GDPR, Article 20(1).

104. "ECLI:NL:RBAMS:2021:1020," *Rechtspraak*, March 11, 2021, https://uitspraken .rechtspraak.nl/details?id=ECLI:NL:RBAMS:2021:1020&showbutton=true&keyword =uber.

105. Ludmila Costhek Abílio, "Uberization: The Periphery as the Future of Work?," in *Platformization and Informality: Dynamics of Virtual Work*, ed. Surie and Huws (Palgrave Macmillan, 2023), 139–60.

106. De Stefano, "The Rise of the 'Just-in-Time Workforce,'" 32.

107. "Cour de cassation Pourvoi no. 19-13.316," Cour de Cassation, Chambre sociale—Formation plénière de chambre, March 4, 2020, https://www.courdecassation .fr/decision/5fca56cd0a790c1ec36ddc07.

108. Paul-Henri Antonmattei and Jean-Christophe Sciberras, *Le Travailleur économiquement dépendant: quelle protection?*, report to the Minister for Labor, Social Relations, the Family and Solidarity, November 2008.

109. Delphine Gardes, "Une définition juridique du travail," *Droit social*, no. 4 (2014): 373–82.

Chapter Eight

1. Christian Fuchs, *Digital Labor and Karl Marx* (New York: Routledge, 2014).

2. Kate Crawford, *Atlas of AI: Power, Politics, and the Planetary Costs of Artificial Intelligence* (New Haven, CT: Yale University Press, 2021).

3. Georges Friedmann, *The Anatomy of Work: Labor, Leisure, and the Implications of Automation* (1956; repr., Free Press, 1961).

4. Friedmann, *The Anatomy of Work*, 143.

5. Annalee Newitz, "Op-ed: Mark Zuckerberg's Manifesto Is a Political Trainwreck," *ArsTechnica*, February 18, 2017, https://arstechnica.com/staff/2017/02/op-ed-mark -zuckerbergs-manifesto-is-a-political-trainwreck/.

6. Sarrah Kassem, *Work and Alienation in the Platform Economy: Amazon and the Power of Organization* (Bristol: Bristol University Press, 2023).

7. Kylie Jarrett, *Feminism, Labour and Digital Media: The Digital Housewife* (London: Routledge, 2015).

8. Mark Graham et al., "Could Online Gig Work Drive Development in Lower-Income Countries?," in *The Future of Work in the Global South*, ed. Hernan Galperin and Andrea Alarcon (Ottawa: IDRC, 2018), 8–11.

9. Katherine K. Chen, "Prosumption: From Parasitic to Prefigurative," *The Sociological Quarterly* 56, no. 3 (2015): 446–59.

10. Carlo Formenti, *Felici e sfruttati. Capitalismo digitale ed eclissi del lavoro* (Milan: EGEA, 2011).

11. Davide Dusi, "Investigating the Exploitative and Empowering Potential of the Prosumption Phenomenon," *Sociology Compass* 11, no. 6 (2017): e12488.

12. Eran Fisher, "How Less Alienation Creates More Exploitation? Audience Labor on Social Network Sites," *tripleC* 10, no. 2 (2012): 171–83 (175).

13. Marie-Anne Dujarier, "The Activity of the Consumer: Strengthening, Trans-

forming, or Contesting Capitalism?," *The Sociological Quarterly* 56, no. 3 (2015): 460–71.

14. Richard Barbrook, *The Class of the New* (London: Openmute.org, 2006).

15. Friedrich Engels, *The Condition of the Working-Class in England* (New York: Engels, 1892).

16. Vladimir I. Lenin, *Imperialism, the Highest Stage of Capitalism* (1917).

17. William Morris and E. Belfort Bax, *Socialism: Its Growth and Outcome* (Swan Sonnenschein, 1893), 275.

18. Antonio Gramsci, *Prison Notebooks* [1929–1935], trans. and ed. Joseph Anthony Buttigieg II, vol. 3 (New York: Columbia University Press, 1991–2011).

19. Thorstein Veblen, *The Engineers and the Price System* (1919).

20. William Whyte, *The Organization Man* (New York: Simon & Schuster, 1956).

21. Serge Mallet, *La Nouvelle Classe ouvrière* (Paris: Seuil, 1964).

22. Arthur Kroker and Michael A. Weinstein, *Data Trash: The Theory of the Virtual Class* (Montreal: New World Perspectives, 1994).

23. Richard Florida, *The Rise of the Creative Class, and How It's Transforming Work, Leisure, Community and Everyday Life* (New York: Basic Books, 2002).

24. Ian Angell, *The New Barbarian Manifesto: How to Survive the Information Age* (London: Kogan Page Ltd., 2000).

25. Barbrook, *The Class of the New*; Christian Fuchs, "Labor in Informational Capitalism and on the Internet," *The Information Society* 26, no. 3 (2010): 179–96.

26. Mike Wayne, *Marxism and Media Studies: Key Concepts and Contemporary Trends* (London: Pluto, 2003).

27. André Gorz, *Adieux au prolétariat: Au-dela du Socialisme* (Paris: Galilée, 1980), 61.

28. Alvin Toffler, *Previews and Premises* (New York: William Morrow, 1983).

29. Franco Berardi (Bifo), *La fabbrica dell'infelicità: New economy e movimento del cognitariato* (Rome: Derive Approdi, 2002).

30. Nick Dyer-Witheford, *Cyber-Marx: Cycles and Circuits of Struggle in High-Technology Capitalism* (Urbana: University of Illinois Press, 1999).

31. Ursula Huws, *The Making of a Cybertariat: Virtual Work in a Real World* (New York: Monthly Review Press, 2003), 19.

32. Guy Standing, *The Precariat: The New Dangerous Class* (New York: Bloomsbury Academic, 2011).

33. Manuel Castells, "An Introduction to the Information Age," in *The Information Society Reader*, ed. Frank Webster et al. (New York: Routledge, 2001), 138–49.

34. McKenzie Wark, "The Vectoralist Class," *e-flux Journal* 65 (2015).

35. In spite of Amazon having over 600 retail stores (85 percent of which are Whole Foods), they've only opened a handful of bookstores since 2015. In 2022, the platform announced that all locations would close.

36. Wark, "The Vectoralist Class."

37. McKenzie Wark, *Telesthesia: Communication, Culture and Class* (Polity Press, 2012), 164.

38. Wark, *Telesthesia*.

39. Wark, "The Vectoralist Class."

40. Lilly Irani et al., "Postcolonial Computing: A Lens on Design and Development," in *Proceedings of the SIGCHI Conference on Human Factors in Computing Systems*, Atlanta, GA, Association for Computing Machinery, 2010, 1311–20.

41. Syed Mustafa Ali, "A Brief Introduction to Decolonial Computing," *Crossroads* 22, no. 4 (June 13, 2016): 16–21.

42. Antonio A. Casilli, "Digital Labor Studies Go Global: Towards a Digital Decolonial Turn," *International Journal of Communication* 11 (2017): 3934–54.

43. Roberto Casati, *Contre le colonialisme numérique: Manifeste pour continuer à lire* (Paris: Albin Michel, 2013).

44. Dmytri Kleiner, "Mr. Peel Goes to Cyberspace: Resisting Digital Colonization," communication at the *Digital Bauhaus Summit 384 2016—Luxury Communism*, Neufert-Box & Deutsches Nationaltheater, Weimar, 2016.

45. Jim Thatcher, David O'Sullivan, and Dillon Mahmoudi, "Data Colonialism through Accumulation by Dispossession: New Metaphors for Daily Data," *Environment and Planning D: Society and Space* 34, no. 6 (July 26, 2016): 990–1006, https://doi.org/10.1177/0263775816633195.

46. Petar Jandrić and Ana Kuzmanić, "Digital Postcolonialism," *IADIS: International Journal on WWW/Internet* 13, no. 2 (2015): 34–51.

47. Adrienne LaFrance, "Facebook and the New Colonialism," *The Atlantic*, February 11, 2016, https://www.theatlantic.com/technology/archive/2016/02/facebook-and-the-new-colonialism/462393/.

48. Gordon Bell, "The Colonization of Cyberspace," seminar held at the Institut d'Estudis Catalans, Barcelona, 1999.

49. Howard Rheingold, *Virtual Community: Homesteading on the Electronic Frontier* (Reading, UK: Addison-Wesley, 1993).

50. Marc Prensky, "Digital Natives, Digital Immigrants," *On the Horizon* 9, no. 5 (2001): 1–6.

51. Nick Could and Ulises Ali Mejias, *The Costs of Connection: How Data Is Colonizing Human Life and Appropriating It for Capitalism* (Stanford University Press, 2019).

52. Nick Couldry and Ulises Ali Mejias, "The Decolonial Turn in Data and Technology Research: What Is at Stake and Where Is It Heading?," *Information, Communication & Society* 26, no. 4 (2023): 786–802.

53. The Tierra Común Network, *Resisting Data Colonialism: A Practical Intervention* (Amsterdam: Institute of Network Cultures, 2023); James Muldoon and Boxi A Wu, "Artificial Intelligence in the Colonial Matrix of Power," *Philosophy & Technology* 36 (2023): art. no. 80, https://link.springer.com/article/10.1007/s13347-023-00687-8. See also the project by Kate Crawford and Vladan Joler, "Anatomy of an AI System: The Amazon Echo as an Anatomical Map of Human Labor, Data and Planetary Resources," AI Now Institute and Share Lab, September 7, 2018, https://anatomyof.ai.

54. Christian Fuchs, "Digital Labor and Imperialism," *Monthly Review* 67, no. 8 (2016): 14–24.

55. Jack Linchuan Qiu, *Goodbye iSlave: A Manifesto for Digital Abolition* (Urbana: University of Illinois Press, 2017).

56. Jack Linchuan Qiu, *Working-Class Network Society: Communication Technology and the Information Have-Less in Urban China* (Cambridge, MA: MIT Press, 2009).

57. Jack Linchuan Qiu and Lin Lin, "Foxconn: The Disruption of iSlavery," *Asiascape: Digital Asia* 4, no. 1–2 (2017): 103–28.

58. Qiu, *Goodbye iSlave*.

59. Vili Lehdonvirta et al., "Online Labour Markets: Levelling the Playing Field for International Service Markets?," IPP2014: Crowdsourcing for Politics and Policy Conference, Oxford Internet Institute, 2014.

60. Clément Le Ludec, Maxime Cornet, and Antonio A. Casilli, "The Problem with Annotation: Human Labour and Outsourcing between France and Madagascar," *Big Data & Society* 10, no. 2 (July 1, 2023), https://journals.sagepub.com/doi/10.1177/20539517231188723.

61. Couldry and Mejias, "The Decolonial Turn."

62. Payal Arora, "Bottom of the Data Pyramid: Big Data and the Global South," *International Journal of Communication* 10, no. 1 (2014): 1681-99.

63. OECD, "Crisis Squeezes Income and Puts Pressure on Inequality and Poverty," 2013, http://archives.strategie.gouv.fr/cas/system/files/oecd2013-inequality-and-poverty-8p.pdf.

64. International Monetary Fund, *Jobs and Growth: Analytical and Operational Considerations for the Fund*, Paris, March 14, 2013, https://www.imf.org/external/np/pp/eng/2013/031413.pdf.

65. Andrew Ross, *Fast Boat to China: Corporate Flight and the Consequences of Free Trade: Lessons from Shanghai* (New York: Random House, 2006).

66. Niels Van Doorn, Fabian Ferrari, and Mark Graham, "Migration and Migrant Labour in the Gig Economy: An Intervention," *Work, Employment and Society* 37, no. 4 (2023): 1099-111.

67. Mark Graham, Isis Hjorth, and Vili Lehdonvirta, "Digital Labour and Development: Impacts of Global Digital Labour Platforms and the Gig Economy on Worker Livelihoods," *Transfer: European Review of Labour and Research* 23 (2017): 135-62.

68. Ayhan Aytes, "Return of the Crowds: Mechanical Turk and Neo-Liberal States of Exception," in *Digital Labor: The Internet as Playground and Factory*, ed. Trebor Scholz (New York: Routledge, 2012), 79-97.

69. *Global Wage Report 2016/17: Wage Inequality in the Workplace* (Geneva: ILO, 2016); *Global Wage Report 2022-23: The Impact of COVID-19 and Inflation on Wages and Purchasing Power* (Geneva: ILO, 2022).

70. Brendan M. Sullivan et al., "Socioeconomic Group Classification Based on User Features," Patent Application US15/221,587, filed July 27, 2016, priority date July 27, 2016, assigned to Meta Platforms Inc, published February 1, 2018, granted and published on March 31, 2020, with adjusted expiration on October 5, 2038.

71. Sullivan, "Socioeconomic Group Classification."

72. Sullivan, "Socioeconomic Group Classification."

73. Edward Andrew, "Class in Itself and Class against Capital: Karl Marx and His Classifiers," *Canadian Journal of Political Science* 16, no. 3 (1983): 577-84.

74. Roberto Ciccarelli, *Labour Power: Virtual and Actual in Digital Production* (Springer, 2021).

75. Bruce J. Berman, "Artificial Intelligence and the Ideology of Capitalist Reconstruction," *AI & Society* 6, no. 2 (1992): 103-14.

76. Edward A. Feigenbaum, "Some Challenges and Grand Challenges for Computational Intelligence," *Journal of the ACM* 50, no. 1 (2003): 39.

77. Feigenbaum, "Some Challenges."

78. Brian Hayes, "The Manifest Destiny of Artificial Intelligence," *American Scientist* 100, no. 4 (2012): 282-87.

79. Lilly Irani, "The Cultural Work of Microwork," *New Media & Society* 17, no. 5 (2015): 720-39.

80. Hamid R. Ekbia and Bonnie A. Nardi, *Heteromation, and Other Stories of Computing and Capitalism* (Cambridge, MA: MIT Press, 2017).

81. Irani, "The Cultural Work of Microwork."

82. Hope Reese and Nick Heath, "Inside Amazon's Clickworker Platform: How Half a Million People Are Being Paid Pennies to Train AI," *Tech Republic*, December 16, 2016, http://www.techrepublic.com/article/inside-amazons-clickworker-platform-how-half -a-million-people-are-training-ai-for-pennies-per-task/.

83. OpenAI, "AI and Compute," *OpenAI Blog*, May 16, 2018, https://blog.openai.com /ai-and-compute/.

84. Jeremias Prassl, *Humans as a Service: The Promise and Perils of Work in the Gig Economy* (Oxford, UK: Oxford University Press, 2018).

85. Paul N. Edwards, *The Closed World: Computers and the Politics of Discourse in Cold War America* (Cambridge, MA: MIT Press, 1996).

86. Danièle Linhart, *La Comédie humaine du travail. De la déshumanisation taylo-rienne à la sur-humanisation managériale* (Toulouse: Érès, 2015).

87. Berman, "Artificial Intelligence."

88. Jonas Oppenlaender, "The Creativity of Text-to-Image Generation," in *Proceedings of the 25th International Academic Mindtrek Conference*, Tampere, Finland, Association for Computing Machinery, 2022, 192–202.

89. Marc Cheong et al., "Investigating Gender and Racial Biases in DALL-E Mini Images," *Proceedings of the National Academy of Sciences* 30, no. 2 (2023): 1–16.

90. Jan Smits and Tijn Borghuis, "Generative AI and Intellectual Property Rights," in *Law and Artificial Intelligence. Information Technology and Law Series*, vol. 35, ed. Bart Custers and Eduard Fosch-Villaronga (T. M. C. Asser Press, 2022), 323–44.

91. Pawel Korzynski et al., "Artificial Intelligence Prompt Engineering as a New Dig-ital Competence: Analysis of Generative AI Technologies such as ChatGPT," *Entrepreneurial Business and Economics Review* 11, no. 3 (2023): 25–38.

92. Hayes, "The Manifest Destiny."

93. Jennings Brown, "IBM Watson Reportedly Recommended Cancer Treatments That Were 'Unsafe and Incorrect,'" *Gizmodo*, July 25, 2018, https://gizmodo.com/ibm -watson-reportedly-recommended-cancer-treatments-tha-1827868882.

94. Bernie Hogan, "From Invisible Algorithms to Interactive Affordances: Data after the Ideology of Machine Learning," in *Roles, Trust, and Reputation in Social Media Knowledge Markets: Theory and Methods*, ed. Elisa Bertino and Sorin Adam Matei (Springer, 2014), 103–19.

95. Tianyu Wu et al., "A Brief Overview of ChatGPT: The History, Status Quo and Potential Future Development," in *IEEE/CAA Journal of Automatica Sinica* 10, no. 5 (2023): 1122–36.

96. "Earn Free Hours," MidJourney, August 3, 2023, https://docs.midjourney.com /docs/free-hours.

97. Jean-Gabriel Ganascia, *Le Mythe de la Singularité: Faut-il craindre l'intelligence artificielle?* (Paris: Seuil, 2017), 51.

98. Ganascia, *Le Mythe de la Singularité*.

99. Davide Castelvecchi, "Can We Open the Black Box of AI?," *Nature* 538, no. 7623 (2016): 21–23.

100. Aravindh Mahendran and Andrea Vedaldi, "Understanding Deep Image Rep-resentations by Inverting Them," in *Computer Vision and Pattern Recognition (CVPR '15)*, Boston, MA, IEEE, 2015, 5188–96.

101. "New International Consortium Formed to Create Trustworthy and Reli-

able Generative AI Models for Science," Argonne National Laboratory, November 10, 2023, https://www.anl.gov/article/new-international-consortium-formed-to-create-trustworthy-and-reliable-generative-ai-models-for.

102. Adrian Mackenzie, *Machine Learners: Archeology of a Data Practice* (Cambridge, MA: MIT Press, 2017).

103. Boris Van Breugel, Zhaozhi Qian, and Mihaela Van Der Schaar, "Synthetic Data, Real Errors: How (Not) to Publish and Use Synthetic Data," in *Proceedings of the 40th International Conference on Machine Learning* (ICML'23) 2023, JMLR.org, Article 1448, 34793–808.

104. Philip N. Johnson-Laird and Marco Ragni, "What Should Replace the Turing Test?," *Intelligent Computing* 2 (2023), https://doi.org/10.34133/icomputing.0064.

105. Nils J. Nilsson, "Human-Level Artificial Intelligence? Be Serious!," *AI Magazine*, 2005, 73 (emphasis mine).

106. Ernest Mandel, "Marx, the Present Crisis and the Future of Labour," *Socialist Register* (1985/1986), 436–44. This paper was initially delivered at a colloquium, The Future of Human Labour, organized by the Institute for Marxist Studies at the Vrise University, Brussels, February 14-16, 1985.

Conclusion

1. Axel Honneth, *The Struggle for Recognition. The Moral Grammar of Social Conflicts* (Cambridge, MA: Polity Press, 2015 [1992]).

2. Apart from the examples cited, there is also the class action brought against Google by a group of American citizens who seek to have users of reCAPTCHA integrated into the firm as employees. See Gabriela Rojas-Lozano v. Google Inc., United States District Court for the District of Massachusetts, case 3:15-cv-10160-MGM, 2015.

3. For example, Selina Wang, "Uber Drivers Strike to Protest Fare Cuts in New York City," *Bloomberg*, February 1, 2016, https://www.bloomberg.com/news/articles/2016-02-01/uber-drivers-plan-strike-to-protest-fare-cuts-in-new-york-city; TNN, "Ola, Uber Drivers on Hunger Strike from Today," *The Times of India*, May 18, 2018, https://timesofindia.indiatimes.com/city/jaipur/ola-uber-drivers-on-hunger-strike-from-today/articleshow/64214992.cms; Hilary Osborne and Sean Farrell, "Deliveroo Workers Strike Again over New Pay Structure," *The Guardian*, August, 15, 2016, https://www.theguardian.com/business/2016/aug/15/deliveroo-workers-strike-again-over-new-pay-structure; Karen Cheung, "Hong Kong Riders for Takeaway App Deliveroo Go on Strike over New Work Arrangements," *Hong Kong Free Press*, January 23, 2018, https://www.hongkongfp.com/2018/01/23/hong-kong-riders-takeaway-app-deliveroo-go-strike-new-work-arrangements/.

4. The court cases are accessible via the EuroFund platform economy database. "Platform Economy Database," EuroFund, 2023, https://apps.eurofound.europa.eu/platformeconomydb/.

5. "Platform Workers: Council Confirms Agreement on New Rules to Improve Their Working Conditions," Council of the EU, press release, March 11, 2024, https://www.consilium.europa.eu/en/press/press-releases/2024/03/11/platform-workers-council-confirms-agreement-on-new-rules-to-improve-their-working-conditions/.

6. Antonio Aloisi and Valerio De Stefano, "'Gig' Workers in Europe: The New

Platform of Rights," *Social Europe*, March 16, 2024, https://www.socialeurope.eu/gig
-workers-in-europe-the-new-platform-of-rights.

7. "Spain: Supreme Court Decision on the Employment Status of Workers for a Delivery Company and Social Dialogue Process on a 'Riders Law,'" Industrial Relations and Labor Law Newsletter, October 2020, https://industrialrelationsnews.ioe-emp.org/industrial-relations-and-labour-law-october-2020/news/article/spain-supreme-court-decision-on-the-employment-status-of-workers-for-a-delivery-company-and-social-dialogue-process-on-a-riders-law.

8. Anna Ilsøe, "The Hilfr Agreement: Negotiating the Platform Economy in Denmark" (FAOS, 2020), 176.

9. Nicola Coutouris and Valerio De Stefano, "Collective-Bargaining Rights for Platform Workers," *Social Europe*, October 6, 2020, https://www.socialeurope.eu/collective-bargaining-rights-for-platform-workers.

10. In addition to Amazon Mechanical Turk's microworkers, who attempted to organize under the aegis of WeAreDynamo (see chap. 4), later initiatives in organized protest include those of Upwork's users. See Alex Wood, "Variable Geographies of Protest among Online Gig Workers," Oxford Internet Institute Blog, February 13, 2017, https://www.oii.ox.ac.uk/blog/variable-geographies-of-protest-among-online-gig-workers/.

11. Grasielle Castro, "TRT2 reconhece vínculo empregatício de terceirizados em ambiente virtual," JOTA, March 30, 2023, https://www.jota.info/tributos-e-empresas/trabalho/trt2-reconhece-vinculo-empregaticio-de-terceirizados-em-ambiente-virtual-30032023.

12. Court of cassation, *Contrôle et contentieux—Distinction entre "Clic and walk" et emploi salarié*, judgment handed down on April 5, 2022, by the Criminal Division of the French Supreme Court, https://www.dalloz-actualite.fr/sites/dalloz-actualite.fr/files/resources/2022/04/20-81.775.pdf.

13. Fair Crowd Work, "Fair Crowd Work Shedding Light on the Real Work of Crowd-, Platform-, and App-Based Work," 2016, http://faircrowd.work/platform-reviews/platform-review-information/.

14. See, for example, Aitor Riveiro, "Los usuarios de Tuenti se declaran en huelga contra los términos de uso," *El Pais*, March 5, 2009, https://elpais.com/tecnologia/2009/03/05/actualidad/1236245278_850215.html; Kim LaCapria, "Great Reddit Clackout of 2015 (a.k.a. AMAgeddon)," *Snopes*, July 3, 2015, https://www.snopes.com/news/2015/07/03/reddit-blackout-2015/; Nicola Slawson, "Faceblock Campaign Urges Users to Boycott Facebook for a Day," *The Guardian*, April 7, 2018, https://www.theguardian.com/technology/2018/apr/07/faceblock-campaign-urges-users-boycott-facebook-for-one-day-protest-cambridge-analytica-scandal; Sheera Frenkel et al., "Facebook Employees Stage Virtual Walkout to Protest Trump Posts," *New York Times*, October 10, 2020, https://www.nytimes.com/2020/06/01/technology/facebook-employee-protest-trump.html; Guillemette Faure, "Should We Boycott Twitter?," *Le Monde*, December 4, 2022, https://www.lemonde.fr/en/m-le-mag/article/2022/12/04/should-we-boycott-twitter_6006525_117.html; Atanu Biswas, "View: Delete Account to Boycott Social Media?," *Economic Times*, September 3, 2022, https://economictimes.indiatimes.com/opinion/et-commentary/view-delete-account-to-boycott-social-media/articleshow/93973288.cms.

15. For example, the Dutch union Datavakbond, which protects the interests of users of major platforms like Google and Facebook by sending their reps to the union's platform: https://datavakbond.nl/.

16. Kwame Opam, "Uber Drivers in California Join with Teamsters Union to Fight for Better Benefits," *The Verge*, April 24, 2016, https://www.theverge.com/2016/4/24/11497842/uber-drivers-teamsters-partnership-wages-benefits.

17. Sarah Butler, "Gig Economy Union Seeks to Raise £50,000 to Fund Deliveroo Fight," *The Guardian*, May 16, 2018, https://www.theguardian.com/business/2018/may/16/gig-economy-union-seeking-to-raise-cash-to-fund-deliveroo-fight-legal-action.

18. Raquel Pascual Cortés, "CC OO y UGT acusan a CEOE de someterse a lobbies explotadores en la regulación de plataformas digitales," *El País*, December 4, 2020, https://cincodias.elpais.com/cincodias/2020/12/04/economia/1607103727_243182.html.

19. Billy Perrigo, "150 African Workers for ChatGPT, TikTok and Facebook Vote to Unionize at Landmark Nairobi Meeting," *TIME*, May 1, 2023, https://time.com/6275995/chatgpt-facebook-african-workers-union/.

20. For example, Fairwork, which works with the ILO and the UN Conference on Trade and Development to develop equitable standards for digital labor. Fairwork, "Fairwork Homepage," 2024, https://fair.work/en/fw/homepage/.

21. "Frankfurt Paper on Platform-Based Work," Crowdwork IG Metall, December 6, 2016, http://crowdwork-igmetall.de.

22. Lukas Sonnenberg et al., "Irresponsible Technologies: Who Is Accountable for the Workers in the AI Supply Chains?," keynote panel, INDL "Digital Labor in the Wake of Pandemic Times," Weizenbaum Institut, Berlin, 2023, https://www.youtube.com/live/RWYg6b1CWDs; Antonio A. Casilli and Intérêt à Agir, "IA: trois mesures urgentes pour protéger les travailleurs de la donnée," *Libération*, November 1, 2023, https://www.liberation.fr/idees-et-debats/tribunes/ia-trois-mesures-urgentes-pour-proteger-les-travailleurs-de-la-donnee-20231101_OG4HDG7RLNAFLGDLUXQGKYY6JA/.

23. Jaron Lanier, *Who Owns the Future?* (New York: Simon & Schuster, 2013).

24. Génération Libre, "Rapport: Mes data sont à moi: Pour une patrimonialité des données personnelles," January 2018, https://www.generationlibre.eu/wp-content/uploads/2018/01/2018-01-generationlibre-patrimonialite-des-donnees.pdf.

25. "Data laborers could organize a 'data labor union' that would collectively bargain with siren servers. While no individual user has much bargaining power, a union that filters platform access to user data could credibly call a powerful strike. Such a union could be an access gateway, making a strike easy to enforce and on a social network, where users would be pressured by friends not to break a strike, this might be particularly effective" (Imanol Arrieta-Ibarra et al., "Should We Treat Data as Labor? Moving beyond 'Free,'" *AEA Papers and Proceedings* 108 [2018]: 40).

26. See on this subject the series of "Platform cooperativism" international conferences organized by the consortium coordinated by Trebor Scholz and Nathan Schneider, "Supporting the Platform Co-Op Ecosystem: Events," Platform Cooperativism Consortium, 2018, https://platform.coop/events.

27. Niels Van Doorn, "Analysis: Platform Cooperativism and the Problem of the Outside," *Culture Digitally*, February 7, 2017, http://culturedigitally.org/2017/02/platform-cooperativism-and-the-problem-of-the-outside/.

28. Trebor Scholz and Nathan Schneider, *Ours to Hack and to Own: The Rise of Platform Cooperativism: A New Vision for the Future of Work and a Fairer Internet* (New York: OR Books, 2017).

29. Trebor Scholz, *Platform Cooperativism: Challenging the Corporate Sharing Economy* (New York: Rosa Luxemburg Stiftung, 2016), 18-19.

30. James Sullivan, "Home Cleaning Co-Ops in USA Get Their Own 'Uber,'" *Cooperative News*, January 28, 2016, https://www.thenews.coop/101205/sector/worker-coops/home-cleaning-coops-usa-get-uber/.

31. Trebor Scholz, *Uberworked and Underpaid: How Workers Are Disrupting the Digital Economy* (New York: Polity Press 2016).

32. "We Socialize Bike Delivery," CoopCycle, 2024, https://coopcycle.org/en/.

33. "Illustration & Animation Highlights," Stocksy, 2024, https://www.stocksy.com; "A Music Platform We Can All Control. No, Really," Resonate, 2024, https://resonate.coop.

34. Rafael Grohmann, "Dead Platform Co-Ops: Archiving Worker-Owned Experiences in the Delivery Sector," paper presented at *International Network on Digital Labor* INDL-6, "Digital Labor in the Wake of Pandemic Times," Weizenbaum Institut, Berlin, October 9-11, 2023, https://www.indl.network/indl-6-program/.

35. Trebor Scholz, *Uberworked and Underpaid*.

36. Marisol Sandoval, "Entrepreneurial Activism? Platform Cooperativism Between Subversion and Co-optation," *Critical Sociology* 46, no. 6 (2020): 801-17.

37. In May 2018, the Platform Cooperativism Consortium received a million-dollar grant from Google. See Robert Raymond, "Platform Cooperativism Consortium Awarded $1 Million Google.org Grant," *Shareable*, May 31, 2018, https://www.shareable.net/blog/the-platform-cooperativism-consortium-awarded-1-million-googleorg-grant.

38. Platform Cooperativism Consortium, "Mission," New School, 2018, http://platformcoop.newschool.edu/index.php/mission/.

39. For a definition of this concept, see Lionel Maurel, "La reconnaissance du 'domaine commun informationnel': tirer les enseignements d'un échec législatif," in *Vers une république des biens communs*, ed. Nicole Alix (Paris: Les liens qui libèrent, 2018), 133-41.

40. On the concept of a bundle of rights, see Fabienne Orsi, "Elinor Ostrom et les faisceaux de droits: l'ouverture d'un nouvel espace pour penser la propriété commune," *Revue de la régulation. Capitalisme, institutions, pouvoirs* 14, no. 2 (2013), http://regulation.revues.org/10471.

41. See Antonio Casilli and Paola Tubaro, "Notre vie privée, un concept négociable," *Le Monde*, January 24, 2018; and Lionel Maurel and Laura Aufrère, "Pour une protection sociale des données personnelles," *S.I.Lex*, February 5, 2018, https://scinfolex.com/2018/02/05/pour-une-protection-sociale-des-donnees-personnelles/.

42. Karl H. Metz, "Pauperism to Social Policy: Towards a Historical Theory of Social Policy," *International Review of Social History* 37, no. 3 (1992): 329-49; Robert Castel, "Emergence and Transformations of Social Property," *Constellations* 9, no. 3 (2002): 318-34.

43. James Muldoon, "Data-Owning Democracy or Digital Socialism?," *Critical Review of International Social and Political Philosophy* (2022), https://doi.org/10.1080/13698230.2022.2120737.

44. The association Europe-v-Facebook.org took a pioneering class action to the CJEU in 2014 (see chapter 5), and in 2018, La Quadrature du Net, an organization that fights for digital freedoms, took one class action against each of the Tech Giants (Alphabet [Google], Amazon, Apple, Meta [Facebook], and Microsoft). See https://gafam .laquadrature.net.

45. Antonio A. Casilli, "Four Theses on Digital Mass Surveillance and the Negotiation of Privacy," presented at the 8th Annual Privacy Law Scholar Congress, Berkeley Center for Law & Technology, June 2015, Berkeley, CA, https://shs.hal.science/halshs-01147832.

46. Valerio De Stefano, "'Negotiating the Algorithm': Automation, Artificial Intelligence and Labour Protection" working paper no. 246 (Geneva: ILO, 2018).

47. Mariya Vyalykh, "IG Metall continues its successful cooperation with the YouTubers Union by founding the FairTube e. V. association," Fair Crowd Work, http://faircrowd.work/2021/02/03/ig-metall-continues-its-successful-cooperation-with-the-youtubers-union-by-founding-the-fairtube-e-v-association/.

48. Maurel and Aufrère, "Pour une protection sociale des données personnelles."

49. Aarian Marshall, "Uber Makes Peace with Cities by Spilling Its Secrets," Wired, April 16, 2018, https://www.wired.com/story/uber-nacto-data-sharing; "Two Years Later: The City Portal," Airbnb, January 20, 2023, https://news.airbnb.com/two-years -later-the-city-portal/.

50. Martin Anderson "A Cartel of Influential Datasets Is Dominating Machine Learning Research, New Study Suggests," Unite.Ai, December 9, 2022, https://www.unite.ai/a-cartel-of-influential-datasets-are-dominating-machine-learning-research-new-study-suggests/.

51. Cletus Gregor Barié, "Nuevas narrativas constitucionales en Bolivia y Ecuador: el buen vivir y los derechos de la naturaleza," Latinoamérica. Revista de Estudios Latinoamericanos, no. 59 (2014): 9-40.

52. Sofía Beatriz Scasserra and Paola Ricaurte Quijano, "La Pachamama, descentralización y ciudades inteligentes en América Latina," Friedrich-Ebert-Stiftung Union Project, October 3, 2023, https://sindical.fes.de/detalle/la-pachamama-descentralizacion -y-ciudades-inteligentes-en-america-latina.

53. Nicolas Colin and Pierre Collin, Rapport relatif à la fiscalité du secteur numérique (Paris: La Documentation française, 2013).

54. Vili Lehdonvirta et al., Data Financing for Global Good: A Feasibility Study (Oxford: Oxford Internet Institute, 2016).

55. The Conseil national du numérique français (the French Digital Affairs Council) came down against establishing private property rights over personal data in 2014. Given the asymmetrical power relations between consumers and platforms, the text read, the sale of personal data as private property would, first, only generate "insignificant sums" and, second, it would increase the inequalities between citizens. Francis Jutand et al., "Neutralité des plateformes: réunir les conditions d'un environnement numérique ouvert et soutenable," Conseil national du numérique, May 2014, http://www.ladocumentationfrancaise.fr/var/storage/rapports-publics/144000332.pdf.

56. Tera Allas et al., "An Experiment to Inform Universal Basic Income, McKinsey & Co.," September 15, 2020, https://www.mckinsey.com/industries/social-sector/our -insights/an-experiment-to-inform-universal-basic-income.

57. Alvise Armellini and Giuseppe Fonte, "Protests Erupt in Italy over Cuts to Pov-

erty Relief Scheme," Reuters, July 31, 2023, https://www.reuters.com/world/europe/protests-erupt-italy-over-cuts-poverty-relief-scheme-2023-07-31/.

58. Ryan Abbott and Bret Bogenschneider, "Should Robots Pay Taxes? Tax Policy in the Age of Automation," *Harvard Law & Policy Review* 12, no. 1 (2018): 145–75.

59. Nick Srnicek and Alex Williams, *Inventing the Future: Postcapitalism and a World without Work* (London: Verso, 2015).

60. Jean-Marie Monnier and Carlo Vercellone, "Le Financement du Revenu Social Garanti comme Revenu Primaire: Approche Méthodologique," *Mouvements* 73, no. 1 (2013): 44–53.

61. Yuri Biondi and I proposed this option long before digital platforms took off in our contribution. Yuri Biondi and Antonio A. Casilli, "Reddito universale di cittadinanza e riforma della moneta. Una proposta di portafoglio elettronico di moneta di cittadinanza," in *Tute bianche. Disoccupazione di massa e reddito di cittadinanza*, ed. Andrea Fumagalli and Maurizio Lazzarato (Rome: DeriveApprodi, 1999), 63–73.

62. Monnier and Vercellone, "Le Financement du Revenu Social," 47.

INDEX

Accenture, 139, 173
Achilles and the tortoise paradox, 212
ACMU (African Content Moderator
    Union), 217
Acxiom, 135
Admiral, 50
Adorno, Theodor, 156
Ad Rank, 44
ad sales brokers, 131
Adsense, 131
Advanced Technologies Group, 74
Adwords, 131
Aerosolve, 50
Africa, xiv-xv, 61, 81, 92, 100-103, 139,
    200, 206
African Americans, 65
African Content Moderator Union
    (ACMU), 217
agency (sociological notion), xii, 79, 117,
    169, 171, 189, 200, 208, 223, 272n56
AGI (Artificial General Intelligence), xiii,
    11, 24, 33
Agosti, Claudio, 70
AI. *See* artificial intelligence (AI)
AI Index, 3
Airbnb, 5, 37, 49, 60, 62, 65, 78, 186, 195,
    197, 219, 222
AlexNet, 206
Algeria, 146
algorithms, xiii, xv, 18, 31, 34, 44, 88, 104,
    107, 133, 140, 188; algorithmic wage
    discrimination, 65-66; as artificial
    objects, 91

alienation, 191-92; exploitation, 114;
    false consciousness, 116
AllBnB, 219
Alphabet Inc., 46, 105-7, 223, 288-89n44.
    See also Google
Amazon, 3, 8, 11, 38, 41-42, 74, 87, 90,
    110, 120, 123, 179, 195, 288-89n44;
    Alexa, xii, 31; alienation, 191; Amazon
    Web Services, 37; "artificial artificial
    intelligence," 80; autonomous tech-
    nology, experimenting with, 88; big
    data strategy, 80; bookstores, closing
    of, 281n35; cloud storage, 80; commis-
    sions, 85; data, reselling of, 88; data
    security services, 80; drone and robot
    delivery programs, 88; Goodreads, 46
Amazon Mechanical Turk (MTurk), 5,
    34, 78, 93-95, 99, 104, 108-9, 138,
    140, 143, 163, 169, 171, 177-78, 185-
    86, 191, 206-7; application program-
    ming interface (API), 89; as commer-
    cial operation, 83; commissions, 85;
    ethical alternative to, 87; exploitative
    practices, 84; gamification of, 83-84;
    GroundTruth interface, 91; human
    intelligence task (HIT) scrapers, 83;
    human intelligence tasks (HITs), 85,
    89, 138; as intermediary, 85-86; labor
    of automation, 88; microtasks, as re-
    wards, 84; microworkers, use of, 89; as
    neutral, 86; participation agreement,
    84-85; in presidential campaigns,
    82-83; requesters, 80; requesters and

Amazon Mechanical Turk (*continued*) microworkers, as competing against each other, 86; sociability and play, 83; software-as-a-service, 80; as third-party beneficiary, 86; unpaid, 84; value of automation, as dehumanizing process, 91; WeAreDynamo, 87–88, 286n10. *See also* Mechanical Turk; Turkers; Turkopticon

American Association for Artificial Intelligence, 211

America Online (AOL), 121; class action lawsuit, 120; Community Leader Program, 120

Anatomy of Work, The (Friedmann), 190

Andreessen, Marc, 198

Angell, Ian, 193

Anglicus, Bartholomaeus, 40

AOL. *See* America Online (AOL)

API (application programming interface), 73

Appen, 87, 96

Apple, 35, 74, 93–94, 120, 288–89n44; Apple Maps, 94; iPhone, 199; Siri, xii, 31; TryRating, use of, 94

application programming interface (API), 73

Arab states, 101

Artificial General Intelligence (AGI), xiii, 11, 24, 33

artificial intelligence (AI), vi, ix, xi, xiv–xvi, 1, 11, 13, 17, 22, 33, 36, 46–47, 49, 94–95, 104–5, 122, 155, 179, 189–90, 201, 212, 227; affective turbulence, 154; affects and exploitation, link between, 154; "AI washing," 3; algorithms, xiii, 2; "artificial artificial intelligence," 80, 206–7; automated learning, relying on human "supervision" and "reinforcement," 32; biases of, 154; colonization, as metaphor, 196–97; destruction of jobs, 25; through digital labor, 196; display of intelligence, 18; expert predictions, bad track record of, 24; fake, xiii; generative, 142, 154, 207–8, 211; ghost work, 164; human input, 31; "humans in the loop," 91, 93, 139; imitations of,

93; learning process of, 32; linguistic, as doubly exploited, 154; machine learning, 19; misleading advertising, 2; narrow or weak smart technologies, resort to, 207; platforms, as crucial to, 151; political debate on, 204–5; progress in, 210–11; as prophecy, 206–7; recAPTCHA system, 107–8; recognition, 214; reinforcement phase, 154; reward system for, 32; solutions, 31–32, 90, 93, 137, 154, 164, 183, 214; speed of human work, benefit from, 206; as "super brain," 31; synthetic data, 211; training, 32, 91–92, 138, 213; unsupervised learning, 209–10; verification, xiii, 92; woke bias, accusation of, 20

artistic critique (sociological notion), 157–58

Arvidsson, Adam, 114–15

Asia, 6, 48, 101, 124, 145, 173, 180–81, 206, 215

"assetized work," 119

Association for the Advancement of Artificial Intelligence, 211

Atlas of AI (Crawford), 189

ATMs (automatic teller machines), 26–27

atomized labor, 108

attention capture, 179

attention economy, and audience economics, 155

auctoramentum, 204

audience labor, 159, 161; activity of watching, 155; creation of meaning, 155; vs. digital labor, 156; and economics, 155; visibility, data that measures and organizes, 155

Audience Network for Meta, 131

Aurora Innovation, 74, 77, 94, 167

Australia, 100

automatic teller machines (ATMs), 26–27

automation, xii, 5, 9–10, 30, 34–36, 68, 77, 93, 107, 159–60, 163–64, 202, 210, 212–14, 227; AI solutions, 31–32; animals, 21; content moderation, 140; data generation, 136; digitalization of human tasks, 26–27; digital labor, 19, 28, 33; "end of work," 4, 11; fantasy of complete automation, 205–6; frictionless, 23; "great technological replacement" theory, 4, 24; "humans in the loop," 3; humans stealing jobs from robots, 2; human work, reliance on, 138; labor negotiations, 23; low-cost labor, 21; "manifest destiny," 11; as method of squashing conflict, 21; nature of work, altering of, 11; outsourcing of tasks, 27; replacing humans, 23–26; standardization of tasks, 27; task interdependence, 22; unemployment rates, 24–25; value, 183; work offshoring and concealment, 12

autonomous taxis, 74. *See also* autonomous vehicles (AVs); robotaxis; self-driving cars

autonomous vehicles (AVs), 74, 76; car experiments, failure of, 75; fatal accidents, 75; human input, 75; microworkers, 77; "safety drivers," 75. *See also* autonomous taxis; robotaxis; self-driving cars

Autor, David, 26

Babbage, Charles, 79
Bacon, Francis, 40
Badger, Adam, 50
Baidu, 74
bait-clicks, 146
Baker, Dean, 24
Baker, Paul, 106
Bali, 145
Bangladesh, 100, 143
Barbrook, Richard, "new class," 192
Bardella, Jordan, 20
Bauman, Zygmunt, 168
Beck, Ulrich, 168
Beckett, Samuel, vii
Bell, Daniel, viii, 21
Benjamin, Walter, 34, 79
Benkler, Yochai, "wealth of networks," 115
Berardi, Franco "Bifo," 13, 36, 194
Berman, Bruce, 204–5
Betterworks, 184
Bezos, Jeff, 80, 88, 93; "Christmas cards" to, 86–87

BHV, 50
Bidet, Alexandra, 28–29
Big Brother, 184
big business, 42
big data, 6, 22, 80; as commodity, 197
Big Tech, 20, 94; autonomous vehicles (AVs), investing in, 74; impact sourcing (IS), 102; social leverage, 102
Bilibili, 140
Bilić, Paško, 107
Bing, 38, 127; quality raters, 93
bin Laden, Osama, 92
Biondi, Yuri, 290n61
BlaBlaCar, 51
Black Lives Matter (BLM), xi
blended workforce, 61
blogging, 48, 110
Bogost, Ian, 158, 165–66; hyperemployment, 164
Bolivia, xv, 224
Boltanski, Luc, 157–58
bond of subordination, 177
Bossware apps and platforms, 181
bots, 25–26. *See also* robots
brands, 110–11; "non-organic" content, 145
Braverman, Harry, vii
Brazil, 97, 145–46, 226; Constitution for the Net, 200
Bruckner, Caroline, 60
bucklige Zwerge ("humpbacked dwarfs"), 34
bureaucratization, 47
business law, 10
Buttigieg, Pete, 82–83
buybacks, 42
BYD, 74

California, 62, 125; Prop 22 minimum-wage bill, 64
California Consumer Privacy Act, 186
Callon, Michel, 54; qualification and requalification of goods, 52–53
Cambridge Analytica scandal, 139
Cambridge Platform of the Puritan Congregationalist Churches of New England, 40
Cameroon, 102

Camus, Renaud, 20
Canada, 100, 143
Capgemini, 173
capitalism, 193, 197, 213, 219; cognitive, 13, 159, 161, 226; colonialism, 198; consumption-based, 111; industrial, 20, 153, 195, 198; information, 194; just-in-time lives and sleeplessness, 158; platform, 198; pre-internet, 112; production-based, 111; surveillance, 183; "vectoralist class," 194–95
CAPTCHAs, 128
Casati, Roberto, 197
Castel, Robert: "social property," 221; workers' autonomy, 171
Castells, Manuel, vii, 194; networked society, 22
CCM. See commercial content moderation (CCM)
CCOO (Workers' Commission, Spanish trade union), 217
cellular organizations, 50
Charles I, 41
chatbots, 32, 137–38, 209
ChatGPT, vii, xii–xiii, 82, 102, 122, 137, 174, 206; affective turbulence, 154; priming of, 31–32; reinforcement learning, 209
"cherry blossoming," 97
Chiapello, Ève, 157–58
Chile, 190
China, 3, 6, 25, 35, 74, 110, 111, 127, 180–81, 201; anti-lockdown riots, xi; WeChat, 200
Christensen, Clayton, 235n1
Christian anarchist movements, 40–41
Churchill, Winston, 41
Ciccarelli, Roberto, 204
circular economy, 58
citizen journalists, 130
class action lawsuits, 120–21, 285n2, 288–89n44
Clearview AI, 136
CleverControl app, 181
Clic and Walk app, 216
click farms, 7–8, 110, 127, 132, 133, 146–47, 160, 198; "click on content," 142;

data sweatshops, 199; as parasite platforms, 145; as slavery, 143
Clickworker, 78, 94, 96
climate crisis, xi
CloudFactory, 138
cloud platforms, 37
"cloudwork," 78
Coase, Ronald, 43
Codealphabet, 173
Codementor, 173
"cognitariat," 194
Cohen, Julie, "surveillance-innovation complex," 183
Cohen, Nicole, 48
Coinbase Earn, 128
collaborative economy, 57–58, 151, 154, 219. See also gig economy
collective labor structures, eroding of, 43
collective production and reproduction, 13
collectivizing, 222; digital platforms, 223
Colleoni, Elanor, 114–15
Colombia, 100–101
colonialism: colonization, 196–97; and digital labor, 196; and industrial capitalism, 198
Columbia School of Journalism, 130
Comcast, 48
commercial content moderation (CCM), 140; in-house in big-tech companies, 141; on microwork platforms, 141; offshore in boutique local firms, 141; in remote call centers, 141. See also content moderation
CommonCrawl, 137, 206
commons, 219, 221, 223; global, 220
Competition and Consumer Authority (Denmark), 216
conspicuous consumption, 163
conspicuous tasks, 172
consumer labor, 27, 161
consumer work, 159; customization, 152; digital labor, resemblance to, 151–52
content creators, 7–8, 109, 117; collaboration, 118
content filtering, 140
content mills, 142

content moderation, 141; automation, furthering cause of, 140; offshoring of, 140; visibility, standards of, 140
content producers, 110, 114, 124, 126–27, 130–33, 156, 160; as "fantrepreneurs," 119; monetization, 177; payment, 121
CoopCycle, 219
"coopetition," 35
Coopify app, 219
Cornerstone OnDemand, 183
Corrigan, Thomas, 123
Côte d'Ivoire, 102–3
cottage industry, 176
Cottereau, Alain, principle of bilateral free will, 175
Couldry, Nick, 198
couriers, 65, 69–70, 86, 123, 160, 162–63, 169, 171, 178–79, 182, 184–85, 219; delivery, xi, 8, 62; demanding conditions, 64; piecework rates, 45; work accidents and occupational diseases, 63–64
Court of Cassation (Paris), 187
COVID-19 pandemic, xi, 3, 22, 58, 65, 69–70, 81, 161, 166, 168, 179, 181, 202; remote work, 95, 164
Craigslist, 41–42, 49
Crary, Jonathan, 159; just-in-time lives and sleeplessness, 158
Crawford, Kate, xii, 189
Cromwell, Oliver, 40
Crowdin, 120
Crowdsource, 96, 105
crowdsourcing, 95, 105
crowdwork, xi, 78, 87–88, 163
Cruz, Ted, 82–83
cryptocurrency, 128
cultural capital, 112
cybercriminals, 146
cybernetics, 19, 207

Daemo, 87–88
data, 68; algorithmic control, 66; brokers, 135; collection, 66; colonialism, and historic colonialism, 198; discrimination, 66; economy, 131; extraction, 4; laborers, 287n25; personal, value of, 132; protection laws, 70–71; qualifica-tion, monetization, and automation, 133; sharing, 222–23; surveillance, 66; tracking of, 62; use, 63; value capture, 66, 72, 130–31; value of, 130–31, 133; work, 78
datafication, 19, 151, 207
Datalogix, 135
Davos, Switzerland, 23
deep labor, 94
deep learning, 135–36, 138; "algorithmic black box," 210
Deleuze, Gilles, 160
Deliverance Milano, 70
Deliveroo, 5, 57–58, 65, 69, 86, 177–78
delivery platforms, contactless delivery, 69–70, 195
Deloitte, 173
Denis, Jérôme, 162
DeviantArt, 130
Diggers, 40; abolition of private property, 41; abolition of wage labor, 41; pooling of productive resources, 41
digital age, 3, 27, 34, 193
digital autonomy, 223
digital colonialism, 197
digital commons, 220
digital divide, 125
digital economy, 39, 41, 44–45, 189; colonial nature of, 199; taxation, 10; value chain, 226
digital game labor, 156–58
digital immigrants, 125
digital labor, ix, 7, 9, 20, 34, 67, 83, 105, 127, 138–39, 160, 173, 179, 183, 225, 227; AI, 196; as assembly line, 169; vs. audience labor, 156; automation, 19, 21, 33; boom in, 187; care work, 153; certification principles, 218; characteristics of, 161; chatbots, 137; class conflict, 12, 202–4; co-determined labor, 219; cogs in the machine, 190; collaborating, 8; collective redistributive income, 10; coloniality, 196; commercial content moderation (CCM), 140; commons, access to, 220–21; computing, tied to, 211; consumer work, resemblance to, 151–52; as continuous-time occupa-

digital labor (*continued*)
tion, 158; on continuum, 188; datafication, 19; and "dead labor," 28; as de-
territorialized work, 202; digital labor
platforms, 104; digital proletariat, 192;
digital sociability, as interrelated, 29;
domestic and parental work, similarity to, 153–54; emails, 165; exploitation and alienation, 214; exploitation
and empowerment, 192; flexibility
of, 167; Fordist era, comparison to,
190; as foreigners at work, 202; form
of contract, 187; fragmentation, 30,
189; free, 141; gendered distinction
between private and public spheres,
blurring of, 154; gray zone of, 161,
186; "housewife," modern incarnation of, 153; human contributions, and
accuracy, 206; immaterial labor, 159;
inconspicuous, 164–65; inequalities,
100; informality, association with, 153,
187; labor of monetization, 74; labor
of qualification, 73; labor protections,
214–15; lack of recognition, 162; logical impossibility, 212; mental health
and well-being, 140; microtasks, 78;
microwork, 54, 108; monetization,
159; offshoring, 140; on-demand
work, 54, 57–59; outsourcing of tasks,
28, 30; of platforms, 43, 57; pleasure
approach, 154; and PTSD, 140; as real
work, 218; recruiting and managing of,
174; remote exploitation without local
protection, 202; sociability, 57; social
media work, 54, 108, 132; standardization of, tasks, 28; subordination, 178;
supervised learning, 209–10; surveillance, 182, 187; taskification, 19, 30,
151; as temporary, 212; as term, 37, 113,
122; time, distortion of, 159; as unpaid
labor, 153–55; users' actions, forms of
labor embedded in social relations,
53; visibility and invisibility, oscillating between, 163, 167; as voluntary
servitude, 184; as work, 8; work and
leisure, interweaving of, 156; workerist approach, 13; work for others and

work for oneself, tensions between, 12;
working conditions, 170
digital natives, 9
digital oligopolies, 35–36
Digital Personal Data Protection Act (India), 186
Digital Platform Labor (DiPLab), xiv xv,
229
digital platforms, 9–11, 35, 42, 44, 95, 156,
158, 163–64, 172, 175–77, 181, 201–2,
227; collectivizing, 223; consumer work,
152; data collection, 66; discrimination, 66; economy, 199; ecosystems of,
46; informal workers, 153; legislation,
153; loopholes, 64; as multisided, 37,
45, 155; owners and users, 7; regulatory
capture, 51; side-hustle culture, 169;
surveillance, 182–83; Taylorism of, 27;
technical, economic, and systemic coordination, 46; two-sided markets, 37,
45; users, as work and apart from work,
151; value creation, 51–52; visibility, 162
digital proletariat, 8–9, 192, 194, 204, 212
digital sociability, and digital labor, as
interrelated, 29
digital socialism, 221
digital studies: decolonial computing,
196; decolonial turn, 198; postcolonial
computing, 196
digital sweatshops, 205
digital use, race and ethnicity, 125
digital utopianism, 171
digitus, 13, 160
DiPLab (Digital Platform Labor), xiv–xv,
229
Discord, 140, 209
discrimination, 65–66
disruptive digital technologies, 26
domestic labor, 27, 159, 167; invisibility
of, 161
DoorDash, 58, 62
dot-com boom, 170
DoubleClick for Google, 131
Dowd, Maureen, 73
driverless cars, 6. *See also* autonomous
vehicles (AV)
drones, 80

dual value production, 50
Dubai, United Arab Emirates, 145
Dubal, Veena, 65-66
Düsseldorf, Germany, 50
Dyer-Witheford, Nick, 194

eBay, 41-42
e-books, 45-46
Echazú, Luis Alberto, 224
Egypt, xv, 146
El Alto, Bolivia, 225
electronic frontier, 170, 198, 278nn75-76,
    282n49
Electronic Frontier Foundation, 181
Elerding, Carolyn, 73, 243n75
email, 52, 166; hyperemployment, 165; as
    unending lists of tasks, 165
empowerment, 9; collective, 192; eco-
    nomic, 103; users', 191
energy companies, 50
Engels, Friedrich, 192
English Civil War, 40
English colonies, 40
Ennius, 20
entrepreneurship, 168, 170-71, 175
Epsilon, 135
Epweike, 96
Etsy, 9, 38, 60, 62-63
ETUI Foresight Unit, 70
EuroFund, 285n4
Europe, ix, xiv-xv, 1, 6, 37, 59-61, 68, 81,
    92-93, 100-101, 120, 139, 173-74, 187,
    201, 215-16, 220, 222, 226; "Frankfurt
    Declaration," 218; yellow vests, xi
European Commission, 60
European Network on Digital Labour, x
European Union (EU), ix, 60, 71, 168;
    Court of Justice, 120
"Europe vs. Facebook," 120, 288-89n44
evaluation data, 66
"exalted" workers, 170-74, 192
"Exercising Workers' Rights in Algorith-
    mic Management Systems" (report), 70
exploitation, 129-30, 176, 189-91, 198,
    219; empowerment, balancing of, 192
Exxon Mobil, 42
EY, 173

Facebook, 3, 5, 41, 48, 51, 93, 102-3, 109-
    11, 117, 120, 122, 125-29, 131, 144, 171,
    172, 213, 288-89n44; algorithmic fac-
    tory, 134; artificial and organic traffic,
    as blurred, 147; automated multilin-
    gual versions of posts, 137; digital liter-
    acy, 135; Facebook Editor, 134; Face-
    book Leaks, 139; Facebook Reactions,
    134; facial recognition, 136; fact check-
    ing, 139; FAIR (Facebook Artificial In-
    telligence Research), 135, 257n70; fake
    clicks, 143; fake news, fight against,
    139; Free Basics, 198; "likes," value
    of, 132-33, 146-47; "M" chatbot, 32,
    138; organic brand posts, 146; paid
    media, 147; paid reach, 146-47; "real
    name" policy, 179; targeting, 147;
    terms of use, 113-14; transformation,
    from website to "factory," 7; triggers,
    178; user classification patent, 202-4;
    "viewability" measure, call for, 147;
    visibility, selling of, 147; Workplace,
    183. See also Meta
Facebook Diaries, The (reality TV series), 48
facial recognition, 136, 211
Fair Crowd Work, 216
fair pay, 10
FairTube campaign, 222
fake clicks, 143, 145, 146
fake followers, 142-44, 147
fake news, 136, 139-40, 145
"fauxtomation," 11
Federal Trade Commission, 135
Federici, Silvia, 154-55
Feigenbaum, Edward, 205
Ferraris, Maurizio, 179
Figure Eight, 95
financialization, 164; value production,
    42-43
Finland, 180, 226
firms, 47, 175; acquisitions, mergers, and
    optimization, 43; price mechanism,
    43; primary objective, betrayal of, 44;
    theory of, 43; transaction costs, 43;
    undermining of, 43; value production,
    collapse of, 42-43
Fisher, Eran, 192

Fiverr, 9, 96, 143, 152, 173
flexibility, 12, 51, 60, 101-2, 161, 167-70, 172; imperative of, 173; on-demand labor, 57, 61-62
"flexicurity," 170
Flichy, Patrice, society of amateurs, as democratic, 115
Flickr, 48, 111, 219
Florida, Richard, 193
Fogg, BJ, 178
Foodinho, 65, 215; hidden rating, 70
Foodora, 219
Ford, Henry, 190
Fordism, 190; reduction of heterogeneity, 47
Ford Motor Company, 74
formal employment, end of, 174
Forman, Jonathan, 60
Fortunati, Leopoldina, 13
Foursquare, 49-50
Foxconn, 199
Foxglove, 217
France, xv, 40, 50, 63, 92, 201, 216, 218, 276n39; *marchandage* system, abolishing of, 176; minimum wage, 64
Freelancer, 95-96, 143
freelancing, 95, 98-99, 151, 175, 180, 215; and microwork, overlapping of, 96-97
free speech, 142
free time, 156
free work, 29, 166; as exploitation, 129
French Digital Affairs Council, 289n55
Frey, Carl Benedikt, 22-23, 25
Friedmann, Georges, 190
Friendster, 48
Fuchs, Christian, 114, 198
functional economy, 58
Future of Jobs (World Economic Forum reports), 23

gambling, 158
gamers, 156-57
game theory, 207
gamification, 106, 157; competition, 84; as "exploitationware," 158; microtask platforms, 158; of microwork, 105; time, changing relationship to, 158

gaming platforms, 130
Ganascia, Jean-Gabriel, 209
Garcia, Ignacio, 106
Gazier, Bernard, 170
Gehl, Robert, 129
General Data Protection Regulation (GDPR), 62, 120, 186
General Motors (GM), 74; Cruise robotaxi, 75
Germany, xv, 25, 37, 50, 59, 160, 187, 216, 222
GetPaidForLikes, 145-46
Ghana, 74
ghost work, 164
gig economy, 58-59, 63, 167, 175, 180; collaborative, 119; on-demand labor, 57. *See also* collaborative economy
Gigster, 173
Gilles, Laurent, 132
Gillespie, Tarleton, 39-40
globalization, 35
Global North, 6, 100, 104, 173, 196-97, 199-200, 202, 217-18, 224
global outsourcing chains, 105
Global South, 61, 97, 100-101, 103-4, 189, 194, 197-200, 202, 217-18, 223-24
global value chains, 104
Globetech, 94
Glorious Revolution, 40
Glovo, 69-70, 215
GM. *See* General Motors (GM)
Gmail, 52
gold farming, 97, 180-81
Goodreads, 46
Google, 3, 11, 41, 44, 46, 51, 74, 108, 120, 140, 171, 288n37, 288-89n44; class action lawsuit against, 285n2; Google Ads, 37; Google Books, 7, 107; Google Data Liberation Front, 186; Google DeepMind, 88, 157; Google Images, 7, 107; Google Knowledge Graph, 122; Google Maps, 49, 105, 197; Google Marketing Platform, 38; GoogLeNet, 17; Google Neural Machine Translation, 106; Google Opinion Reward application, 127; Google Play, 127; Google Street View, 49,

77; Google Translate, 105–6; Google Trekker, making users work for free, trend toward, 105–6; manual digital labor program, 49; Partner Program, 117; Raterhub (formerly EWOQ), 93–94, 97, 123; "real name" policy, 179; reviews on, 143; search engines, 106; Waymo, 7, 76–77; Workspace, 183. See also Alphabet Inc.

Gorz, André, 193, 274n4

GPT, 90; biases of, 154; software powered by, 22

Gramsci, Antonio, 193

Grande, Ariana, 71

Gray, Mary L., xii, 30, 164

"great replacement" theory, critique of, xiv, 20–21, 24, 36, 108, 212

Grohmann, Rafael, 220

Grok (chatbot), 137

Grossman, Lev, 110

Grundrisse (Marx), 115

Guattari, Félix, 160

Guyer, Jane, 38–39

Hall, Jonathan, 71

Handy app, 59

Hastings, Reed, 158

HEC (French business school), 64

Helpling, 59

heterogeneity, 47

"heteromation," 11

Hilfr, 216

HITs. See human intelligence tasks (HITs)

"hiving off," 167

honeypots, 146

Hong Kong, 147

Honneth, Axel, 214

"hope labor," 123

Howard, Dorothy, 122

Howtank, 185

Huffington Post, 121, 123

"human-assisted virtual agents" (HAVA), 32

human-based computation approach, 88

human intelligence tasks (HITs), 85, 88–89, 138; scrapers, 83

human labor: banking sector, 26; obsolescence of, as difficult to prove, 26

"human robots," 6, 77

hustling, 61

Huws, Ursula, unpaid labor of consumption work, 27

Hyderabad (India), 197

hyperemployment, 164; email, 165; societal celebration of, 166

hypertext organizations, 50

IAC/InterActiveCorp, 46

IBM, 49, 93, 173; Deep Blue supercomputer, 17–19; Watson, 94, 208

Idaho, 181

IG Metall (German union), 216, 222

Illich, Ivan, 163–64

ImageNet, 206

IMDb, reviews on, 143

immaterial labor, 160–61; affective labor, 119; as intellectual labor, 159

impact sourcing, 104; impact sourcing platforms, 102; microwork and microfinance, at crossroads of, 103; as "philanthrocapitalism," 103

imperialism, 192, 198

impersonation of AI, 33, 92–93

inconspicuous production, 163

Independent Workers Union of Great Britain (IWGB), 217

India, 6, 81, 97, 99–100, 140, 143–44, 146, 160, 173, 176, 186, 198; Aadhaar digital identification system, 200

Indonesia, 97, 100, 143, 145, 176

industrialism, 20, 153

Industrial Revolution, vii, 180, 198; "exalted workers" of, 171; first, 27, 111, 170

influencers, 31, 109–10, 117, 119, 176, 186, 188, 197

informal economy, 153, 187

informal labor, 61

informal work, 29, 101

informational commons, 220

information management systems, 162

information theory, 207

Infosys, 173

insecure jobs, 30

Instacart, 58, 62
Instagram, 7–8, 109, 111, 118, 125, 127, 133, 135–37, 140, 143, 166, 171, 197, 209; influencers, 176
instant messaging, 138
Institut Polytechnique de Paris, xiv
insurance sectors, 50
Intérêt à Agir, 218
International Brotherhood of Teamsters, 217
International Federation of Robotics, 24
International Labour Office, 84
International Labour Organization (ILO), 60, 168, 187; World Employment and Social Outlook, 100
International Network on Digital Labour (INDL), x, xv
International Platform Association, 40
internet, 130, 154–55; "cognitariat," 194; "cyber-proletariat," 194; internet culture, sleep-deprived humanity, 158; internet multinationals, 35; "precariat," 194
internet studies, 160
Iraq, 146
Isahit, 102–3
Italy, 60, 70, 187, 226; caporalato system, banning of, 176
IT devices, 32; hidden "helpers," 33
Ixia, xi, 216

Japan, 6, 25, 140, 176; cell phone novels (keitai shōsetsu), 124
Jarrett, Kylie, xii
Jefferson, Thomas, xv–xvi, 205
Jehel, Sophie, 125
Jenkins, Henry, vii
Jeopardy (TV show), 94
jobs gap rate, 100
JPay, 181
Jumia, 200
Just Eat, 215

Kakao, 48, 111
Kalanick, Travis, 71, 75
Kasparov, Garry, 17
Kassem, Sarrah, 191

Kelly, Kevin, 45
Kempelen, Wolfgang von, 79
Kendrick, Cory, 71
Kenya, 7, 59, 102, 107, 138, 140, 160, 217
Kleiner, Dmytri, 197
Kroker, Arthur, 193
Kücklich, Julian, 157; "playbor," 156
Kuehn, Kathleen, 123
Kurzweil, Ray, 32–33

labor law, 187; reforms, 61
labor of automation, 88; self-driving cars, 74
labor of monetization, 74, 88
labor of qualification, 73, 88
labor studies, 36
Laitenberger, Ulrich, xiv
La Paz, Bolivia, 224–25
La Quadrature du Net, 120, 288–89n44
Latin America, xiv–xv, 61, 63, 100, 145, 174, 220
Law of Freedom, The (Winstanley), 40–41
"Layers of Silence, Arenas of Voice" (Star and Strauss), 161
Lazzarato, Maurizio, 159; machinic enslavement, 272n56
LDTalentWork, 173
LeadGenius, 87
Leapforce, 94
LeCun, Yann, 135
Lenin, Vladimir, 192
Leroi-Gourhan, André, 25, 29
Levandowski, Anthony, 77
Lévy, Maurice, 235n2
libertarianism, 171
LibraryThing, 45–46
Lieberman, Jethro, 164; inconspicuous production, 163
linear optimization, 207
LinkedIn, 90, 166, 178
Lionbridge, 94
LiveJournal, 48
Lyft, 60, 64, 74

machine learning, 22, 34, 49–50, 70, 77, 79–80, 95, 108, 130–31, 156, 203, 206, 211; algorithms and computing power,

91; data annotation, 136; generative AI, 207–8; interchangeable specialists, use of, 88–89; political beliefs, rooted in, 208; reinforcement learning, 209; streamlining and refining data, difficulty of, 90; supervised learning, 209, 212; translation, focus on, 137; unsupervised learning, 209–10

machines, 31; humans, as interdependent, 36; humans, replaced by, 21; "machine that thinks," 18; replacing jobs, myth of, xiii–xiv; as threat, 20–21

Madagascar, xv, 93, 200–201

Madison Square Garden, 71–72

Magic Island portal, 124

Mallet, Serge, 193

Mandel, Ernest, 213

marchandage, 176–77, 180

Marx, Karl, general intellect, 13, 114–15, 159–60

Marxism, 12–13, 28, 105, 130, 192, 204; exploitation and theft, 189; technical composition of capital, 115

Massachusetts Bay Colony, 40

materialism, 34

mCent, 143–44

Mechanical Turk, 79–80; as philosophical metaphor, 34. *See also* Amazon Mechanical Turk (MTurk)

Méda, Dominique, 231n7

media industry, and platformization, 51

Mejias, Ulises, 198

Messenger, 135; "M," 138

Meta, 7–8, 52, 102, 128, 130, 135, 139–40, 145, 147, 155, 196, 217, 223, 288–89n44; AI model, 137; "Face CAPTCHA," 136; face recognition feature, 136; false clicks, 146; fetishized work culture, promotion of, 191; "Generative AI Data Subject Rights," 137; Meta Business Partners, 135; natural language processing, 137; News Feed algorithm, 146. *See also* Facebook; Instagram; Messenger; Quest; WhatsApp

metadata, 131–32, 135

Mexico, 97; #Niunrepartidormenos ("Not one courier less") initiative, 63

microroyalties, 218

Microsoft, 11, 94, 120, 223, 288–89n44; Edge browser, 127; GitHub, 174; Microsoft Teams, 166, 181; Tay (conversational bot), 137–38; 365 (app), 183; Universal Human Relevance System (UHRS), 93, 97, 179

MicroSourcing, 180

microtasking, 7, 8, 86, 95, 101, 128, 132, 162–63, 171; automated work and work delegated to, blurring line between, 90; certifications, 188; code of ethics, 87–88; commissions on, 85; fake profiles and fake links, 143; gamifying of work, 83, 158; geography of, 100; low entry requirement, 79; "Master Turkers," 85; micro-offshoring, 100; multimedia content, managing of, 83; nuanced thinking, 82; participation agreement, 84–85; poverty trap, 99; reintermediation, 98–99; as requesters, 78, 82; rewards, 84–85; self-employed, 103; sentiment analysis, 82; software units, 89; translation software, 90–91

microwork, 30, 77, 79, 81, 83, 86–87, 89–91, 93, 95, 99–100, 104, 107, 138–40, 143, 154–55, 159, 169, 176–78, 200, 210, 213, 216–17; commercial content moderation (CCM), 141; digital labor, on social media, 108; flexibility, 101–2; free digital labor, 141; freelancing, overlapping of, 96–97; gamification of, 105, 157; "hitters," 103; "human computation" of, 108; impact outsourcing, 102; impact sourcing, 103; income, 129; microtasks, execution of, 78; people-as-a-service, 80; platformization, 105; precarity of, 101–2; as reflection of unemployment, 101; social media labor, 116; standard employment, as alternative to, 101; underemployment, 166; unpaid, 84; unskilled, viewed as, 82; verifiers, 92; women, and entrepreneurial values, 103

Microworkers (platform), 96, 144

microwork platforms, 94, 97, 101, 103, 175; adaptability, 104; as hubs for protest movements, 88; human layer, 91; mobile apps, 89–90; teaching machines, 90

Middle Ages, 180

Middle East, 100

MidJourney, 209

MightyAI, 94

Milan, Italy, 68–69

mindwork, viii, 6

Minecraft, 52

Minsky, Marvin, 205

MinuteWorkers, 143

misinformation, 139

moderators, xii, 8, 44, 102, 110, 120, 127, 140–41, 160, 163, 169–71, 185–86, 190, 196, 200, 215, 217; free speech, 142; paradoxical situations of, 142

Moderators, The (documentary), 141

monetization, 5, 31, 48, 86, 88, 109, 119, 133, 135, 159, 181, 215–17; of content, and collective ownership, 53; labor of, 74, 85; value, 183

Monnier, Jean-Marie, 226–27

Morocco, 146

Morris, William, 193

Mortimer, Thomas, 20

Moscow, Russia, 144

MTurk. See Amazon Mechanical Turk (MTurk)

MTurk Crowd, 86, 138

Muldoon, James, xiv

multitasking, 164

Murthy, Vivek, 126

Musk, Elon, 47, 74–75, 110, 137

mutualist movement, 219

MyEasyTask, 143

Myspace, 48, 111; Data Portability, 186

Nairobi, Kenya, 102, 197

Nakamura, Lisa, 46

Napster, 48

National Writers Union, 121

Neff, Gina, 170

Negroponte, Nicholas, vii

neoliberalism, 28

Nepal, 138, 143

Netflix, 158

New Spirit of Capitalism, The (Boltanski and Chiapello), 157–58

New Zealand, 147

N-form corporations, 50

Nigeria, xv, 97, 102

Nilsson, Nils, 212; employment test, 211

nonemployees, 60

None of Your Business (Austrian not-for-profit organization), 120

nonstandard employment, 43; concealed work, 60–61; growth of, 61; labor law reforms, 61; multi-employer work, 60; outsourcing, 61; part-time work, 60; project-based labor, 60; service economy, 61; temporary jobs, 60; women, growing presence of, 61

nonwork, 30, 151

North America, 40, 61, 100–101, 146, 218, 226

Norway, 100–101

Nosko, Chris, 71

Nuance Communications, 32

occasional work, 29

Oceania, 81

OECD (Organisation for Economic Co-operation and Development), 25, 168

offshoring: of content moderation, 140; digital labor, 140; and platformization, close connection between, 104

Ola, 65

oligopolies, 34–35; digital, 218; digital colonialism, 197

on-demand applications, 65

on-demand economy, 78–79; job insecurity, 60

on-demand labor, 61, 63, 109, 178, 213, 217–18; digital, 57, 119; underemployment of, 166

on-demand platforms, 65–66, 173; algorithmic management rules, 86; health risks, acknowledging of, 70; labor of automation, 74; labor of monetization, 74; labor of qualification, 73

on-demand workers' rights, 188

OneForma, 94
online communities, 190
online employment resources, 124
online games, 52, 158
online microwork, 99–100
online outsourcing, 97–98
online video games, 97
online workers, 96
OnlyFans, 119; subscription intimacy, 118
OpenAI, 102, 130, 137, 217; chatbots, 209; and games, 157; generative models, 174; GPT models, 154
open work, 169
operations research, 207
Oracle, 49
O'Reilly, Tim, "Government 2.0," 41
Organisation for Economic Co-operation and Development (OECD), 25, 168
Orkut, 48, 111
Osborne, Michael, 22–23, 25
Ostrom, Elinor, 223
outsourcing, 30, 35, 43, 61, 95, 104, 108, 164; extensive vs. intensive, 196; platforms, 94
Overblog, 48
Oxford Internet Institute: Connectivity, Inclusion, and Inequality group, 98; Online Labour Index, 95–97
"Oxford Paper" (Frey and Osborne), 22, 25–26; task-based approach, 23

PageRank, 44
Pakistan, 97, 100, 140, 143
Partnership on AI, 88
Patreon, 118
personhood, 110
Pfizer, 42
"philanthrocapitalism," 103
Philippines, 97, 100, 107, 140, 145–46, 180
piecework, 94–95, 108, 151, 176; digital, 177
Pittsburgh, Pennsylvania, 74
platform capitalism, 38, 57, 220–22, 226; fragmentation of work, 207
platform cooperativism, 10, 219; class action lawsuits, 222, 285n2; dead

platform co-ops, 220; failure of, 220; global commons, protecting and conserving of, 220; principles of, 219; protective system, based on resource allocation, 221
Platform Cooperativism Consortium, 220, 288n37
platform economy, 100, 169, 172, 198, 200; algorithmic management, rooted in, 182; alienation, 191; collective mobilization, difficulty of, 88; "downstairs" workers and "upstairs" workers, 170; labor, as geographically dispersed, 201; legal presumption of employment, 215–16; as uneven and polarized, 199; unpaid digital labor, 125; wage labor, 168
platformization, 9–10, 37, 189, 201, 212, 219–20; infrastructural, 51; institutional, 51; microwork, 105; offshoring, close connection between, 104; prisons, as intertwined, 181; soft surveillance, 184
platforms, xiv, 7, 13, 35–36, 74, 76–77, 83, 86, 88, 91, 98, 102, 132, 166–67, 177, 180, 183–84, 189, 219, 221, 224, 227; access arrangements, 179; AI, crucial to, 151; algorithmic matching, 38; algorithmic methods, 4–5; algorithmic wage discrimination, 65–66; algorithms, 44; amateurs, 117–18; "artificial artificial intelligence," 80; blogging, 48; characteristics of, 37–38; class conflict, 71; cloud, 37; commission, paying of, 127–28; computer architecture, borrowing from, 39; concept of, 40; consumers, outsourcing of work to, 105; consumers, tracking of, 50; content creators, 117–18; content production, 49, 51; content tagging, 136; contributors, 12; cultural goods, 48; data use, tracking of, 62; detractors of, 59; in digital contexts, 41; digital labor of, 4, 53, 57–60; as "doubly free," 127–29; dual value production, 50; ecosystems, 46, 52; as firm-market hybrids, 54; greater access, promise

platforms (*continued*)
of, 153; higher wages, promises of, 64; "hope workers," 123; horizontality of, 39; industrial, 37; informal labor of, 61; instability of tasks and payment, 153; instrumentalization, lending itself to, 39; as intermediaries, 39, 44, 46, 85, 117; and labor law, 65; labor market inequalities, perpetuating of, 65; lack of choice, 170; lean, 37; as market-company hybrids, 38; matching mechanisms, 134; microwork, 78–79, 93, 96–97; mobile workers and digital nomads, 181; monetization, 5, 135; as multisided infrastructures, 44–45, 51–52, 128; neutrality of, 39, 84; nonstandard occupations, 60; notion of, 54; on-demand code development, 173; on-demand labor, 61, 65; as paradigm, 50; participation agreements, 84–85; payments, 45; personal data, harvesting of, 70; philosophical and political foundations of, 39–42; pricing structure, 129; product, 37; qualification, 5, 53, 135; "real name" policy, 136; reintermediation, 99; software, 39; as specific types, 37; standardization, 46–47; surveillance, 63; targeting, 146; taskification of, 47; task uncertainty, 47; task visibility, 188; and tax laws, 81; technical subordination, 178, 188; as term, 39–41; time, distortion of, 158–59; tracking and extracting data, 62, 66–67; unpaid labor, glamorization of, 125; user activity, continuous monitoring of, 188; user data, 12, 38, 44, 47–53, 62; user participation, 123–24; value, representing of, 52; value capital vs. value creation, 47–50; value capture, 53, 66, 135, 152; value creation, 38, 42, 51–52; value production, 155, 187; virtual supply chains, 196; vision of society, 40; vision of wealth, 52; workers, as independent contractors, 85; working conditions, 62, 188; "work of the future," 9. *See also* digital platforms; platformization

platte-fourme, 40
playbor, 8, 46, 156–57, 159, 169; adaptation of artistic critique of work, 158; ambiguity of, 161
Poe, Edgar Allan, analytical machines, 79
political economy of insecurity, 168
porn, 119, 158; DIY, 118; revenge, 136
PornHub, 118
Porta, Jérôme, 28–29
post-Fordism, viii
Postmates, 5, 64
post-work, 168
"Potemkin AI," 11
Potts, Amanda, 106
precarity, 102
price discrimination, 66
Princeton University, 223
prison labor, 180–81
privacy, collective negotiation, 183
"pro am," 115
production standardization, 35
"produsers," 111–12, 115. *See also* "prosumers"
professionalization, 161
proletariat, 194; and bourgeoisie, 193
"prosumers," 111. *See also* "produsers"
protected labor employment, 215
Ptak, Laurel, 154–55
publishing: "crisis of the book," 45; curator services, 45–46; e-books, 45–46
Puritans, 40
Putin, Vladimir, 110
"putting out system," 176

Qiu, Jack Linchuan, 198–99
QQ, 48
qualification value, 183
Quest, 136
Quick, Draw! (game app), 106

reCAPTCHA system, 7, 108, 215, 285n2; as concealed digital labor, 107
recognized work, 171–72
Reddit, 51, 120, 133, 137, 206
reintermediation, 98–99, 176
relation of production, outsourcing work directly to consumers, 105

relation of subordination, 187
Remotask, 140
remote work, 95, 97, 164, 181; migration, 201–2
remuneration, 187–88
Republic of the Congo, 103
Resonate, 219
retail sectors, 50
Rheingold, Howard, theory of smart mobs, 115
Ricardo, David, 20, 23
Ricaurte Quijano, Paola, "La Pacha-mama" (Earth Mother) symbol, 223
ride-hailing economy, 58–59, 62, 68
Rifkin, Jeremy, 21, 231n7
right to privacy, 10
Ring Twice, 64
Risam, Roopika, 73
Roberts, Sarah, xii; commercial content moderation (CCM), 140–41
Robinson, Laura, 125
robotaxis, 75. See also autonomous taxis; autonomous vehicles (AVs); self-driving cars
robotization, 25
robots, ix, 30, 33, 36; productivity growth, 24; replacing humans, 23; unemploy-ment rates, 25. See also bots
Rockefeller Foundation, 102
Russia, 111, 127, 143, 180, 200

SageMaker Ground Truth, 138
Salesforce, 49; Salesforce Chatter, 183
Sama, 102, 139–40, 217
San Francisco, California, 75, 99, 276n39
Savoy Declaration, 40
scalability, 47, 206
Scale AI, 89
Scholz, Trebor, 114, 219
Schrems, Max, 131
Screenwise Panel Trend, 127
search engines, 41–42, 106, 209–10; hu-man judges, need for, 93–94; optimi-zation, 93
Seattle, Washington, 65
Second Life, 110
Seinfeld (TV series), vii

self-driving cars, 74–76. See also autono-mous vehicles (AVs)
Self-Driving Coalition, 74
self-employed, 59, 61–62, 100, 177, 215; accident coverage, 63–64; "hacker spirit," 171; microwork, 103
self-service technologies, 27
Senegal, 102
Sennett, Richard, 168
serfdom, 105
service economy, 61
Sevignani, Sebastian, 114
shadow banning, 62
SHARE Lab, 134–35
sharing economy, 57–58
Shirky, Clay, 116; cognitive surplus, 115
ShortTask, 143
Siemens Digital Industries Software, 37
Silicon Valley, 198; universal basic in-come, 10, 225
Simon, Herbert, 155
Simondon, Gilbert, 29
Sina Weibo, 127
slavery, 105, 115, 143; "iSlavery," 198–99
smart mobs, 115
Smythe, Dallas Walker, 157; audience commodity, 155; free time, 156
Snapchat, 125
SNCF, 51
Snopes.com, 139
sociability, 134; digital labor, 57, 156; human-machine collaboration, 115
social capital, 114
social change, and digital labor, 12
social critique, 157–58
socialism: digital, 221; Mandel, 213, 285n106; Morris, 193, 281n17; plat-form, 14, 231n5, 288n4
social media, 8, 35, 49, 52, 83, 122, 131, 166, 176, 178–79, 188; automated learning, reliance on human work, 138; autonomy, sapping of, 126; celeb-rity, 117; conflict, 119–20; content, vs. data, 130; content producers, 124; dig-ital divide, 125; digital labor, 108, 132; economy of links, 132; FOMO (fear of missing out), 125; "free work," 109,

social media (*continued*)
113–14, 116; harm to young people's mental health, 126; influencers, 109–10; informational habitus, 125; lawsuits, 120; need to "disconnect," 126; "90-9-1" empirical rule, 116–17; optimization, 142; organic users, 109–10, 142–43; personal data, value of, 132; pleasure vs. work, 114–16, 123–25; qualification, monetization, and automation, 109, 159; sociability and human-machine collaboration, 115; user-generated content classification, 133; users as workers, debate over, 112–13; virality, 142; wealth, development of, 114–15; younger members, as pathologized, 125

social media content: political victimization, 112; social pessimism, 112; technophobia, 112

social media labor, 140, 180, 213; microwork, 116; work vs. pleasure, 130

social networks, 152

social platforms, 184–85; activity on, as "free," 127; alienation, 114, 116; audiences' "work of being watched," 156; brands, 110–11; for coders, 174; content creation, 111; content producers, 114; deep learning, 138; digital piecework, 177; exploitation, 114; gamers, 158; microcelebrities, 111; monetization, 177; revenue, sources of, 130; sociability, 134, 156; value of data, 130–31; "working for free," 128

social property, 221, 227

social safety net, 169–70

social security systems, 101

soft surveillance, 184

Solow's paradox, 24

South Africa, 81

South America, xv, 48, 81, 139, 144, 222

South Korea, 24–25, 111, 226

Spain, 63–64, 187; riders' law, xi, 59, 216–17

Spare5, 94

Spotify, 37, 190, 219

Sri Lanka, 143

Srnicek, Nick, 37–38

StackOverflow, 174

StaffCop, 181

Standing, Guy, 194

Stanford University, 3, 223

Star, Susan Leigh, 161–62, 164; invisible work and shadow work, 163

"starbursting," 167

StarCraft II, use of to train AI model AlphaStar, 157

Steam gaming platform, 120, 130

Stocksy United, 219

Strauss, Anselm, 161–62, 164; invisible work and shadow work, 163

strikes, 63, 86, 218; user strikes, 216–17

subcontracting, 29

subjection, triggers of, 178

subjectivity, 189, 272n56; collective, 9, 12, 190; common, 202

Supply Chain Act and Due Diligence Provisions (Germany and EU), 218

surge pricing, 68

Suri, Siddharth, xii; ghost work, concept of, 164

surveillance, 63, 219; algorithmic management, 182–83; capitalism, 183; digital labor, 187; panoptic, 184; participatory, 184–85; and production, as inseparable for users, 183; reciprocal disclosure, 184; soft, 184; transparent tracking, 184

surveillance capitalism, 183

survival economy, 61

sustainability, 223

sweatshops, 176–77; data, 199

systems analysis, 207

Taiwan, 190

Taskcn, 96

taskification, 19, 30, 174, 207; of digital labor, 151; of platforms, 47; scalability, goal of, 47

TaskRabbit, 5, 58, 171, 185, 219

tasks, 30; outsourcing of, 27; standardization of, 27

task uncertainty, 47

Tata, 173

Tatoeba, 137
Taxation of the Digital Sector, The (French report), 223
Tay (conversational bot), 137–38
Taylorism, 27, 207
technical subjection, 179
technological innovation, 18, 23, 25–26, 76, 135, 199
tech workers, 170, 173–74; entrepreneurial spirit of, 171
telematics, 21
Teleperformance, 139
telework, 95. *See also* remote work
Telus, 139
Tempe, Arizona, 75
temping, 151
Tencent WeChat, 111
terms of service (TOS), 184, 187, 222; employment contracts, replacing of, 215; labor lock-in, 185–86; protest against, 216–17
Terranova, Tiziana, 113
Tesla, 77; Full Self-Driving system, 31, 74–75
TextBroker, 265–66n190
TGVPop, 51
Thailand, 110, 143
Thousand Plateaus, A (Deleuze and Guattari), 160, 272n56
TikTok, 5, 31, 52, 111, 121, 125, 129, 133, 140, 143, 217
TIME (magazine): "person of the year," 110; social media revolution, 116
time theft, 182
Tinder, 46, 131, 140; swiping right, and ELO score, 133–34
Tirole, Jean, 37
Tocqueville, Alexis de, xi
Toffler, Alvin, 111, 194
TOS. *See* terms of service (TOS)
training AI, 91–92
transhumanist theory, 32–33
translation software, 90–91
transportation platforms, African American passengers, 65
transportation sector, 50–51; on-demand methods in, 66

transport on-demand services, use of lidar (Light Detection and Ranging), 76–77. *See also* autonomous vehicles (AVs)
Trillion Parameter Consortium, 210
Tripadvisor, reviews on, 143
Trump, Donald, 82–83
TryRating, 94
Tsinghua University, 223
Tsu (social platform): advertisers, 128; subscribers, 128
Tubaro, Paola, xiv
Tunisia, 146, 200
Turing, Alan, 17–19; Turing test, 211
Turker Nation, 86
Turkers, 6, 80, 83–84, 96–97, 99, 105, 178, 185–86; classification of, 246n16; demands and grievances, 86; gender equality, 81; human intelligence task (HIT), 82, 86; labor, renting out, 85; "Master Turkers," 85, 86–87; median hourly wage, 81; pay levels, discrepancies in, 82; piecework basis, 108; systemic misrecognition, 82; WeAreDynamo, 87–88. *See also* Amazon Mechanical Turk (MTurk)
TurkerView, 86
Turkopticon, 84, 86–87
Twitch, 118
Twitter, 41, 47, 49, 121, 137, 198; fake followers, 143. *See also* X
2008 subprime crisis, 42

Uber, 3, 6, 11, 37, 57, 58–60, 64–65, 76–78, 84, 86, 88, 94, 109, 123, 160, 169, 175–77, 182, 186–87, 215, 222; Advanced Technologies Group, 74; algorithmic management system, 73, 172; application programming interface (API), 73; Aurora Innovation, 74, 167; collaborative economy and on-demand digital labor, 5; dead miles, 68; debt trap, 64; digital labor on, 67, 73; drivers, as employees, 276n39; drivers, tracking of, 66–67; fatal accidents, 75; insecurity of workers, 61–62; labor of automation, 74; pricing algorithm, 71–73; rat-

Uber (*continued*)
ing system, 73; Self-Driving Coalition, 74; social aspect of, 67–68; as social network, 68; surge pricing, 68, 195; surveillance, 66–67; value capture, 74; wage theft, avoidance of, 67
Uber Eats, 63, 69, 123, 162–63, 184–85, 209
uberization, 10
Uganda, 102
UGT (General Workers' Confederation), 217
unions, 216, 217–18, 225; fair-washing, 70
United Kingdom, 3, 7, 50, 59, 100, 138, 143, 176, 217; workers in, as unique category, 60
United States, xv, 3, 6–7, 22, 25–26, 40, 42, 59, 74, 81, 85, 100–101, 125, 127, 140, 143–44, 168, 180, 200–201, 206, 215, 217; algorithmic management, 66; anti-Asian sentiments, 65; Black Lives Matter movement, xi; case law, 183; hustling in, 61
universal basic income (UBI), 10, 225–26
universal digital income, 225–26; financing of, 227
unpaid labor, 153–55; glamorization of, 125
unpaid work, 35, 115, 125
unrecognized work: audience labor, 151; conspicuousness of, 151; consumption work, 151; domestic work, 151; gamified work, 151; immaterial labor, 151
unworked hours, 30
Upwork, 95–96, 98, 129, 143, 173, 182, 225, 286n10
Urban Company, 63
Ure, Andrew, 21, 23

valuation studies, 52
value: of automation, 5, 91, 135–38; capture, 47, 51, 53, 66, 74, 131, 227; exchange, 52; manufacturing process vs. circulation of goods and services, 52; of monetization, 5, 132–33, 135; production, 13, 28, 30, 42–43, 50, 88, 113, 155; of qualification, 5, 133–35

value creation, 155, 192; qualification and requalification, 52–53; as work of qualification, 53
Van Dijck, José, 117
van Doorn, Niels, 50
Vatin, François, 36
Veblen, Thorstein, 193; conspicuous consumption, 163
vectoralist class, 194–95, 201, 204; colonialism, 198; outsourcing, 196
Venezuela, xv, 7, 107, 144, 160
venture labor, 170
Vercellone, Carlo, 226–27
verification of AI, xiii, 92
Verizon, 144
videoconferencing, 153–54
video game industry, 156
Vimeo, 130
Vine, 121
virality, 142
Virno, Paolo, 13; mass intellectuality, 159
virtual assistants, 138
virtual class, 9, 193
visible work, 162
VKontakte, 48, 111, 127
voice assistants: Alexa, xii, 31; Siri, xii, 31
Volle, Michel, 160
Volvo, 74
von Ahn, Luis, 107; digitalization of human knowledge, 108

wage labor, 8–9, 41, 105, 157; platform economy, 168; social stigma, 180
wages, 31; as micropayments, 30
"Wages for Facebook" (Ptak), 154–55
"Wages for Housework" (Federici), 154–55
wage theft, 189
Waiting for Godot (Beckett), vii
Walk app, 216
Wark, McKenzie, 195–96; "vectoralist class," 194
WeAreDynamo, 87–88, 286n10
Weber, Max, 171
Webster, Daniel, 40
Web 2.0, 111
Weinstein, Michael, 193

"weisure," 8
WelectGo, 50
WhatsApp, 135, 137
white-collar professionals, 3, 35, 59, 193
Whole Foods, 281n35
Whyte, William, "organization man," 193
Wikimedia Foundation, 121–22
Wikipedia, 41, 111, 139, 206, 257n70; content creation and users, 121–22; contributors and compensation, 121–23
Williams, Raymond, 113
Winstanley, Gerrard, 40–41
Wired (magazine), vii, 45
Wittgenstein, Ludwig, 17, 19; technological innovation, 18
work: altering of, 11; atomization of, viii; changing nature of, 30; degradation of, viii; diversification, 30; "end of work," 4, 11; fragmentation of, 30–31, 190, 207; inconspicuous, 212; invisibility of, 164, 167; invisible and shadow, 163; vs. leisure, as indistinguishable, 169; manual, 21, 192; nonwork, 30; recognition of, 162, 167; regime of dispersion at, 179; side hustle, 169; as term, 28; unskilled chores, 163; virtual captives, 180; visibility of, 161–62; wages, as micropayments, 30; work for hire, 174–75, 180; "work outside of work," 8, 160
workerism, 12–13
workforce: changing demographics of, 20; low-cost labor, 21; migrant, 202; split between hyperspecialized and irreplaceable jobs, 22
working class, 69, 125–26, 193, 199, 203, 213
World Bank, 95, 97
World Economic Forum, 23
World Wide Web, 48

X, 47, 49, 52, 121, 127, 140; fake followers, 143; Grok (chatbot), 137; post on, value of, 133. *See also* Twitter

Yahoo!, 41–42
Yandex (search engine), 200
yellow vests, xi
Yelp, 120, 123
YouTube, 35, 37–38, 48, 111, 118, 123, 125, 129, 133, 137–38, 155, 176, 180, 217, 222; "adpocalypse," 121; contributors, 117; racist and violent content, 121; "superusers," 141
YSense, 144

ZBJ, 96, 98, 129
Ziddio, 48
Zoom, 166
Zuboff, Shoshana, surveillance capitalism, 183
Zuckerberg, Mark, 146, 225–26; "Building Global Community" manifesto, 190–91

Printed and bound by CPI Group (UK) Ltd, Croydon, CR0 4YY

13/04/2025

14656512-0005